Heck's Pictorial Archive of Nature and Science

EDITED BY

J. G. HECK

DOVER PUBLICATIONS
Garden City, New York

Bibliographical Note

This Dover edition, first published in 1994, contains the complete sections "Mathematics and Astronomy," "Physics and Meteorology," "Chemistry, Minerology, and Geology," "Botany," "Anthropology and Surgery" and "Technology" (176 plates) and the accompanying descriptive contents of the plates from the *Iconographic Encyclopœdia of Science, Literature, and Art. Systematically Arranged by J. G. Heck. Translated from the German, with Additions, and Edited by Spencer F. Baird, A.M. M.D., Professor of Natural Sciences in Dickinson College, Carlisle, PA. Illustrated by Five Hundred Steel Plates, Containing Upwards of Twelve Thousand Engravings,* originally published by Rudolph Garrigue, Publisher, New York, in 1851. The table of contents has been slightly rearranged for this edition.

DOVER *Pictorial Archive* SERIES

Library of Congress Cataloging-in-Publication Data

Heck, J. G. (Johann Georg), d. 1857.
[Bilder Atlas zum Conversations Lexicon. English. Selections]
Heck's pictorial archive of natural and science / edited by J.G. Heck.
p. cm. — (Dover pictorial archive series)
A selection of engravings from the 1851 English ed., entitled Iconographic encyclopaedia of science, literature, and art, which was a translation of J.G. Heck's Bilder Atlas zum Conversations Lexicon.
ISBN-13: 978-0-486-28291-6 (pbk.)
ISBN-10: 0-486-28291-0 (pbk.)
1. Scientific illustrations—Handbooks, manuals, etc. 2. Engraving—Handbooks, manuals, etc. 3. Picture dictionaries—Handbooks, manuals, etc. 4. Illustration of books—Handbooks, manuals, etc. I. Title. II. Title: Pictorial archive of nature and science. III. Series.

Q222.H432513 1994
502.2'2—dc20 94-13115
CIP

Printed in Canada
28291018 2025
www.doverpublications.com

Contents

MATHEMATICS AND ASTRONOMY

[Including the translation of German words and phrases in the same.]

GLOSSARY.

Milchstrasse, Milky Way.

GLOSSARY.

a = *der Projection des Bergstriches*, a = Projection of the mountain tract.
b = *der Höhe der horizontalen Schicht*, b = Height of the horizontal stratum.
c = *wahre Länge des Bergstriches*, c = True length of the mountain tract.

GLOSSARY.

Abend, Evening ; — ***stern***, Evening star.
Abnehmender Mond, Decreasing moon.
Aphelium, Aphelion.
Apsidenlinie, Line of Apsides.
Aufsteigender Knoten, Ascending node.
Austritt beim Aufgang der Sonne, Emersion at sunrise ; — ***beim Untergang der Sonne***, Emersion at sunset.
Axe der Ekliptik, Axis of the Ecliptic.
Bahn des Merkur, Orbit of Mercury ; — ***der Venus***, Orbit of Venus.
Colur des Frühlings-Æquinoxiums, Colure of the vernal equinox ; — *des Herbst-Æquinoxiums*, Col. of the Autumnal equinox ; — *der Nachtgleichen*, Col. of the equinoxes ; — *des Sommersolstitiums*, Col. of the Summer solstice ; — *des Wintersolstitiums*, Col. of the Winter solstice.
Comet von 1811 vor seiner Erscheinung. Projectirt auf die Ekliptik am 25 März 1811. Neigung der Bahn 75° 5′, Comet of 1811 before its appearance. Projected on the ecliptic March 25, 1811, Inclination of the orbit 75° 5′. — *von 1811 nach seiner Erscheinung b. am 1 März 1812*, Comet of 1811 after its appearance until March 1, 1812.
Dauer der längsten Nächte, Duration of the longest nights ; — *des längsten Tages*, Duration of the longest day.
Dichteres Medium, Denser medium.
Drei Uhr Morgens, 3 o'clock A. M. ; — *Nachmittag*, 3 o'clock P. M.
Dritter Octant, Third octant.
Dünneres Medium, Rarer medium.
Ebene der Pallas, d. Juno, &c., Planes of the orbits of Pallas, Juno, &c.
Eintritt beim Aufgang der Sonne, Entrance at sunrise ; — *beim Untergang d. Sonne*, Entrance at sunset.
Ekliptik, Ecliptic.
Erdaxe, Axis of the earth.
Erdbahn, Orbit of the earth.
Erde, Earth.
Erster Octant, First octant.
Erstes Viertel, First quarter.
Excentricität, Eccentricity.
Frühling, Spring.
Frühlingsnachtgleiche, Vernal equinox.
Gemässigte Zone, Temperate zone.
Grosse Axe, Axis of the heavens.
Halley'scher Comet v. 1759 u. 1835, Neigung seiner Bahn, Halley's Comet of 1759 and 1835, inclination of its orbit.
Heisse Zone, Torrid zone.
Herbst, Autumn.
Herbstnachtgleiche, Autumnal equinox.
Horizont, Horizon.
Kalte Zone, Frigid zone.
Kometenbahn, Orbit of a comet.
Letztes Viertel, Last quarter.
Millionen Meilen, Millions of miles.
Mittag, Noon.
Mittelkraft, Mean force.
Mitternacht, Midnight.
Monate, Months.
Mond, Moon ; — *-bahn*, Orbit of the moon ; — *-finsterniss*, Eclipse of the moon.
Morgen, Morning ; — *-stern*, Morning star.
Neumond, New moon.
Neun Uhr Abends, 9 P.M. ; — *Morgens*, 9 A.M.
Niedersteigender Knoten, Descending node.
Nord, North ; — *-pol*, Northpole.
Nördliche Declination der Sonne, Northern declination of the sun.
Nördlicher Polarkreis, Arctic circle.
Obere Conjunction, Superior conjunction.
Œstliche Digression, East digression.
Ost, East.
Perihelium, Perihelion.
Polarstern, Polar star.
Polhöhe über dem Horizont, Elevation of the pole above the horizon.
Richtung des Schattens um Mittag, Direction of the shadow at noon.
Rotationsaxe, Axis of rotation.
Scheinbarer Himmelsbogen, Apparent arch of the heavens ; — *Horizont*, visible horizon.
Sechs Uhr Abends, 6 P.M. ; — *Morgens*, 6 A.M.
Solstitial oder Wendepunktlinie, Solstitial colure.
Sommer, Summer ; — *-Sonnenwende*, Summer Solstice.
Sonne, Sun ; — *Sonnen-Æquator*, Sun's equator ; — *-finsterniss*, Eclipse of the sun ; — *-scheibe im Grössenverhältniss zu den Planeten*, The sun's disk ; its size compared to the diameters of the planets.
Stunden, Hours ; — *Entfernung*, Hours' distance ; — *-ring*, Hour-circle.

Süd, South ; — *-pol*, South pole.
Südliche Declination der Sonne, Southern declination of the sun.
Südlicher Polarkreis, Antarctic circle.
Südwestlicher Sonnenrand, Southwestern edge of the sun.
Trabanten des Jupiter ; — des Saturn ; — des Uranus, Satellites of Jupiter, Saturn, and Uranus.
Untere Conjunction, inferior conjunction.
Vierter Octant, Fourth octant.
Vollmond, Full moon.
Vom Pol bis zum Zenith, From pole to zenith ; — *Zenith biz zum Æquator*, From zenith to equator.
Wahrer Horizont, True horizon.
Wendekreis des Krebses, Tropic of Cancer ; — *des Steinbocks*, Tropic of Capricorn.
Westliche Digression, West digression.
Winter-Sonnenwende, Winter solstice.
Zoll, Digit, inch.
Zunehmender Mond, Increasing moon.
Zweites Octant, Second octant.

PLATE 11.

Beer and Mädler's Map of the Moon

PLATES 12, 13.

Map of the Northern Heavens (*Nördliche Sternkarte*), and of the Southern Heavens (*Südliche Sternkarte*)
Names of constellations

PLATE 14.

Figs. 1–11. Relative sizes of the planets with respect to the sun's diameter
" 12–15. Positions of Saturn and its ring with respect to the earth
Figs. 16–35. Apparent size of the planets in their perigee and apogee
" 36–45. Apparent size of the sun as seen from the planets
" 46–54. Comparison of the diameters of the moon and the planets to the diameter of the earth
" 55. Mode of observing a transit of Venus
" 56. Total eclipse of the sun
" 57. Aurora borealis
" 58. Aurora australis
" 59. Mock suns

GLOSSARY.

Anfang bei Sonnenaufgang, Beginning at sunrise ; — *bei Sonnenuntergang*, beginning at sunset.
Arabien, Arabia.
Atlantischer Ocean, Atlantic Ocean.
Azorische Inseln, The Azores.
Berberey, Barbary.
Berührung des Sonnen-, und Mondrandes, Contact of the edges of sun and moon.
Canarische Inseln, Canary Islands.
Capverdische Inseln, Cape Verd Islands.
Centrale oder totale Verfinsterung, Central or total eclipse.
Drei Zoll Verfinsterung, Three digits eclipsed.
Ende bei Sonnenaufgang, End at sunrise ; — *bei Sonnenuntergang*, End at sunset.
Grönland, Greenland.
Grossbritanien, Great Britain.
Grosser Ocean, Pacific Ocean.
Indisches Meer, Indian Sea.
Island, Iceland,
Mittel bie Sonnenaufgang, Middle at sunrise ; — *bei Sonnenuntergang*, Middle at sunset.
Mittelländisches Meer, Mediterranean Sea.
Mongolei, Mongolia.
Neun Zoll Verfinsterung Nine digits eclipsed.
Nordpol, North pole.
Norwegen, Norway.
Nubien, Nubia.
Ost Indien, East Indies.
Russland, Russia.
Sechs Zoll Verfinsterung, Six digits eclipsed.
Sibirien, Siberia.

PLATE 15.

Astronomical Instruments.

Figs. 1. Herschel's reflecting telescope
" 2. Fraunhofer's achromatic refractor at Dorpat
" 3–10. Meridian circle at Pulkowa
" 11, 12. Meridian circle at Hamburg
" 13. Roemer's transit instrument
" 14. Dollond's repeating circle
" 15–17. Repsold's Equatorial at Hamburg
" 18. Troughton's quadrant
" 19. Tycho's mural quadrant
" 20. Clockwork of Fraunhofer's refractor
" 21. Ptolemy's triquetrum
" 22. Portable transit instrument
" 23. Reflecting sextant
" 24. Reflecting sector
" 25–34. Repsold's small transit instrument at St. Petersburg
" 35–37. Ertel's theodolite
" 38. Henderson's planetarium
" 39. Henderson's explanation of the seasons

PHYSICS AND METEOROLOGY

[Including the translation of German words and phrases in the same.]

PLATE 16.

Fig. 1. Parallelogram of forces
" 2. Triangle of forces
" 3. Polygon of forces
" 4. Parallelopipedon of forces
" 5. Centre of gravity of a line
" 6. Do. do. triangle
" 7. Do. do. polygon
" 8. Do. do. parallelogram
" 9. Do. of irregular surfaces
" 10. Do. of a cube
" 11. Do. do. parallelopipedon
" 12. Do. do. pyramid
" 13. Do. of two different bodies
" 14. Illustrating Varignon's funicular machine
" 15, 16. Ill'g the theory of the pulley
" 17, 18. Atwood's machine for demonstrating the laws of freely falling bodies
" 19. The fixed pulley
" 20. Illustrating the oscillation of the pendulum
" 21. The gridiron pendulum
" 22. The mercurial pendulum
" 23. Lever of the first class
" 24. Application of the same in the steelyard
" 25. Lever of the second class
" 26. Lever of the third class
" 27, 28. Illustrating the theory of equilibrium
" 29. Lever and pulley
" 30. Bent lever
" 31. Compound lever
" 32, 33. Wheel and axle
" 34. Endless screw attached to windlass
" 35. Wheel and axle with pulley
" 36–40. Movable pulleys
" 41, 42. White's pulley
" 43. Pulley with cords working obliquely
" 44–46. Inclined plane
" 47, 48. The wedge
" 49–52. Illustrating theory of strength and stress of materials

PLATE 17.

Fig. 1–3. Illustrating the theory of the screw
" 4. Hunter's differential screw
" 5–21. Illustrating the theory of strength and stress of materials
" 22. Ill'g the theory of uniformly varying motion
" 23. Ill'g the theory of unequably accelerated motion
" 24. Ill'g the influence of gravitation on the motion of projectiles
" 25–20. Ill'g the theory of projectiles
" 29. Ill'g the centrifugal force
" 30. Ill'g the lateral pressure of moving liquids
" 31–33. Ill'g the velocity of efflux
" 34. Ill'g the forcing pump
" 35. Prony's swimmer, a contrivance to regulate the velocity of efflux
" 36. Ill'g the hydraulic ram
" 37, 38. Ill'g the ballistic pendulum
" 39, 40. Valve and parachute of the Hampton balloon
" 41. Henson's flying machine
" 42–48. Illustrating the pneumodrome of Partridge

PLATE 18.

Fig. 1–8. Illustrating the pressure of liquids
" 9, 10. Ill'g the law of Archimedes
" 11. The Cartesian devils
" 12–14. Hydrostatic balance
" 15. Nicholson's areometer
" 16. The scale areometer
" 17–19. Illustrating capillary attraction
" 20–30. Ill'g phenomena of attraction of liquids and solids
" 31. Dutrochet's endosmometer
" 32. Common barometer
" 33–35. Gay Lussac's syphon barometer
" 36–38. Fortin's barometer
Fig. 39–41. Arago and Dulong's apparatus for proving Mariotte's law
" 42. Kopp's differential barometer
" 43. Kopp's volumeter
" 44. Gay Lussac's hand air-pump
" 45, 46. Larger air-pump
" 47. Two-cylindered air-pump
" 48–50. Babinet's cock for obtaining the greatest rarefaction
" 51–53. The condensing pump
" 54–56. The syphon
" 57, 58. Apparatus involving the theory of the syphon
" 59. Cup of Tantalus
" 60. Fountain involving the syphon
" 61. Hero's ball,
" 62. Intermitting spring
" 63, 64. Hero's fountain
" 65–71. Hydraulic press
" 72. Apparatus to measure the pressure of falling water
" 73. Ordinary air balloon
" 74. Marey Monge's copper balloon
" 93 see 43. (The figure is marked 93 by mistake of the engraver.)

PLATE 19.

Fig. 1. Gasometer
" 2, 3. Cylindrical blower
" 4. Water regulator
" 5. Wind measurer
" 6. Common bellows
" 7. Mercurial thermometer
" 8. Apparatus for measuring the expansion of solids
" 9–11. Do. for higher temperatures
" 12. Metal, or quadrant thermometer
" 13. Breguet's thermometer
" 14. Air thermometer
" 15. Pneumatic apparatus
" 16. Illustrating the relative expansion of mercury, water, and alcohol
" 17. Ill'g the laws of vaporization
" 18. Ill'g the tension of vapor in unequally heated places

Fig. 15–17. Neef and Wagner's induction coil
" 18. Apparatus to show the production of electrical currents by magnetism
" 19. Induction coil in connexion with magnets
" 20–22. Ettinghausen's magneto-electric rotating machine
" 23–27. Stöhrer's triple magneto-electric machine
" 28. Differential galvanometer
" 29. Apparatus for determining the conducting powers of metals
" 30. App's for determining the requisite qualities of conducting wires
" 31, 32. App's for obtaining an unalterable thermo-electric current
" 33–35. Thermo-electric pile
" 36–38. Stöhrer's improvement of Daniel's electro-magnetic machine
" 39. Ritchie's improved machine
" 40. Wheatstone's telegraph
" 41, 42. Steinheil's telegraph
" 43–47. Illustrating Ampère's theory of the rotation of magnets
" 48, 49. Astatic multiplier
" 50, 51. Weber's apparatus for demonstrating the laws of magnetism
" 52, 53, 53*a*. Arago's apparatus for investigating phenomena in rotation magnetism
" 54. Illustrating the intensity of electricity in combined elements
" 55–58. Clarke's magneto-electric machine
" 59. Henley's quadrant electrometer
" 60. Apparatus for proving that free electricity is confined to the surface of bodies
" 61. Gold leaf electrometer
" 62. Buff's app's for determining the kind of electricity in a body
" 63. Apparatus to show the separation of the combined electricities when an electrified body is brought near
" 64. Apparatus to show that the connexion of two similar electroscopes by an insulated conductor will separate their electricities
" 65. Construction of the electrometer
" 66, 67. Different windings around the electro-magnet
" 68. Rotation of the magnet around its own axis
" 69. The voltaic pile
" 70. Illustrating the action of electrical currents
" 71. Production of electric light in a vacuum
" 72. Electric dance
" 73–75. Eisenlohr's hydro-electric machine

PLATE 23.

Fig. 1. Howard's diagram to elucidate the action of the sun's rays on the earth
" 2. Illustrating the snow line
" 3. Curvature of the Alpine snow line
" 4–23. Forms of snow crystals
" 24–35. Forms of hailstones
" 36–38. Illustrating the amount of dew on trees and herbs
" 39, 40. Ill'g the amount of dew on differently inclined surfaces
" 41–43. Ill'g the greatest deposit of dew on the highest points of an object
" 44–47. Ill'g the increase of dewdrops
" 48. Ill'g that an object rolled up in a ball receives less dew than if spread out flat
" 49–52. Ill'g that gilt paper is bedewed only when its unmetallic side is in contact with wood
" 53, *ab*. Zone of dew deposited in the cavity of a watch-glass
" 54. Regular deposit of dew *around* wafers on a glass plate, not *on* them
" 55. Deposit of dew on a window
Fig. 56, 57. Deposit of dew on spiders' webs
" 58. Perpendiculars from the widest boughs of a tree marking the limits of the formation of frost
" 59. Curious formation of frost on the chiselled star of a tombstone
" 60, 61. Curious formation of frost on leaves
" 62. Illustrating the theory of winds
" 63. Ill'g the theory of whirlwinds
" 64. Rain gauge

GLOSSARY.

Februar, February.
Fische, Pisces.
Frühling 90 *Tage*, Spring 90 days.
Heisseste Temperatur, Hottest temperature.
Herbst 90 *Tage*, Autumn 90 days.
Januar, January.
Juli, July.
Jungfrau, Virgo.
Juni, June.
Kälteste Temperatur, Coldest temperature.
Krebs, Cancer.
Löwe, Leo.
Mai, May.
März, March.
Mittlere Temperatur, Medium temperature.
Schütze, Sagittarius.
Skorpion, Scorpio.
Sommer 93 *Tage*, Summer 93 days.
Steinbock, Capricornus.
Stier, Taurus.
Waage, Libra.
Wassermann, Aquarius.
Widder, Aries.
Winter 93 *Tage*, Winter 93 days.
Zwillinge, Gemini.

PLATE 24.

Fig. 1. Peculiar crystallization on the window of a drug store
" 2. Two trees trained on a wall, only in part exposed to frost
" 3–6. Ice flowers
" 7. Formation of fog over rivers
" 8. Overtopping of trees by fog
" 9. Formation of fog over rivers inclosed by hills
" Ill'g the different appearance of clouds to observers at different places
" 11. Ill'g the difference of shape of clouds according to their elevation
" 12. Ill'g the difference of currents of air at different heights
" 13–18. Different formations of fog near the harbor of Plymouth (England)
" 19. Mirage in Lower Egypt
" 20, 21. Mirages at sea
" 22. Theory of mirages
" 23. Mirage on the lake of Geneva

PLATE 25.

Fig. 1–8. Forms of clouds
" 9. Halo around the sun
" 10, 11. Haloes around the moon
" 12. Halo and colored ring
" 13–20. Mock suns

PLATE 26.

Fig. 1–9. Ill'g phenomena in clouds
" 10–12. Rainbows
" 13. Aurora borealis
" 14. Midnight sun in the polar regions

PLATE 27.

Fig. 1. Whirlwind and waterspout
" 2, 3. The sun behind clouds
" 4. The drawing of water
" 5. Converging in the east, of the shadow of the setting sun
" 6. Illus'g that objects on earth can throw shadows into space
" 7. Shadows thrown on a bank of fog
" 8, 9. Coloring of clouds by the sun
" 10, 11. Phenomena of the sun's apparent contact with clouds
" 12. Apparent contact of clouds
" 13. Zodiacal light
" 14–19. Lightning rod
" 20–26. Magnetic observatory at Greenwich

PLATE 28.

Fig. 1, *ab*. Inclination charts
" 2, *ab*. Isodynamic charts
" 3 (centre). Chart of magnetic declinations
" 4, *ab*. Barometrical wind rosettes
" 5, *ab*. Thermometrical wind rosettes, (For Glossary see after plate 29.)

PLATE 29.

Fig. 1. Map of Humboldt's isothermal lines showing the distribution of heat and rain on the earth
" 2. Diagram of the mean annual temperature at various places
" 3. The mean decennial temperature of various places
" 4. Comparison of temperature

GLOSSARY.

Ægypten, Egypt.
Æquator-Grenze des Schneefalls, Equatorial boundary of the snow region.
Aleuten, Aleutian Islands.
Arabien, Arabia.
Asien, Asia.
Barometer steigt,—fällt, Barometer rises,—falls.
Berberei, Barbary.
Beständiger Regen, Perpetual rain.
C. der guten Hoffnung, Cape of Good Hope.
Canarische Inseln, Canary Islands.
Capverdische Inseln, Cape Verd Islands.
Deutschland, Germany.
Erd-Æquator, Terrestrial equator.
Europa, Europe.
Frankreich, France.
Freundschafts Inseln, Friendly Islands.
Gebiet der Monsun Regen, Region of the Monsoons.
Gesellschafts Inseln, Society Islands.
Grönland, Greenland.
Grossbritanien, Great Britain.
Grosser Ocean, Pacific Ocean.
Herbst und Winter Regen, Autumn and winter rains.
Indisches Meer, Indian Sea.
Italien, Italy.
Kopenhagen, Copenhagen.
Kurve von Leith, Br., Curve of Leith, latitude.
Kurve von Padua, Br., Curve of Padua, latitude.
Linie ohne Inclination, Line without inclination.
Magnetischer Æquator, Magnetic equator.
Manschurei, Mandshoo territory.
Maximum der magnet. Kraft, Maximum of magnetic power.
Mittlere Tageswärme, Medium daily temperature.
Mongolei, Mongolia.
N. O., N. E.
Neufundland, Newfoundland.
Neu Guinea, New Guinea.
Neu Sealand, New Zealand.
Niedrige Inseln, Low Islands.
Nord Amerika, North America.
Nördlicher Gürtel der beständigen Niederschläge, Northern zone of perpetual deposits.
Nördliche Hemisphäre, Northern Hemisphere.
Nördlicher Polarkreis, Arctic Circle.
Nördliches Eismeer, Arctic Sea.
Nordpol, North pole.
Nord See, North sea.
Norwegen, Norway.
Nubien, Nubia.
Oceanien, Oceania.
Ost Indien, The East Indies.
Ostseite, East side.
Persien, Persia.
Provinz des Herbstregens; — des Sommerregens; — des Winterregens, Region of autumnal, of summer, and of winter rains.
Regenloses Gebiet, Rainless region.
Russland, Russia.
Schwacher Sommerregen, Light summer rain.
Sibirien, Siberia.
S. O., S. E.
S. O. Monsun im Apr.–Oct., N. W. Monsun im Oct.–Apr., S. E. Monsoon from April to October, N. W. Monsoon from October to April.
Spanien, Spain.
Süd Amerika, South America.
Südpol, South pole.
Südlicher Continent, Southern continent.
Südlicher Gürtel der beständigen Niederschläge, Southern zone of perpetual deposits.
Südliche Hemisphäre, Southern Hemisphere.
S. W. Monsun im Apr.–Oct., N. O. Monsun im Oct.–Apr., S. W. Monsoon from April to October, N. E. Monsoon from October to April.
Thermometer steigt, — fällt, Thermometer rises, — falls.

Turkei, Turkey.
Vereinigte Staaten, United States.
Wärme-Æquator, Equator of heat.
Wendekreis des Krebses; — des Steinbocks, Tropic of Cancer; — of Capricorn.

West Indien, The West Indies.
Westseite, West side.
Winterregen, Winter rains.
Wüste Schamo oder Gobi, Desert of Shamo, or Gobi.

Zone häufiger, fast beständiger Niederschläge, stets mit electrischen Explosionen, Zone of frequent, nearly perpetual deposits, always accompanied by electrical explosions.

CHEMISTRY, MINERALOGY, AND GEOLOGY

[Including the translation of German words and phrases in the same.]

PLATE 30.

Fig. 1. Ground-plan of the chemical laboratory at Giessen
" 2. Perspective view of part of the analytical laboratory at Giessen
" 3, 3*a*. Portable wind furnace
" 4, 5. Portable distilling apparatus
" 6, 7. Fixed furnace
" 8, 9. Sand bath furnace
" 10. Small table for raising or lowering receivers
" 11, 12. Small alembic
" 13. Pharmaceutical steam apparatus
" 14–20. Do. with its utensils
" 21, 22. Crucible furnace
" 23. Tongs for handling hot crucibles
" 24. Splitting iron
" 25, 26. Retorts
" 27–29. Globes or matrasses
" 30–34. Hessian crucible
" 35, 36. Evaporating dishes
" 37–41. Thermometers
" 42. Separating funnel
" 43. Bent tube with bulb
" 44, 45. Walter's safety tube
" 46. Filtering funnel
" 47. Separating funnel
" 48. Special funnel
" 49. Washing bottle
" 50. Measure
" 51–54. Florentine flasks
" 55. Pneumatic tub
" 56, 57. Wolff apparatus
" 58, 59. Gasometer
" 60, 61. Davy's apparatus for investigating earths
" 62. Apparatus for generating prussic acid
" 63. Apparatus for producing small quantities of illuminating gas from various substances
" 64. Argand's oil lamp
" 65–68. Pharmaceutical extract presses
" 69–73. Chemical balance

PLATE 31.

Fig. 1, 2. Oxy-ethereal lamp
" 3. Distilling apparatus
" 4, 5. Gasometer
" 6, 7. Apparatus for rendering carbonic acid liquid
" 8. App's for obtaining phosphorus
" 9. Furnace for obtaining caustic soda
" 10. Furnace for heating matrasses
" 11. Berzelius lamp
" 12. Triangle of iron wire
" 13. Apparatus for obtaining the gaseous elements
" 14. App's for obtaining chlorine
" 15-23. Drying apparatus
" 24. Air pump
" 25. Calcium tube
" 26. Chloride of calcium tube
" 27–29. Combustion tubes
" 30–32. Combustion furnace
" 33. Apparatus for elementary analysis with application of oxygen
" 34–38. App's for determining the nitrogen in organic substances
" 39. App's for filling the potash apparatus
" 40, 41. App's for determining the specific gravity of vapor
" 42. Muffle furnace
" 43. Cupel
" 44. Muffle
" 45, 46. Apparatus for generating carbonic acid gas
" 47. Two-branched tube for washing precipitates

Fig. 48, 49. Washing bottles
" 50–52. Filtering apparatus
" 53. Suction tube
" 54. Washing bottle
" 55. Galley furnace
" 56. Gas generator
" 57. Crucible
" 58. Apparatus for ascertaining the amount of pure metal in oxydes
" 59. Davy's safety lamp
" 60. Apparatus for procuring oxygen
" 61. Safety tube
" 62, 63. Blowpipe
" 64. Iron pincers for handling heated vessels
" 65. Apparatus for obtaining nitrogenous substances from organic matter
" 66. Eudiometer

PLATE 32.

Fig. 1. Blowpipe
" 2. Attenuated point of flame caused by the blowpipe
" 3. Forceps
" 4. Berzelius blowpipe
" 5. Nicholson areometer
" 6. Magnetic needle
" 7, *ab*. Instrument for detecting electricity in minerals
" 8. Common goniometer
" 9. Gambay's do.
" 10. Wollaston's do.

Forms of Crystallization.

Fig. 11, 12. Regular octahedron
" 13. Octahedron abbreviated to a six sided plate
" 14. Cube
" 15. Relation of the cube and octahedron
" 16, 17. Cubic octahedron
" 18. Rhombic dodecahedron
" 19. Relation of the rhombic dodecahedron and cube
" 20. Pyramidal cube
" 21,24. Cube with dodecahedral faces replacing its edges
" 22. Octahedron with dodecahedral faces
" 23. Octahedron passing into a dodecahedron
" 24. See 21
" 25. Combination of cube faces and those of the pyramidal cube
" 26. Tetrahedron
" 27. Octahedron passing into a tetrahedron
" 28. Cube with its edges replaced by three faces
" 29. Pentagonal dodecahedron
" 30. Trapezohedron
" 31. Octahedron passing into a trapezohedron
" 32. Obtuse square octahedron
" 33. Acute do. do.
" 34. Square octahedron with two corners truncated
" 35. Right square prism
" 36. Regular eight sided prism
" 37. Right square prism with its corners truncated
" 38. Square octahedron with truncated basal edges
" 39. Do. with truncated lateral edges
" 40. Eight sided pyramid
" 41. Twelve sided prism
" 42. Six sided prism with bevelled edges

Fig. 43. Right square prism, with basal edges truncated
" 44. Rectangular octahedron
" 45. Do. truncated
" 46. Square octahedron with bevelled lateral edges
" 47, *a*. Obtuse rhombic octahedror
" *b*. Acute rhombic octahedron
" *c*. Rectangular octahedron
" 48. Rhombic octahedron with two corners bevelled
" 49. Octahedron with the corners of the vertical axis replaced by four plane faces
" 50. Right rhombic prism
" 51. Do. with obtuse edges truncated
" 52. Irregular eight sided prism
" 53–55. Combination of prisms belonging to the trimetric system
" 56. Rectangular prism
" 57, 58. Modified octahedrons
" 59, 60. Oblique six sided prism
" 61. Octahedron with half the edges truncated
" 62. Prism with half the basal edges truncated
" 63–65. Oblique rhomboidal prism
" 66. Oblique six sided prism
" 67. Double six sided pyramid
" 68. Pyramidal six sided prism
" 69. Modified rhombic octahedron
" 70. Regular six sided prism
" 71. Do. with truncated basal edges
" 72. Do. with truncated corners
" 73. Obtuse double six sided pyramid
" 74. Scalene octahedron
" 75. Rhombohedron,
" 76. Six sided prism with four edges bevelled
" 77, 78. Natural situation of two rhombohedrons
" 79. Grouping of crystallizations
" 80–87, 90, 91. Forms of ground jewels
" 88, 89. Crystalline formations
" 90, 91. See 80, &c.
" 92, 93. Crystalline structure

PLATE 33.

Minerals and their Crystalline Forms.

Fig. 1. Meteoric iron
" 2. Iron pyrites
" 3, 4. Sulphur
" 5–10. Native bismuth
" 6–11. Galena
" 7. Weidmannstedtian figures
" 8. Sulphate of lead
" 9. Native silver
" 10. See 5
" 11. See 6
" 12. Arborescent native silver
" 13. Native copper
" 14. Native gold
" 15. Pyrites in calcareous spar
" 16. Filiform native silver
" 17. Galena on quartz
" 18. Arsenical pyrites
" 19. Galena on calcareous spar
" 20. Galena in cubes
" 21. Selenid of mercury
" 22. Crystallization of iron pyrites
" 23. Molybdenite
" 24. Native copper
" 25. Native antimony
" 26–29. Sulphur
" 30. Native silver
" 31. Antimony
" 32–36. Gold, galena, &c.
" 37. Sulphuret of copper
" 38. Antimony

Fig. 39, 47. Antimonial silver
" 40–45. Gold, 40 and 42, Silver 42, 43, Diamond
" 46. Diamond
" 47. See 39
" 48. Copper pyrites
" 49, 50. Glance cobalt
" 51. Diamond, Grey copper ore
" 52, 53, 56, 59. Grey copper ore
" 54, 55. Glance cobalt
" 56. See 52
" 57, 58, 60. Glance cobalt
" 59. See 52
" 60. See 57
" 61–63. Native mercury
" 64. Glance cobalt

PLATE 34.

Minerals and their Crystalline Forms.

Fig. 1. Micaceous iron ore
" 2, 3. Quartz
" 4. Tourmaline
" 5. Hematite
" 6. Datholite
" 7. Pyrolusite
" 8. Tin ore
" 9. Fibrous brown hematite
" 10. Rutile
" 11. Scaly hematite
" 12. Andalusite
" 13. Asbestos
" 14. Stilbite
" 15. Analcime
" 16. Mesotype
" 17. Apophyllite
" 18. Chabazite
" 19. Tin ore in granite
" 20. Reniform hematite
" 21, 22. Silica
" 23, 25. Oxyde of iron
" 24. Manganite
" 25. See 23
" 26. Braunite
" 27. Red copper ore
" 28. Apophyllite
" 29. Hausmannite
" 30. Sapphire
" 31. Oxyde of iron
" 32. Anatase
" 33. Red copper ore
" 34. Tourmaline
" 35. Augite
" 36. Quartz
" 37. Oxyde of tin
" 38. Corundum
" 39, 40. Arsenious acid
" 41. Picrosmine
" 42. Oxyde of tin
" 43. Rutile
" 44. Oxyde of copper
" 45. Nepheline
" 46. Scapolite
" 47. Staurotide
" 48, 53. Idocrase
" 49. Beryl
" 50. Thomsonite
" 51. Silica
" 52. Augite
" 53. See 48
" 54. Chondrotite
" 55. Humboldtite
" 56. Various silicates
" 57. Emerald
" 58. Euclase
" 59. Natrolite
" 60. Dichroite
" 61. Oxyde of copper
" 62. Prehnite
" 63. Epistilbite
" 64. Garnet
" 65. Albite
" 66. Chabazite
" 67. Pyroxene
" 68, 69. Beryl
" 70. Staurotide
" 71. Iolite
" 72. Silica
" 73. Datholite
" 74. Olivine
" 75. Hornblende
" 76. Feldspar

PLATE 35.

Minerals and their Crystalline Forms.

Fig. 1. Spodumene
" 2. Feldspar twin crystal
" 3. Sphærosiderite
" 4. Natrolite
" 5. Harmotome
" 6. Green garnet
" 7. Strontianite
" 8, 10, 11, 14, 16–22, 24, 25, 27–29, 31–37, 40, 41, 43–47, 49. Various forms of arragonite and calcareous spar
" 9. Topaz
" 10, 11. See 8
" 12, 30, 38, 39, 42, 48, Boracite
" 13, 23. Stalactite
" 14. See 8
" 15. Opake beryl
" 16–22. See 8
" 23. See 13
" 24, 25. See 8
" 26. Feldspar
" 27–29. See 8
" 30. See 12
" 31–37. See 8
" 38, 39. See 12
" 40, 41. See 8
" 42. See 12
" 43–47. See 8
" 48. See 12
" 49. See 8

PLATE 36.

Minerals and their Crystalline Forms.

Fig. 1, 2, 4, 6, 8–10, 13. Sulphate of baryta
" 3. Double spar
" 4. See 1
" 5. Wavellite
" 6. See 1
" 7, 20. Strontianite
" 8–10. See 1
" 11. Thorns coated with salt
" 12. Gypsum
" 13. See 1
" 14. Spathic iron
" 15. Wolfram
" 16, 24, 25. Alum
" 17. Sulphate of soda
" 18. Sulphate of ammonia
" 19, 34, 35. Copperas
" 20. See 7
" 21. White lead ore
" 22. Yellow lead ore
" 23. Sulphate of zinc
" 24, 25. See 16
" 26, 36, 41, 43, 47, 48. Saltpetre
" 27. Green vitriol
" 28. Sulphate of potassa
" 29, 37. Calcareous spar
" 30. Rocksalt
" 31. Karstenite
" 32. Fluor spar
" 33, 40. Sulphates of alkalies
" 34, 35. See 19
" 36. See 26
" 37. See 29
" 38, 39, 42, 44, 46, 49. Phosphates of alkalies
" 40. See 33
" 41. See 26
" 42. See 38
" 43. See 26
" 44. See 38
" 45. Sulphate of iron
" 46. See 38
" 47, 48. See 26
" 49. See 38

PLATE 37.

Fig. 1–59. Fossils of the transition slate
" 60–72. Fossils of the carboniferous period
a–f. Buckland's representation of animals of the Lias

PLATE 38.

Fig. 1, 2. Fossils of the Copper Slate Formation
" 3–11. Fossils of the Rock Salt Formation
" 12–67. Fossils of the Jura
" 68. Section of the Wirksworth cave

PLATE 39.

Fig. 1–45. Fossils of the Cretaceous System
" 46–50. Fossils of the Tertiary Period
" 51–80. Skeleton of Hydrarchos harlani
" 81–87. Illustrating the formations of veins of ores

PLATE 40.

Fig. 1–3, 11–22. Fossils of the Tertiary Period
Fig. 4–9. Fossils of the Rock Salt Formation
" 10. Plesiosaurus macrocephalus (Jura)
" 11–22. See 1.

PLATE 41.

Fig. 1–10. Fossil remains of Arachnida &c. (Tertiary)
" 11–27. Fossils of the Jura
" 28. Fossil Irish elk (Tertiary)
" 29. Dinotherium giganteum (Tert.)
" 30. Missourium theristocaulodon (Tert.)
" 31, *ab*. Batrachoid tracks (Rock Salt)
" 32. Birds' tracks in Connecticut (Rock Salt)
" 33. Scull of Hyæna spelæa (Tert.)
" 34. Scull of Ursus spelæus (Tert.)
" 35. Pterodactylus crassirostris (Jura)
" 36. Human skeleton from Guadaloupe (fossil in Limestone)

PLATE 42.

Fig. 1–12. Fossils of the Tertiary
" 13–23. Do. do. Cretaceous
" 24–31. Do. do. Jura
" 32–40. Do. do. Rock Salt
" 41–56. Do. do. Carboniferous
" 57–66. Do. do. Transition Slate

PLATE 43.

Fig. 1–3. Illustrating the formation of valleys
" 4–6. Ill'g the formation of rock beds
" 7–9, 12. Ill'g stratification
" 10, 11. Penetration of various rocks in volcanic regions
" 12. See 7.
" 13–15. Illustrating the relation of stratification to mountain masses
" 16–19. Ill'g subordinate strata
" 20, 21. Ill'g disseminated and imbedded ores
" 22–25. Ill'g the cleavage of rocks
" 26. Table mountain at the Cape of Good Hope

PLATE 44.

Fig. 1. The valley of Pyrmont
" 2–5, 12–14. Illustrating the relation of stratification to mountain masses
" 6, 7. Tertiary basin of Paris, 555, 598
" 8. Dunes on the sea shore
" 9. Section of a volcanic cone
" 10. Mont Blanc chain
" 11. Mountains in the vicinity of Bärschwyl
" 12–14. See 2.
" 15. Fissures in the earth in Calabria
" 16. Circular cavities in the plain of Rosarno
" 17. Crater of the Geyser in Iceland
" 18. Volcano of Eyafiël in Iceland

PLATE 45.

Fig. 1. Illustrating valleys of disruption
" 2. Extinct volcanic district in Auvergne
" 3. Section of Judæa and the Dead Sea
" 4. Vesuvius in the time of Pliny
" 5. Vesuvius as seen at present
" 6. Mount Etna
" 7. Campi Phlegræi near Naples
" 8. Chart of the Bay of Naples
" 9. Volcanic chart of Etna
" 10. Volcanic chart of Iceland
" 11. Map of the vicinity of Lake Titicaca, in Peru
" 12. Section of the mountains around the lake

GLOSSARY.

Basalt Boden, Basaltic soil.
Bugt, Bay.
Busen von Neapel, — von Salerno, Bay of Naples, — of Salerno.
Cyclopen Inseln, Cyclops Islands.
Die Flegreischen Felder, The Phlegræan fields.
Fiorden, The inlet.
Island, Iceland.
Längenthal im Trachyt, Long valley in the trachyte formation.
Lava Ströme, Lava streams.
Neapel, Naples.
Nord Cap, North Cape.
Œ, Island.

Passhöhe, Height of the pass.
Polarkreis, Arctic circle.
See, Lake.
Vesuv, Vesuvius.

PLATE 46.

Special Geognosy.

Fig. 1–3. Silurian system
" 4–6. Devonian system
" 7, 8. Carboniferous group
" 9, 10. Copper slate formation
" 11. Rock salt formation
" 12–16. Jura formation
" 17. Cretaceous system
" 18. Upper tertiary
" 19. Erratic boulder

GLOSSARY.

Aufgeschw. Land, Deposited ground.
Braunschweig, Brunswick.
Bunter Sandstein, Variegated sandstone.
Donau, Danube.
Ebbe Geb., Ebbe Mountains.
Egge Geb., Egge Mountains.
Geb. v. Charolais, Mountains of Charolais.
Harz Gebirge, Hartz Mountains; — *und Thüringer Wald*, Hartz and Thuringian Mountains.
Hochwald, High forest (mountain tract).
Jura Gebirg, Jura Mountains; — *-kalk*, Jurassic Limestone.
Kohlengebirge, Carboniferous formation.
Kohlenkalk, Carboniferous limestone.
Kohlensandstein, Carboniferous sandstone.
Kreide, Chalk.
Laacher See, Lake Laach.
Porphyr, Porphyry.
Quadersandstein, Freestone.
Rhein, Rhine.
Rheinisches Uebergangsgebirg, Rhenish transition rock.
Rothhaar Geb., Rothhaar Mountains.
Schuttland, Conglomerate.
Steink. Gebilde, Carboniferous formation.
Tertiär, Tertiary.
Teutoburger Wald, Teutoburg forest.
Todtliegendes, Red sandstone.
Trias, Rock salt formation.
Trier, Treves.
Uebergangskalk, Transition limestone.
Vogel Geb., Vogel Mountains.
Vogesen sandstein, Vosges sandstone.
Vulcan Gebilde; — *Gerölle*, Volcanic formation; — rubble.
Weser Geb., Weser Mountains.

PLATE 47.

Fig. 1. Volcanic chart of the world (Also chart of the winds glossary at pl. 29.)
" 2. Chart of European volcanoes
" 3. Volcanic chart of Lower Italy
" 4. Crater of Kirauea
" 5. Crater on the Island of Hawaï
Fig. 6. Landscape from the Southern Continent
" 7. Profile of mountains in the Old World
" 8. Profile of mountains in the New World
" 9, 10. Charts of hurricanes

GLOSSARY.

Ægatische In., Ægatian Islands.
Atlantischer Ocean, Atlantic Ocean.
Canal von Malta, Maltese Channel.
Deutschland, Germany.
Frankreich, France.
Grosse Antillen, The larger West India Islands.
Indischer Ocean, Indian Ocean.
Kleine Antillen, The smaller West India Islands.
Liparische In., Liparian Islands.
Lissabon, Lisbon.
Meerb. v. Taranto, Bay of Taranto.
Mittelländisches Meer, Mediterranean Sea.
Mozambique Kanal, Mozambique Channel.
Neapel, Naples.
Nord Amerika, North America.
Sicilien, Sicily.
Spanien, Spain.
Str. von Messina, Strait of Messina.
Türkei, Turkey.
Tyrrhenisches Meer, Tyrrhenian Sea.
Venedig, Venice.
Vesuv, Vesuvius.
Wien, Vienna.

PLATE 48

Fig. 1. *Coup d'œil* of the normal masses
" 2. Illustrating the theory of springs and Artesian wells
" 3. Section of the Artesian well of Grenoble
" 4. Interior of a boring shed
" 5–14. Boring tools

PLATE 49.

Fig. 1. Interior of the crater on the Island of Hawaï
" 2. Coral island in the Pacific Ocean,
" 3. Rock groups on the Faroe Islands
" 4. Natural bridge of Icomonzo in Columbia
" 5. Glacier sources of the Rhine
" 6. Trachyte columns in the Siebengebirge
" 7. Basaltic groups on the Island of Staffa
" 8. Fissured clay slate on the Lahn
" 9. Cleft variegated sandstone near Kreuznach, Germany

PLATE 50.

Fig. 1. The Geyser of Iceland
" 2. Submarine volcanic eruption
" 3. Air volcano of Turbaso in Columbia
Fig. 4. The volcanic Island of Ferdinandea
" 5. Crater of the same
" 6. Crater of the Krabla in Iceland
" 7. Crater of Etna in Sicily
" 8. Niagara Falls
" 9. The Dal Elf Fall in Sweden

PLATE 51.

Fig. 1. Primitive forest of Brazil
" 2. Lake of Keswick, Cumberland England
" 3. Arch of rock on Cape Parry
" 4. The fresh water cave, Isle of Wight, England
" 5. The Blackgang cave on the Isle of Wight
" 6. Basalt rocks on the Island of Tahiti
" 7. The Grotto del Cane near Naples,
" 8. The grotto of Jupiter on the Island of Naxos, Greece

PLATE 52.

Fig. 1. Icebergs of the antarctic regions
" 2. Sandstone cave in Sicily
" 3. Lava arch in Iceland
" 4. Burning mountain near Duttweiler
" 5. Cascade of Teverone near Tivoli
" 6. Basalt columns on the Island of Staffa
" 7. Upright tree stems found in carboniferous formations
" 8. Island of Staffa with Fingal's cave
" 9. Section of the cavern of Gailenreuth in the Hartz

PLATE 53.

Fig. 1. The Grotte des Demoiselles in France
" 2. The Dunold Mill Cave, Lancashire England
" 3. The cave of Montserrat in Catalonia, Spain
" 4. Porphyry mountain near Kreuznach, Germany
" 5. The Lurley rock on the Rhine
" 6. The Rhonetrichter (*Perte du Rhone*), France
" 7. Serpentine rocks near Lizard Point Cornwall, England
" 8. Clay slate strata on the coast of Scotland
" 9. Submarine section of the Straits of Gibraltar
" 10. Do. do. between Tarifa and Alcazar, Spain
" 11–13. Illustrating the cleavage of rocks
" 14. Ill'g subordinate beds

BOTANY

[Including the translation of German words and phrases on pl. 73.]

PLATE 54.

Figs. 1–15. Jussieu's Classification of Plants, types of classes
" 16. Ustilago segetum
" 17. Uredo phaseoli
" 18. Tuber cibarium, truffle
" 19. Bovista gigantea
" 20. Morchella esculenta, morel
" 21. Polyporus perennis
" 22*a*. Boletus umbellatus
" 22*b*. Boletus edulis, edible mushroom
" 23. Hydnum auriscalpium
" 24. Cantharellus cibarius
" 25. Agaricus fimetarius
" 26. Agaricus campestris, champignon
" 27. Agaricus procerus
" 28. Agaricus muscarius, toadstool
" 29. Cladonia pyxidata
" 30. Parmelia parietina
" 31. Cetraria islandica, Iceland moss
" 32. Rochella tinctoria
" 33. Usnea florida
Fig. 34. *a*, Conferva bombycina; *b*, C. rivularis; *c*, C. flaccida; *d*, C. glomerata
" 35. Sphærococcus coronopifolius
" 36. Laminaria digitata
" 37. Laminaria saccharina
" 38. Fucus vesiculosus
" 39. Spagnum acutifolium, Peat-moss
" 40. Polytrichum juniperinum
" 41. Climacium dendroides
" 42. Anthoceros punctatus
" 43. Gymnoscyphus repens
" 44. Marchantia polymorpha
" 45. Pilularia globulifera
" 46. Salvinia natans
" 47. Marsilea quadrifolia
" 48. Ceterach leptophylla
" 49*a*. Polypodium vulgare
" 49*b*. Ceterach officinarum
" 50. Aspidium filix mas
" 51. Asplenium trichomanes
" 52. Scolopendrium officinarum
" 53. Lomaria spicans
" 54. Adiantum capillus veneris
Fig. 55. Osmunda regalis
" 56. Botrychium lunaris
" 57. Ophioglossum vulgatum
" 58. Lycopodium clavatum
" 59. Equisetum limosum

PLATE 55.

Fig. 1. Potamogeton natans
" 2. Arum maculatum, wake robin
" 3. Zostera marina
" 4. Acorus calamus
" 5. Typha latifolia, cat-tail
" 6. Eriophorum angustifolium
" 7. Cyperus officinalis
" 8. Saccharum officinale, sugar cane
" 9. Oryza sativa, rice
" 10. Bambusa arundinacea, bamboo-cane
Centre figure: Musa sapientum, banana tree (German: Banana oder Feigenbaum)

PLATE 56.

Fig. 1. Cocos nucifera, cocoa palm
" 2. Phœnix dactylifera, date palm
" 3. Sagus farinifera, sago palm
" 4. Cycas circinalis, sago tree
" 5. Dracæna braziliensis
" 6. Aloë arborescens
" 7. Areca catechu, areca palm
" 8. Zamia elliptica
" 9. Papyrus antiquorum

PLATE 57.

Fig. 1. Dracæna draco, dragon's blood
" 2. Luzula pilosa, woodrush
" 3. Commelyna tuberosa
" 4. Colchicum autumnale
" 5. Veratrum album, sneezewort
" 6. Fritillaria imperialis, crown imperial
" 7. Erythronium dens canis, dog's tooth
" 8. Agave americana, century plant
" 9. Phormium tenax, New Zealand flax
" 10. Scilla maritima, squill

PLATE 58, 59.

Fig. 1. Amaryllis formosissima, daffodil lily
" 2. Narcissus pseudo-narcissus, daffodil
" 3. Crocus sativus, saffron
" 4. Iris germanica, iris
" 5. Musa paradisiaca, plantain
" 6. Curcuma zedoaria, turmeric
" 7. Zingiber officinale, ginger
" 8. Vanilla aromatica, vanilla
" 9. Nymphæa lotus
" 10. Aristolochia clematis, birthwort
" 11. Aristolochia sipho, Dutchman's pipe
" 12. Aristolochia serpentaria, Virginia snake root
" 13. Elæagnus angustifolia
" 14. Daphne mezereum, mezereon
" 15. Camphora officinarum, camphor tree
" 16. Cinnamomum zeylanicum, cinnamon tree
" 17. Laurus nobilis, victor's laurel
" 18. Cinnamomum cassia, cassia tree

PLATE 60, 61.

Fig. 1. Protea speciosa, sugar-bush
" 2. Banksia serrata
" 3. Myristica moschata, nutmeg
" 4. Rheum palmatum, rhubarb
" 5. Beta vulgaris, common beet
" 6. Celosia cristata, cock's comb
" 7. Plantago major, common plantain
" 8. Mirabilis longifolia
" 9. Plumbago europæa, toothwort
" 10. Anagallis arvensis, pimpernel
" 11. Dodecatheon integrifolium, American cowslip
" 12. Cyclamen europæum, sow-bread
" 13. Lysimachia vulgaris, loose-strife
" 14. Veronica officinalis, speedwell
" 15. Pedicularis palustris, lousewort
" 16. Lathræa squamaria, tooth-wort

PLATE 62.

Fig. 1. Ruellia formosa
" 2. Acanthus mollis, bear's claw
" 3. Olea europæa, olive
" 4. Jasminum officinale
" 5. Vitex agnus castus
" 6. Betonica officinalis, betony
" 7. Galeopsis tetrahit, hemp nettle
" 8. Digitalis purpurea, purple foxglove
" 9. Calceolaria corymbosa, slipperwort
" 10. Verbascum thapsus, mullein
" 11. Nicotiana tabacum, tobacco

PLATE 63.

Fig. 1. Hyoscyamus niger, henbane
" 2. Datura stramonium, jimson weed
" 3. Atropa belladonna, belladonna
" 4. Solanum dulcamara, climbing nightshade
" 5. Capsicum annuum, cayenne pepper
" 6. Borago officinalis, borage
" 7. Exogonium purga, jalap plant
" 8. Polemonium cœruleum, Jacob's ladder
" 9. Jacaranda tomentosa
" 10. Bignonia leucoxylon
" 11. Gentiana pneumonanthe, common gentian
" 12. Spigelia marilandica, pinkroot
" 13. Nerium oleander, oleander

PLATE 64.

Fig. 1. Cynanchum vincetoxicum, celandine
" 2. Asclepias cornuti, swallowwort
" 3. Mimusops dissecta
" 4. Styrax benzoin, benzoin tree
" 5. Ledum palustre, marsh tea
" 6. Erica filamentosa, cape heath
" 7. Campanula trachelium
" 8. Lobelia fulgens
" 9. Lactuca virosa, poison lettuce
" 10. Cynara scolymus, artichoke
" 11. Carthamus tinctorius, safflower
" 12. Serratula tinctoria, saw wort
" 13. Tanacetum vulgare, tansy

PLATE 65.

Fig. 1. Artemisia absinthum, wormwood.
" 2. Dipsacus fallonum, cardoon
" 3. Rubia tinctoria, madder
" 4. Coffea arabica, coffee
" 5. Lonicera caprifolium, woodbine
" 6. Aralia nudicaulis, sarsaparilla
" 7. Œnanthe fistulosa, dead tongue
" 8. Cicuta virosa, water hemlock
" 9. Æthusa cynapium, fool's parsley
" 10. Chærophyllum temulum, cow parsley
" 11. Conium maculatum, hemlock
" 12. Sium latifolium, water parsley

PLATE 66.

Fig. 1. Clematis erecta, creeping climber
" 2. Anemone hortensis, garden anemone
" 3. Pulsatilla fratensis, pasque flower
" 4. Ranunculus acris, crowfoot
" 5. Adonis vernalis, bird's eye
" 6. Helleborus niger, black hellebore
" 7. Aquilegia vulgaris, columbine
" 8. Aconitum stœrkianum, helmet flower
" 9. Papaver somniferum, poppy
" 10. Chelidonium majus, celandine
" 11. Sinapis alba, white mustard
" 12. Capparis spinosa, caper plant
" 13. Reseda luteola, dyer's rocket
" 14. Paullinia pinnata, paulinia
" 15. Æsculus pavia, small buckeye

PLATE 67.

Fig. 1. Acer pseudo-platanus, white maple
" 2. Malpighia urens
" 3. Hypericum perforatum
" 4. Garcinia cambogia
" 5. Citrus medica, citron
" 6. Thea chinensis, tea plant
" 7. Swietenia mahogoni, mahogany tree
" 8. Vitis vinifera, wine grape
" 9. Geranium sanguineum, cranesbill
" 10. Oxalis acetocella, wood-sorrel
" 11. Gossipium herbaceum, cotton plant
" 12. Theobroma cacao, cacao tree

PLATE 68.

Fig. 1. Magnolia grandiflora
" 2. Illicium anisatum, star anise
" 3. Anona squamosa, anona
" 4. Cocculus lacunosus
" 5. Berberis vulgaris, barberry
" 6. Tilia grandiflora, lime tree
" 7. Bixa orellana, arnotto tree
" 8. Helianthemum vulgaris
" 9. Guaiacum officinale
" 10. Diptamus albus, white dittany
" 11. Dianthus caryophyllus, carnation
" 12. Saponaria officinalis, soapwort
" 13. Camellia japonica, camelia

PLATE 69.

Fig. 1. Sedum acre, stone crop
" 2. Saxifraga granulata, stone-break
" 3. Cereus hexagonus, six-edged cactus
" 4. Tamarix germanica, tamarisk
" 5. Mesembryanthemum rubrocinctum, fig-marigold
" 6. Œnothera biennis, evening primrose
" 7. Epilobium angustifolium, willow-herb
" 8. Santalum myrtifolium
" 9. Melaleuca cajeput, cajeput tree
" 10. Melaleuca fulgens, fiery cajeput
" 11. Eugenia pimenta, allspice
" 12. Caryophyllus aromaticus, clove-tree
" 13. Melastoma malabathricum

PLATE 70.

Fig. 1. Lythrum salicaria
" 2. Mespilus germanicus, medlar
" 3. Rosa moscata, musk rose
" 4. Potentilla anserina, silver weed
" 5. Amygdalus communis, sweet almond
" 6. Acacia vera, gum arabic tree
" 7. Cassia lanceolata, senna
" 8. Tamarindus indica, tamarind tree
" 9. Hæmatoxylon campechianum logwood
" 10. Coronilla varia, sicklewort
" 11. Phaseolus vulgaris, kidney bean
" 12. Genista tinctoria, dyer's greenwood
" 13. Indigofera anil, anil
" 14. Glycyrrhiza glabra

PLATE 71.

Fig. 1. Anacardium occidentale, cashew nut
" 2. Rhus cotinus, smoke tree
" 3. Pistacia terebinthus, turpentine tree
" 4. Juglans regia, English walnut
" 5. Euonymus europæus, spindle tree
" 6. Ilex aquifolium, European holly
" 7. Rhamnus catharticus, buckthorn
" 8. Euphorbia cyparissias, wolf's milk
" 9. Euphorbia officinarum, spurge
" 10. Siphonia elastica, caoutchouc tree
" 11. Cucumis citrullus, water melon
" 12. Momordica balsamina, balsam apple
" 13. Bryonia alba, white bryony
" 14. Carica papaya, West India pawpaw
" 15. Cucumis melo

PLATE 72.

Fig. 1. Ficus carica, fig tree
" 2. Artocarpus incisa, breadfruit
" 3. Morus nigra, black mulberry
" 4. Humulus lupulus, hops
" 5. Cannabis sativa, hemp
" 6. Piper nigrum, black pepper
" 7. Castanea vesca, chestnut
" 8. Quercus tinctoria, black oak
" 9. Liquidambar styraciflua, sweet gum
" 10. Taxus baccata, European yew
" 11. Juniperus communis, common juniper
" 12. Cupressus sempervirens, European cypress
" 13. Larix cedrus, European cedar

PLATE 73.

Fig. 1. Botanical chart
" 2. Chart of the Chinese and Indian cotton region
" 3. Chart of the American region of the sugar cane, coffee, &c.
" 4. Chart of the *vertical* distribution of plants in Asia
" 5. The same of America
" 6. The same of Europe
" 7. The same of the Canary Islands
" 8. The same of the frigid zone of Europe

GLOSSARY.

Ægypten, Egypt.
Aleuten, Aleutian Islands.
Algerien, Algeria.
Amazonenstrom, Amazon River.
Anden Gebirge, the Andes.
Arabien, Arabia.
Aral S., Lake Aral.
Asien, Asia.

Atlantischer Ocean, Atlantic Ocean.
Australien, Australia.
Azorische In., the Azores.
Bahama In., Bahama Islands.
Baumwolle, Cotton.
Bonin In., Bonin Islands.
Brasilien, Brazil.
Californien, California.
Canarische In., Canary Islands.
Cap der guten Hoffnung, Cape of Good Hope.
Cap Horn, Cape Horn.
Cap Verdische In., Cape Verde Islands.
Caraibisches M., Caribbean Sea.
Carolinen In., Caroline Islands.
Caspisches M., Caspian Sea.
Chinawälder, Bathbark forests.
Deutschland, Germany.
Falklands In., Falkland Islands.
Feuerland, Terra del Fuego.
Frankreich, France.
Freundschafts In., Friendly Islands.
Gallopagos In., Gallopagos Islands.
Gerste u. Hafer, barley and oats.
Gewürznelken, cloves.
Grönland, Greenland.
Grossbritannien, Great Britain.
Grosser oder Stiller Ocean, Pacific Ocean.
Habesch, Habesh,
Himmalaya Geb., Himmalaya Mountains.
I. Island, Iceland.
I. Karafta oder Sachalin, Island of Karafta or Sachalin.
Indisches Meer, Indian Ocean.
Irland, Ireland.
Kaffee, coffee.
Kaschmir, Cashmere,
Kurilen, Kurile Islands.
Ladronen In., Marian Islands.
Mais, Indian corn.
Malediven In., Maldive Islands.
Mandschurei, Manchooria.
Marañon od. Amazonenstrom, Amazon River.
Marianen od. Ladronen In., Marian Islands.
Mb. v. Mexico, Gulf of Mexico.
Mendana's Arch., Mendana's Archipelago.
Mittel Amer, Central America.
Mittelländisches Meer, Mediterranean Sea.
Mongolei, Mongolia.
Moskau, Moscow.
Muscatbaum, nutmeg tree.
N. Gran., New Grenada.
Neufundland, Newfoundland.
Neu Guinea, New Guinea.
Neu Seeland, New Zealand.
Nord Amerika, North America.
Norwegen, Norway.
Nubien, Nubia.
Ost Indien, East India.
Patagonien, Patagonia.
Pfeffer, pepper.
Reis, rice.
Roggen, rye.
Russisches Amerika, Russian America.
Russland, Russia.
Sandwichs In., Sandwich Islands.
Schiffer In., Navigators' Islands.
Schwarzes M., Black Sea.
Schweden, Sweden.
Sibirien, Siberia.
Sklaven S., Slave Lake.
Spanien, Spain.
Süd Amerika, South America.
Thee, tea.
Toisen, toises, (1t.=6 feet.)
Türkei, Turkey.
Vanille, vanilla.
Vereinigte Staaten, United States.
Vulc. v. Aconcagua, Volcano of Aconcagua.
Weitzen, wheat.
West Indien, West Indies.
Wüste Sahara, Desert of Sahara.
Wüste Schamo oder Gobi, Desert of Shamo or Gobi.
Zeichenerklärung für fig. 1, 2, *u.* 3., Explanation of the marks in figs. 1, 2, and 3.
Zimmt, cinnamon.
Zucker, sugar.

ZOOLOGY

[Including Glossary to plate 105 *b*.]

PLATE 74.

Classification.

Fig. 1. Monas, animalcule
" 2. Monocerca, do.
" 3. Spongia, sponge
" 4. Hydra, naked polypi
" 5. Diphies, medusa
" 6. Velella, nautilus
" 7. Sipunculus, tube worm
" 8. Holothuria, sea jelly
" 9. Ligula, strapworm
" 10. Nemertes, cordworm
" 11. Botryllus, sea grape
" 12. Ostrea, oyster
" 13. Chiton, sea cockroach
" 14. Fissurella, fissure shell
" 15. Vermetus, worm shell
" 16. Strombus, screw shell
" 17. Carina, keel shell
" 18. Bulla, wood digger
" 19. Phyllidium, leaf shell
" 20. Tritonium, triton
" 21. Limnæus, mud shell
" 22. Clio, whale louse
" 23. Octopus, polypus
" 24. Hirudo, leech
" 25. Aphrodite, sea mouse
" 26. Amphitrite
" 27. Anatifa, duck muscle
" 28. Cancer, crab
" 29. Squilla, shrimp
" 30. Talitrus, sea flea
" 31. Cyamus, whale louse
" 32. Oniscus, cheslip
" 33. Cyclops, water flea
" 34. Caligus, fish louse
" 35. Aranea, spider
" 36. Chelifer,
" 37. Scolopendra
" 38. Lepisma, book worm
" 39. Pediculus, louse
" 40. Pulex, flea
" 41. Carabus, ground beetle
" 42. Forficula, earwig
" 43. Cimex, bedbug
" 44. Libellula, dragon fly
" 45. Tenthredo, tailed wasp
" 46. Vanessa, butterfly
" 47. Stomoxys, autumn fly
" 48. Petromyzon, lampereel
" 49. Squalus, shark
" 50. Raja, ray
" 51. Acipenser, sturgeon
" 52. Orthagoriscus, moon fish
" 53. Hippocampus, sea horse
" 54. Anguilla, eel
" 55. Pleuronectes, sole
Fig. 56. Merlangus, whiting
" 57. Cyprinus, chub
" 58. Xiphias, swordfish
" 59. Salamandra, salamander
" 60. Rana, frog
" 61. Vipera, viper
" 62. Boa, boa constrictor
" 63. Anguis, adder
" 64. Ophisaurus
" 65. Chirotes
" 66, 67. Chalcides
" 68. Bipes
" 69. Anolis
" 70. Scincus
" 71. Tilicua
" 72. Chamæleo, chameleon
" 73. Ptyodactylus
" 74. Basiliscus, basilisk
" 75. Iguana, guana
" 76. Draco, dragon
" 77. Agama
" 78. Stellio, stellion
" 79. Lacerta, lizard
" 80. Tejus
" 81. Crocodilus, crocodile
" 82. Plesiosaurus
" 83. Ichthyosaurus
" 84. Chelonia, sea tortoise
" 85. Testudo, land tortoise
" 86. Anas, duck
" 87. Sula, dodo
" 88. Pelecanus, pelican
" 89. Procellaria, petrel
" 90. Podiceps, diver
" 91. Phœnicopterus, flamingo
" 92. Rallus, water rail
" 93. Scolopax, snipe
" 94. Ardea, heron
" 95. Grus, crane
" 96. Otis, bustard
" 97. Struthio, ostrich
" 98. Gallus, cock
" 99. Crax, curassow
" 100. Columba, pigeon
" 101. Psittacus, parrot
" 102. Picus, woodpecker
" 103. Buceros, rhinoceros bird
" 104. Merops, bee-eater
" 105. Sitta, nut-pecker
" 106. Alauda, lark
" 107. Cypselus, sea swallow
" 108. Pica, magpie
" 109. Otus, owl
" 110. Vultur, vulture
" 111. Milvus, kite
" 112. Balæna, whale
" 113. Cervus, deer
Fig. 114. Bos, ox
" 115. Camelus, camel
" 116. Equus, horse
" 117. Sus, hog
" 118. Elephas, elephant
" 119. Ornithorhynchus, duckbilled platypus
" 120. Manis, pangolin
" 121. Bradypus, sloth
" 122. Lepus, rabbit
" 123. Sciurus, squirrel
" 124. Castor, beaver
" 125. Mus, mouse
" 126. Didelphys, kangaroo
" 127. Phoca, seal
" 128. Mustela, marten
" 129. Viverra, civet
" 130. Felis, cat
" 131. Hyæna, hyena
" 132. Canis, dog
" 133. Ursus, bear
" 134. Erinaceus, hedgehog
" 135. Vespertilio, bat
" 136. Pteropus, rousset
" 137. Lemur, maki
" 138. Hapale, ouistiti
" 139. Simia, monkey
" 140. Homo, man

PLATE 75.

Fig. 1. Monas lens
" 2. Proteus diffluens
" 3. Bursaria vesiculosa
" 4. Cryptomonas ovata
" 5. Trachelocerca olor
" 6. Trachelocerca viridis
" 7. Vibrio anguillula
" 8. Cyclidium glaucoma
" 9. Paramecium compressum
" 10. Chilodon cucullatus
" 11. Gonium pectorale
" 12. Bursaria truncatella
" 13. Urocentrum turbo
" 14. Trichodina cometa
" 15. Volvox globator
" 16. Rotifer vulgaris
" 17. Opercularia articulata
" 18. Stentor mylleri
" 19. Melicerta ringens
" 20. Carchesium polypinum
" 21. Hydra grisea
" 22. Hydra fusca
" 23. Hydra viridis
" 24. Virgularia juncea
" 25. Pennatula granulosa
" 26. Pennatula phosphorea
" 27. Pennatula grisea

Figs. 28, 29. Tubularia coronata
" 30. Sertularia polyzonalis
" 31. Sertularia falcata
" 32. Thuaria thuia
" 33. Sertularia abietina
" 34. Sertularia operculata
" 35. Corallina rubens
" 36. Flabellaria opuntia
" 37. Penecillus penecillus
" 38. Corallina officinalis
" 39. Tubularia sultana
" 40. Tubularia campanularia
" 41. Acetabulum mediterraneum
" 42. Tubularia indivisa
" 43. Flustra foliacea
" 44. Spongia fistularis
" 45. Spongia officinalis
" 46. Alcyonium ficiforme
" 47. Alcyonium palmatum
" 48. Gorgonia flabellum
" 49. Gorgonia verrucosa
" 50. Gorgonia cerataphyta
" 51. Antipathes spiralis
" 52. Coralium nobile
" 53. Isis hippuris
" 54. Flustra
" 55. Pocillopora polymorpha
" 56. Retepora
" 57. Eschara
" 58. Oculina virginea
" 59. Oculina gemmascens
" 60. Madrepora prolifera
" 61. Porites porites
" 62. Astrea astroites
" 63. Explanaria ananas
" 64. Meandrina labyrinthica
" 65. Fungia fungites
" 66. Tubipora musica
" 67. Teredo navalis
" 68. Sabella ventilabrum
" 69. Vermetus lumbricalis
" 70. Serpula glomerata
" 71. Aspergillum javanum
" 72. Serpula arenaria
" 73. Dentalium elephantinum
" 74. Dentalium politum
" 75. Dentalium entalis
" 76. Patella granatina
" 77. Patella saccharina
" 78. Calyptræa sinensis
" 79. Patella laciniosa
" 80. Fissurella græca
" 81. Ancylus lacustris
" 82. Emarginula fissura
" 83. Pileopsis hungarica (Trochidæ)
" 84. Neritina crepidularia
" 85. Patella vulgata
" 86. Haliotis tuberculata
" 87. Neritina fluviatilis
" 88. Natica canrena
" 89. Sigaretus haliotideus
" 90. Melania amarula
" 91. Limnea auricularia
" 92. Limnea stagnalis
" 93. Bulimus decollatus fasciatus
" 94. Bulimus decollatus albus
" 95. Paludina vivipara
" 96. Janthina ianthina
" 97. Helix nemoralis
" 98. Helix pomathia
" 99. Turbo nautileus
" 100. Clausilia perversa
" 101. Scalaria dathrus
" 102. Scalaria scalaris
" 103. Turbo cochlus
" 104. Delphinula delphinus
" 105. Telescopium indicator
" 106. Trochus solaris
" 107. Trochus magus
" 108. Solarium perspectivum
" 109. Cerithium vertagus
" 110. Pleurotoma babylonia
" 111. Murex ramosa
" 112. Murex haustellum
" 113. Murex tribulus
" 114. Strombus lentiginosus
" 115. Pterocera chiragra
" 116. Rostellaria rectirostris
" 117. Terebra maculata
" 118. Purpura lapillus
" 119. Harpa ventricosa
" 120. Baccinum undatum
" 121. Cassidoria echinophora
" 122. Oliva ispidula
" 123. Mitra episcopalis
" 124. Mitra papalis
" 125. Oliva porphyria
" 126. Oliva maura

PLATE 76.

Fig. 1. Auricula midæ
" 2. Ovula volva
" 3. Bulla physis
" 4. Ovula ovum
" 5. Cypræa moneta
" 6. Cypræa mauritiana
" 7. Cypræa arabica
" 8. Conus aurantiacus
" 9. Conus summus
" 10. Conus cedo nu
" 11. Conus textillis
" 12. Conus marmore
" 13. Spirula spirula
" 14. Nautilus beccari
" 15. Nautilus calcar
" 16. Nautilus pompilius
" 17. Argonauta argo
" 18. Pinna obeliscus
" 19. Pinna rudis
" 20. Avicula margaritifera
" 21. Mytilus cygneus
" 22. Mytilus bidens
" 23. Terebrata caput serpentis
" 24. Crania craniolaris
" 25. Ostrea cristigalli
" 26. Malleus malleus
" 27. Pecten corallinus
" 28. Pecten pallium ducale
" 29. Pecten maximus
" 30. Perna ephippium
" 31. Arca senilis
" 32. Arca noæ
" 33. Tridacna gigas
" 34. Isocardia corallina
" 35. Spondylus gæderopus
" 36. Venus mercenaria
" 37. Cytherea dione
" 38. Donax rugosa
" 39. Donax scripta
" 40. Mactra solidissima
" 41. Hemicardium cardissa
" 42. Cardium echinatum
" 43. Cyclas
" 44. Tellina radiata
" 45. Solen vagina
" 46. Solen siliqua
" 47. Unio complanatus
" 48. Mya pictorum
" 49. Pholas pusillus
" 50. Pholas dactylus
" 51. Pollicipes mitella
" 52. Lepas anatifa
" 53. Coronula
" 54. Balanus psittacus
" 55. Chiton squamosus
" 56 Holopus rangii
" 57. Encrinus radiatus
" 58. Pentacrinus osteria
" 59. Oreaster turritus
" 60. Asterias aurantiaca
" 61. Astrogonium granulare
" 62. Ophiolepis scolopendrina
" 63. Astrophyton caput medusæ
" 64. Ophiurus asterias
" 65. Stellonia rubens
" 66. Solaster papposa
" 67. Spatangus purpureus
" 68. Clypeaster rosacea
" 69. Echinus cidaris
" 70. Cidaris diadema
" 71. Cidaris esculentus
" 72. Lucernaria quadricornis
" 73. Thaumantias cymbaloidea
" 74. Aurelia aurita
" 75. Octopus octopodius
" 76. Loligo loligo
" 77. Sepia officinalis
" 78. Clio borealis
" 79. Scyllæa pelagica (Tritoniidæ)
" 80. Lernæa branchialis
" 81. Lernæa cyprinacea
" 82. Terebella conchilega
" 83. Porpita nuda
" 84. Velella spirans
" 85. Cucumaria frondosa

PLATE 77

Fig. 1. Physalia physalis
" 2. Thetis fimbria
" 3. Salpa maxima
" 4. Pedicellaria
" 5. Actinia undata
" 6. Actinia senilis
" 7. Ascidia lepadiformis
" 8. Ascidia venosa
" 9. Nais serpentina
Fig. 10. Nais proboscidea
" 11. Phyllodoce stellifera
" 12. Nereis tubicola
" 13. Amphitrite reniformis
" 14. Spio filicornis
" 15. Aphrodite aculeata
" 16. Doris papillosa
" 17. Doris argo
" 18. Aplysia depilans
" 19. Limax agrestis
" 20. Limax empiricorum
" 21. Limax fuscus
" 22. Malacobdella grossa
" 23. Nephelis octoculata (Hirudinidæ)
" 24. Clepsina complanata
" 25. Hæmopis vorax
" 26. Hirudo officinalis
" 27. Sipunculus saccatus
" 28. Sipunculus nudus
" 29. Echinococcus veterinorum suis
" 30. Cysticercus cellulosæ
" 31. Tænia cateniformis
" 32. Bothriocephalus latus
" 33. Tænia solium
" 34. Ligula cingulum
" 35. Planaria cornuta
" 36. Distoma hepaticum
" 37. Lumbricus variegatus
" 38. Lumbricus terrestris
" 39. Echinorhynchus gigas
" 40. Trichocephalus dispar
" 41. Ascaris lumbricoides
" 42. Oxyuris vermicolaris
" 43. Filaria medinensis
" 44. Lumbricus aquaticus
" 45. Filaria papillosa
" 46. Argas fischeri
" 47. Argas savignii
" 48. Chelifer beauvoisii
" 49. Galeodes phalangium
" 50. Galeodes araneoides
" 51. Scorpio europæus
" 52. Buthus afer
" 53. Epeira imperialis
" 54. Tetragnatha argyra
" 55. Theridion denticulatum
" 56. Argyroneta aquatica
" 57*a*. Uloborus walcnærius
" 57*b*. Gasteracantha armata
" 58. Mygale avicularia
" 59. Epeira diadema
" 60. Pallene brevirostris (Pycnogonidæ)
" 61. Chelifer cancroides
" 62. Phalangium opilio
" 63. Phalangium ægyptiacum
" 64. Hydrachna abstergens.
" 65. Hydrachna despiciens
" 66. Ixodes annulatus
" 67. Ixodes americanus
" 68. Sarcoptes scabiei
" 69. Gamasus coleopterorum
" 70. Acarus siro
" 71. Ixodes orbiculatus
" 72. Pulex penetrans
" 73. Pulex irritans
" 74, 75. Nirmidæ
" 76. Phthirius pubis
" 77. Pediculus capitis
" 78. Podura nivicola
" 79. Podura villosa
" 80. Smynthurus fimetarius
" 81, 84. Machilus polypoda
" 82, 83. Lepisma saccharina
" 84. See 81.
" 85. Melophagus ovis
" 86. Hippobosca equina
" 87. Bombylius major
" 88. Dioctria ater
" 89. Asilus cabroniformis
" 90. Conops macrocephala
" 91. Stomoxis calcitrans
" 92, 93. Empidæ
" 94. Anopheles bifurcatus
" 95. Culex nemorosus
" 96. Culex pipiens
" 97. Tabanus tropicus
" 98. Tabanus bovinus
" 99—109. Muscidæ
" 110. Clitellaria ephippium
" 111. Volucella pellucens
" 112. Scæva pyrastri
" 113. Crysotoxum vespiformis
" 114. Eristalis tenax
" 115. Helophilus pendulus
" 116. Leptis vermileo
" 117. Stratiomys chamæleon
" 118. Chironomus plumosus

Fig. 119. Anisomera nigra
" 120. Ctenophora elegans
" 121. Psychoda phalænoides
" 122. Mycetophila cericea
" 123. Bibio marci
" 124. Mycetophila mirabilis
" 125. Cephalemyia ovis
" 126, 127. Gasterophilus equi
" 128. Œstrus tarandi
" 129. Œstrus bovis

PLATE 78.

Fig. 1, 2. Cancerinæ
" 3. Thalamita natator
" 4. Gecarcinus lateralis
" 5. Gelasimus annulipes
" 6. Ocypoda arenaria
" 7. Philyra scabriuscula
" 8. Dorippe sima
" 9. Ranina serrata
" 10. Pagurus diogenes
" 11. Palinurus guttatus
" 12. Scyllarus æquinoctialis
" 13. Stenopus hispidus
" 14. Callianassa uncinata
" 15. Squilla maculata
" 16. Gonodactylus stylifer
" 17. Phyllosoma stylicornis
" 18. Orchestia fischeri
" 19. Ancylomera hunteri
" 20. Cymadocea armata
" 21. Sphæroma serratum
" 22. Porcellio granulatus
" 23. Caprella acuminifera
" 24. Cyamus ovalis
" 25. Aprus cancriformis
" 26. Branchipus pisciformis
" 27. Cyclops communis
" 28. Phyllophora cornuta
" 29. Ergasilus sieboldi
" 30. Lernæa polycolpus
" 31. Achtheres percarum
" 32. Pycnogonum litterale
" 33. Nymphon gracile
" 34. Limullus moluccanus
" 35. Nemesia cellicola
" 36. Segestria perfida
" 37. Lycosa tarentula
" 38. Lycosa melanogaster
" 39. Hersilia caudata
" 40. Chersis savinii
" 41. Salticus formicarius
" 42. Eripus heterogaster
" 43. Arcys lanceolarius
" 44. Latrodectus malmignatus
" 45. Nyssa timida
" 46. Tegenaria domestica
" 47. Lachesis perversa
" 48. Uloborus walcnærius
" 49. Argyroneta aquatica

PLATE 79.

Fig. 1. Mutilla europæa
" 2. Apterogyna occidentalis
" 3—9. Formicidæ
" 10. Bombus lapidarius
" 11. Bombus muscorum
" 12. Megachile sementaria
" 13. Bombus terrestris
" 14. Xylocopa violacea
" 15. Nomada variegata
" 16. Eucera longicornis
" 17. Megachile centuncularis
" 18. Apis mellifica
" 19. Vespa maculata
" 20. Polistes paritum
" 21. Vespa vulgaris
" 22. Vespa crabro
" 23. Leucospis dorsigera
" 24. Chrysis cyanea
" 25. Chrysis aurata
" 26. Chrysis ignita
" 27. Pompilus cœruleus
" 28. Pompilus viaticus
" 29. Crabro cribrarius
" 30. Pelopæus spirifex
" 31. Ammophila sabulosa
" 32, 33, 36—39. Ichneumonidæ
" 34, 35. Chalcididæ
" 36—39. See 32
" 40. Rhyssa persuasoria (Ichneumonidæ)
" 41. Urocerus spectrum
" 42. Urocerus gigas
" 43. Lophyrus juniperi
" 44. Nematus capreæ
" 45. Cimbex americana
Fig. 46. Tenthredinidæ
" 47. Cimbex variabilis
" 48—53. Cinipidæ
" 54. Mantispa pagana
" 55. Raphidia ophiopsis
" 56. Termes fatalis
" 57. Bittacus tipularia
" 58. Panorpa communis
" 59. Ascalaphus barbarus
" 60. Myrmeleon libelluloides
" 61. Myrmeleon formicarius
" 62—64. Hemerobiidæ
" 65. Hydropsyche plumosa
" 66. Phryganea striata
" 67. Perla bicaudata
" 68. Limnophilus rhombica
" 69. Phryganea grandis
" 70—72. Ephemeridæ
" 73. Agrion puella
" 74. Calepteryx virgo
" 75. Libellula depressa
" 76. Æschna grandis
" 77. Pterophorus pentadactyla
" 78. Coccyx resinosa
" 79. Carpocapsa pomonella
" 80. Hercyna paliotalis
" 81. Tinea granella
" 82. Plutella xylostella
" 83. Gallerea cereana
" 84. Tinea pellionella
" 85. Lemmatophila salicella
" 86. Hyponomeuta evonymella
" 87. Hyponomeuta pedella
" 88. Hypena rostralis
" 89. Botys verticalis
" 90. Sciaphila literata
" 91. Tortrix viridana
" 92. Halias prasinana
" 93—102. Geometridæ
" 103—123, 129, 136. Noctuidæ
" 124. Callimorpha jacobææ
" 125. Arctia fuliginosa
" 126. Arctia matronula
" 127. Arctia dominula
" 128. Pygæra bucephala
" 129. See 103
" 130—135, 137—151. Bombycidæ
" 136. See 103

PLATE 80.

Fig. 1. Lasiocampa quercus
" 2. Dendrolimus pini
" 3. Odonestis potatoria
" 4—7. Bombycidæ
" 8. Zygæna filipendula
" 9. Sesia fuciformis
" 10—21. Sphingidæ
" 22. Hesperia malvarum
" 23—50. Papilionidæ
" 51. Thrips phisapus
" 52. Lecanium illicis
" 53. Lecanium hesperidum
" 54. Coccus cacti
" 55. Psylla alni
" 56. Dorthesia urticea
" 57, 59. Aphidæ
" 58. Chermes ulmi
" 59. See 57
" 60. Pyrrhocoris apterus
" 61. Coryzus hyoscyami
" 62. Cydnus biguttatus
" 63. Pentatoma juniperinus
" 64. Pentatoma baccarum
" 65. Acanthosoma hæmorrhoidalis
" 66. Cimex rufipes
" 67. Acanthia lectularia
" 68. Ranatra linearis
" 69. Nepa cinerea
" 70. Neucoris cimicoides
" 71. Belostoma grandis
" 72. Corixa striata
" 73. Notonecta glauca
" 74. Aphrophora spumaria
" 75. Cicada fraxini
" 76. Fulgora candelaria
" 77. Fulgoria laternaria
" 78. Œdipoda stridula
" 79. Œdipoda cærulescens
" 80. Œdipoda migratoria
" 81. Acridium cristatum
" 82—85. Locustidæ
" 86—88. Gryllidæ
" 89. Phyllium siccifolium
" 90. Mantis religiosa
" 91. Empusa gongylodes
" 92. Cyphocrana gigas
" 93. Blatta orientalis

PLATE 81.

Upper Division.

Fig. 1—3. Forficulidæ
" 4—6. Brachelytra
" 7. Meloë proscarabæus
" 8. Lytta vesicatoria
" 9. Blaps mortisaga
" 10. Tenebrio molitor
" 11—16. Carabidæ
" 17. Dyticus marginalis
" 18. Hydrophilus piceus
" 19—21. Mordellidæ
" 22. Buprestis mariana
" 23. Euchroma gigas
" 24. Buprestis chrysostigma
" 25. Elaphrus riparius (Carabidæ)
" 26—29. Cicindela
" 30—33. Elateridæ
" 34. Malachius æneus
" 35—38. Cantharis
" 39. Lycus sanguineus.
" 40. Pyrochroa coccinea
" 41, 42. Lampyridæ
" 43. Necydalis cærulea
" 44—61. Longicornia
" 62—64. Clerida
" 65—76. Rhincophora
" 77—79. Bruchus
" 80—101. Trimera
" 102, 103. Nitidulidæ
" 104. Silpha thoracica
" 105, 106. Necrophorus
" 107, 108. Anthrenus
" 109. Byrrhus pilula (Byrrhidæ)
" 110. Gyrinus natator
" 111. Ptinus fur.
" 112. Anobium pertinax
" 113. Apate capucinus
" 114. Hylesinus piniperda
" 115, 116. Bostrichus
" 117, 118. Dermestes
" 119—123. Hister
" 124. Lucanus
" 125—127. Cetoniadæ
" 128—131. Melolonthidæ
" 132, 136—144. Scarabæidæ
" 133—135. Aphodiidæ
" 136—144. See 132
" 145—148. Dynastidæ

Lower Division.

Fig. 1. Alosa vulgaris, shad
" 2. Clupea harengus, herring
" 3. Engraulis encrasicholus, anchovy
" 4. Harengula sprattus, sprat
" 5. Thymallus vulgaris, greyling
" 6. Cobitis fossilis, mudfish
" 7. Exoglossum barbatula, groundling
" 8. Rhombus vulgaris, brill
" 9. Platessa flesus, fluke
" 10. Zeus faber, dory
" 11. Lota vulgaris, burbot
" 12. Trachinus draco, sea cat
" 13. Xiphias gladius, swordfish
" 14. Amodytes tobianus, sand-eel
" 15. Syngnathus ophidion, sea serpent
" 16. Syngnathus acus, sea-needle
" 17. Pegasus draco, sea-dragon
" 18. Lumpus anglorum, lump sucker
" 19. Diodon punctatus, sticklebag
" 20. Tetraodon lagocephalus, star belly
" 21. Centriscus scolopax, trumpet fish
" 22. Aluteres monoceros, unicorn fish
" 23. Lophius piscatorius, widegap
" 24. Acipenser ruthenus, caviar sturgeon
" 25. Acipenser sturio, sturgeon
" 26. Sphyrna malleus, hammer-headed shark
" 27. Spinax acanthias, thorn-hound
" 28. Petromyzon fluviatilis, lamprey
" 29. Seps chalcides, bronze colored lizard
" 30. Scincus cyanurus, scink
" 31. Plestiodon aldrovandii
" 32. Triton cristatum, eft
" 33. Draco dandini, flying dragon
" 34. Rana temporaria, yellow frog
" 35. Hyla viridis, tree or green frog
" 36. Bufo calamita, toad
" 37. Bombinator igneus, glistering toad
" 38. Pelobatus fuscus, water toad
" 39. Testudo geometrica, tortoise
" 40. Testudo græca, mosaic tortoise
" 41. Cistudo europæa, European tortoise

Fig. 11. Trochilus minimus, smallest humming bird
" 12. Trochilus cristatus, crested humming bird
" 13. Trochilus macrourus, swallow tailed humming bird
" 14. Buphaga africana, beefeater
" 15. Sturnella ludoviciana, American lark
" 16. Icterus baltimore, Baltimore oriole

PLATE 100.

Fig. 1. Sturnus vulgaris, starling
" 2. Loxia pityopsittacus, parrot crossbill
" 3. Coccothraustes vulgaris, hawfinch
" 4*ab*. Fringilla canaria, canary bird
" 5. Fringilla cannabina, red poll
" 6. Fringilla domestica, house-sparrow
" 7. Sylvia hippolais, willow wren
" 8. Oriolus galbula, oriole
" 9. Merula vulgaris, blackbird
" 10. Merula saxatilis, rock thrush
" 11. Pastor roseus, rose colored starling
" 12. Merula viscivorus, missel thrush
" 13. Merula iliaca, redwing
" 14. Merula musica, song thrush

PLATE 101.

Fig. 1. Trochilus granatinus, garnet humming bird
" 2. Pyrrhula europea, bullfinch
" 3. Spermophila crassirostris, thick-billed finch
" 4. Coccothraustes chloris, green finch
" 5. Fringilla senegala, senegal finch
" 6. Fringilla amaduva, amaduvat
" 7. Vidua regia, king widow bird
" 8. Vidua erythrorhynchus, red-billed widow bird
" 9. Carduelis spinus, siskin
" 10. Carduelis elegans, goldfinch
" 11. Calliste tatao, seven-colored tanager
" 12. Fringilla cœlebs, chaffinch
" 13. Emberiza hortulana, ortolan
" 14. Emberiza schœniculus, reed bunting
" 15. Emberiza citrinella, yellowhammer
" 16. Rupicola aurantia, cock of the rock
" 17. Muscicapa grisola, grey flycatcher
" 18. Muscicapa albicollis, white-necked flycatcher
" 19. Muscicapa regia, king of the flycatchers
" 20. Ampelis carnifex, scarlet-crested chatterer
" 21. Ampelis cortinga, banded chatterer

PLATE 102.

Fig. 1. Trochilus albicollis, white-necked humming bird
" 2. Certhia familiaris, brown creeper
" 3. Parus cristatus, crested tit
" 4. Parus major, titmouse
" 5. Alauda calandria, field lark
" 6. Alauda cristata, crested lark
" 7. Alauda arvensis, skylark
" 8. Acanthiza campestris, hedge warbler
" 9. Cæreba cyanea, blue creeper
" 10, 11. Phœnicura, redstarts
" 12. Phœnicura suecica, blue throated Swedish redstart
" 13. Erythaca rubecula, robin
" 14. Menura superba, lyre bird
" 15. Dicrurus forficatus, fork-tailed drongo
" 16. Cracticus varius, black and white shrike
" 17. Vauga curvirostris, hooked-bill shrike
" 18. Laniarius barbarus, Barbary shrike
" 19. Laniarius collurio, red-backed shrike
" 20. Laniarius exubitor, butcher bird

PLATE 103.

Fig. 1. Buceros erythrorhynchus, red-billed hornbill
Fig. 2. Buceros rhinoceros, rhinoceros hornbill
" 3. Halcyon atricapilla, black capped kingfisher
" 4. Ceryle javanicus, Java kingfisher
" 5. Alcido ispida, common kingfisher
" 6. Caprimulgus europæus, goatsucker
" 7. Hirundo rustica, swallow
" 8. Cypselus melba, black martin
" 9. Tyrannus severus, Cayenne flycatcher
" 10. Milvulus forficatus, South Amer. flycatcher
" 11. Parus ater, pine tit
" 12. Regulus auricapillus, golden crested wren
" 13. Philomela luscinia, nightingale
" 14. Curruca hortensis, garden warbler
" 15. Curruca atricapilla, blackcap warbler
" 16. Phœnicura ruticilla, common redstart
" 17. Motacilla boarula, grey wagtail
" 18. Saxicola œnanthe, wheatear

PLATE 104.

Fig. 1. Strix flammea, European barn owl
" 2. Otus brachyotus, short eared owl
" 3. Otus wilsonianus, long-eared owl
" 4. Bubo maximus, grand duke
" 5. Falco peregrinus, wandering falcon
" 6. Falco œsalon, merlin
" 7. Milvus regalis, kite
" 8. Buteo vulgaris, common buzzard
" 9. Gyps fulvus, griffin vulture
" 10. Sarcoramphus papa, king vulture
" 11. Sarcoramphus gryphus, condor

PLATE 105.

Fig. 1. Aquila chrysaëtos, golden eagle
" 2. Archibuteo niger, black buzzard
" 3. Pandion ossifragus, osprey
" 4. Haliaëtus albicilla, sea eagle
" 5. Astur palumbarius, goshawk
" 6. Accipiter nisus, sparrow hawk
" 7. Falco subutes, hobby
" 8. Falco alaudarius, kestril

PLATE 105, *b*.

Chart of the migrations of fishes and birds

(*Weltkarte in Mercator's Projection zur Uebersicht der jährlichen Wanderungen der hauptsächlichsten Gattungen der Fische und Vögel*)

GLOSSARY.

Aleuten In., Aleutian Islands.
Amazonenstrom, Amazon River.
Arabien, Arabia.
Arabisches M., Arabian Sea.
Asien, Asia.
Atlantischer Ocean, Atlantic Ocean.
Behringsstrasse, Behring's Straits.
Baffins Meer, Baffin's Bay.
Californien, California.
Canarische In., Canary Islands.
Cap der guten Hoffnung, Cape of Good Hope.
Capstadt, Capetown.
Cap Verds In., Cape Verde Islands.
Caspisches Meer, Caspian Sea.
Donau, Danube.
Felsen Geb., Rocky Mountains.
Fensterschwalben, Domestic swallows.
Feuerland, Terra del Fuego.
Freundschafts In., Friendly Islands.
Gesellschafts In., Society Islands.
Gr. Bären See, Great Bear Lake.
Grönland, Greenland.
Grossbritannien, Great Britain.
Grosser Ocean, Pacific.
Häringe, Herrings.
Hudsons Meer, Hudson's Bay.
I. Melville, Melville Island.
Indisches Meer, Indian Ocean.
Lissabon, Lisbon.
Makrelen, Mackerel.
Meerb. v. Bengalen, Bay of Bengal.
Meerb. v. Mexico, Bay of Mexico.
Mongolei, Mongolia.
Neuseeland, New Zealand.
Neu Sibirien, New Siberia.
Nord Amerika, North America.
Nördliches Eismeer, Arctic Sea.
Patagonien, Patagonia.
Raben u. Kräken, Ravens and crows.
Rauchschwalben, Barn swallows.
Russisch Amerika, Russian America.
Schiffer In., Navigators' Islands.
Schleiereulen, Barn owls.
Schwarzes Meer, Black Sea.
Sibirien, Siberia.
Sklavensee, Slave Lake.
Staare u. Amseln, Starlings and blackbirds.
Süd Amerika, South America.
Uferschwalben, Bank swallows.
Vereinigte Staaten, United States.
Versammlungs- u. Abzugspunkt, Place of meeting and departure.
Wachteln, Quails.
Warschau, Warsaw.
West Indien, West Indies.
Wien, Vienna.
Wüste Sahara, Desert of Sahara.
Zeichenerklärung, Explanation of the marks.

PLATE 106.

Fig. 1. Balæna mysticetus, Greenland whale
" 2, 3. Physeter macrocephalus, sperm whale
" 4. Delphinus delphis, dolphin

PLATE 107.

Fig. 1, 2. Cervus elaphus, stag
" 3. Cervus dama, fallow deer
" 4. Cervus capreolus, roebuck
" 5. Lepus timidus, hare
" 6. Lepus cuniculus, rabit
" 7, 8. Bos taurus, common ox
" 9. Ovis aries, sheep
" 10. Capra hircus, goat
" 11. Sciurus vulgaris, squirrel

PLATE 108.

Fig. 1. Camelus bactrianus, two-humped camel
" 2. Camelus dromedarius, dromedary
" 3. Camelopardalis girafa, giraffe
" 4. Moschus moschiferus, musk
" 5. Antilope dorcas, gazelle
" 6. Antilope redunca, antilope of Senegal

PLATE 109.

Fig. 1. Bison americanus, buffalo
" 2. Bos bubalus, Indian buffalo
" 3. Rupicapra tragus, chamois
" 4. Antilope scripta, Cape-elk
" 5. Auchenia alpaca, paco
" 6. Auchenia lama, lama
" 7. Auchenia vicunna, vicunna
" 8, 9. Tarandus furcifer, reindeer
" 10. Cervus alces, eland
" 11. Siberian horse

PLATE 110.

Fig. 1. Equus zebra, zebra
" 2. Equus asinus, ass
" 3. Equus mulus, mule
" 4. Equus caballus, horse
" 5. Norman team horse
" 6. Arabian mare and colt
" 7. Arabian stallion

PLATE 111.

Fig. 1. Tapir indicus, tapir
" 2. Hyrax syriacus, daman
" 3. Dicotyle labiatus, peccary
" 4. Porcus babyrussa, Asiatic hog
" 5. Sus domesticus, domestic hog
" 6. Sus scropha, wild hog
" 7. Phacochœrus æthiopicus, wart hog
" 8. Hippopotamus amphibius, hippopotamus
" 9. Elephas indicus, Indian elephant

PLATE 112.

Fig. 1. Ornithorhynchus anatinus, duck-billed platypus
" 2. Echidna aculeata, porcupine ant-eater
" 3. Manis pentadactyla, pangolin
" 4. Myrmecophaga didactyla, two-toed ant-eater
" 5. Myrmecophaga jubata, great ant-eater
" 6. Chlamydophorus truncatus, Chili armadillo

Fig. 7. Dasypus sexcinctus, six-girdled armadillo.
" 8. Dasypus novemcinctus girdled armadillo,
" 9. Bradypus didactylus, sloth
" 10. Bradypus tridactylus, aï
" 11*ab*. Halmaturus laniger, rufous kangaroo
" 12*ab*. Halmaturus dorsalis, ashy kangaroo
" 13. Didelphys murina, Brazilian opossum
" 14. Didelphys virginiana, Virginia opossum
" 15. Nasua rufa, brown coati

PLATE 113.

Fig. 1. Chinchilla lanigera, chinchilla
" 2*a*. Cavia cobaya, Guinea pig
" 2*b*. Lagomys alpinus, pika
" 3*a*. Cercolabes villosus, Brazil porcupine
" 3*b*. See 16 (mistake of the engraver's).
" 4. Hystrix cristatus, European porcupine
" 5. Halamys caffer, Cape jerboa
" 6. Mus sylvaticus, field mouse
" 7. Mus rattus, black rat
" 8. Mus musculus, common mouse
" 9. Cricetus vulgaris, hamster
" 10. Myoxus glis, dormouse
" 11. Arctomys alpinus, marmot
" 12, 13. Pteromys volucella, flying squirrel
" 14. Sciurus carolinensis, grey squirrel
" 15. Sciurus maximus, Malabar squirrel
" 16. Lutra vulgaris, European otter
" 17. Talpa europæa, European mole
" 18. Erinaceus europæus, European hedgehog

PLATE 114.

Fig. 1. Castor fiber americanus, beaver
" 2. Phoca vitulina, seal
" 3. Trichechus rosmarus, walrus
" 4. Felis domesticus, cat
" 5. Felis domesticus angorensis Angora cat
" 6. Hyæna striata, striped hyena
" 7. Canis aureus, jackal
" 8. Canis lupus, wolf
" 9. Canis familiaris pastoreus, shepherd's dog
" 10. Lutra canadensis, Canada otter
" 11. Ursus maritimus, white polar bear

PLATE 115.

Fig. 1. Felis leo, lion
" 2. Felis tigris, tiger
" 3. Felis leopardus, leopard
" 4. Felis onca, jaguar
" 5. Felis pardus, panther
" 6. Lynx europæus, lynx

PLATE 116.

Fig. 1. Vulpes fulvus, red fox
" 2. Meles vulgaris, badger
" 3. Mustela martes, marten
" 4. Mustela foina, beech marten
" 5. Mustela putorius, polecat
" 6. Putorius furo, ferret
" 7. Putorius vulgaris, weasel
" 8. Putorius erminea, ermine
" 9. Ursus arctos, brown bear
" 10. Ursus americanus, black bear
" 11. Canis familiaris sibericus, Siberian dog
" 12. Canis familiaris, domestic dog
" 13. Canis familiaris molossus, bull dog
" 14. Canis familiaris leporarius, greyhound

Fig. 15. Canis familiaris normanus, chase dog
" 16. Canis familiaris vertagus, badger dog

PLATE 117.

Fig. 1. Galeopithecus rufus, flying lemur
" 2. Plecotus timoriensis, long-eared bat
" 3. Vespertilio noctula, common bat
" 4. Vespertilio serotinus, serotine bat
" 5. Vespertilio murinus, European bat
" 6. Rhinolophus ferrum equinum horse-shoe bat
" 7. Megaderma lyra, broadwinged bat
" 8. Vampyrus spectrum, vampire
" 9. Pteropus vulgaris, roussette
" 10. Otolicnus senegalensis, galago
" 11. Lemur pusillus, fox-nosed maki
" 12. Lemur macao, maki
" 13. Cynocephalus maimon, mandrill
" 14. Cercopithecus diana, holoway
" 15. Hylobates agilis, agile gibbon

PLATE 118.

Fig. 1. Rhinocerus indicus, Indian rhinoceros
" 2, 3. Cebus capucinus, capucin
" 4. Macacus cynomolgus, hare-lipped monkey
" 5. Cercopithecus ruber, patras
" 6. Cercopithecus griseus, grey patras
" 7. Macacus silenus, maned macaque
" 8. Cynocephalus sphinx, baboon
" 9. Inuus ecaudatus, Barbary ape
" 10. Troglodytes niger, chimpanse
" 11. Pithecus satyrus, orang outang

ANTHROPOLOGY AND SURGERY

[Including the translation of the German words on plate 119.]

PLATE 119.

Figs. 1—9. Varieties of Mankind
" 10—12. Facial angles
Centre figure. Chart of the five races of Blumenbach.

GLOSSARY.

I. *Kaukasische oder Weisse Race*, Caucasian or White Race.
II. *Mongolische oder Gelbe Race*, Mongolian or Yellow Race.
III. *Æthiopische oder Schwarze Race*, Ethiopian or Black Race.
IV. *Amerikanische oder Kupferfarbige Race*, American or Copper-colored Race.
V. *Malayische oder Olivenfarbige Race*, Malay or Olive-colored Race.
F. 1 *bis* 4. *Grundzüge (Typus) der Kaukasischen Race*, Figs. 1—4. Types of the Caucasian Race.
Figur 5. *Grundzüge der Mongolischen Race*, Fig. 5. Type of the Mongolian Race.
Figur 6. *Grundzüge der Æthiopischen Race*, Fig. 6. Type of the Ethiopian Race.
F. 7, 8. *Grundzüge der Kupferfarbigen Race*, Figs. 7, 8. Type of the Copper-colored Race.
Figur 9. *Grundzüge der Olivenfarbigen Race*, Fig 9. Type of the Olive-colored Race.

Afghanen, Afghans.
Algonkiren, Algonkins.
Araber, Arabs.
Azteken, Azteks.
Berbern, Berbers.
Birmanen, Birmans.
Bucharen, Bucharians.
Caffern, Caffres.
Californier, Californians.
Canadier, Canadians.
Caraiben, Caribbeans.
Celten, Celts.
Chaldäer, Chaldæans.
Chinesen, Chinese.
Cinbebassen, Cinbebasses.
Colombier, Colombians.
Eleuten, Aleutians.
Eskimos, Esquimaux.
Ethiopier, Ethiopians.
Fellatas, Fellatah.
Finnen, Finns.
Germanen, Germans.
Gothen, Goths.
Hindus, Hindoos.
Hottentotten, Hottentots.
Iberier, Iberians.
Jakuten, Yacoots.
Japanesen, Japanese.
Jukaghiren, Youkaghirs.
Kalmücken, Calmucks.
Kamtschadalen, Kamskadales.
Kaukasier, Caucasians.
Kirghisen, Kirghese.
Koriälen, Koreans.
Kosaken, Cossacks.
Kriks, Creek Indians.
Lappen, Laplanders.
Letten, Lethonians.
Madagassen, Madagassees.
Mandschus, Manchoos.
Mauren, Moors.
Mongolen, Mongols.
Neger, Negroes.
Negritos, Negritoes.
Neuseeländer, New Zealanders.
Osagen, Osages.
Osmanen, Osmanli.
Ostiaken, Ostiaks.
Patagonen, Patagonians.
Pelasger, Pelasgians.
Perser, Persians.
Samojeden, Samoeids.
Sius, Sioux Indians.
Slaven, Slavonians.
Soyten, Soyetes.
Tibetaner, Thibetans.
Tschucktschen, Tchoukches.
Tungusen, Tungus.
Wogulen, Voguls.

PLATE 120.

Figs. 1—21. Illustrating the psychological relations of the brain (Phrenology)

PLATE 121.

Figs. 1—19. The bones of the head

PLATE 122.

Figs. 1—31. Anatomy of the bones

PLATE 123.

Figs. 1—50. Anatomy of the bones

PLATE 124.

Figs. 1—14. Anatomy of the ligaments
" 15—17. Anatomy of the muscles

PLATE 125.

Figs. 1—18. Anatomy of the ligaments
" 19, 20. Anatomy of the muscles

PLATE 126.

Figs. 1—19. Anatomy of the ligaments
" 20, 21. Anatomy of the muscles

PLATE 127.

Figs. 1—15. Anatomy of the muscles

PLATE 128.

Figs. 1—19. Anatomy of the muscles

PLATE 129.

Figs. 1—22. Anatomy of the fasciæ
" 23—26. Anatomy of the integuments
" 27—30. Organs of mastication and deglutition
" 31—38. Organs of respiration and voice

PLATE 130.

Figs. 1—21. Anatomy of the organs of digestion

PLATE 131.

Figs. 1—45. Anatomy of the eye

PLATE 132.

Figs. 1—25. Anatomy of the ear
" 26—45. Anatomy of the nose

PLATE 133.

Figs. 1—13. Anatomy of the vascular system

PLATE 134.

Figs. 1—13. Anatomy of the vascular system

PLATE 135.

Figs. 1—17. Anatomy of the vascular system

PLATE 136.

Figs. 1—16. Anatomy of the vascular system

PLATE 137.

Figs. 1—21. Anatomy of the brain and nerves

PLATE 138.

Figs. 1—14. Anatomy of the nerves

PLATE 139.

Figs. 1—35. Various surgical operations

PLATE 140.

Figs. 1—91. Various surgical instruments

TECHNOLOGY

PLATE 141.

Figs. 1-27. Illustrating the construction of streets and roads
" 28-34. Illustrating the Thames tunnel in London

PLATE 142.

Figs. 1-54. Illustrating the construction of railroads

PLATE 143.

Figs. 1-29. Illustrating the construction of railroads
" 30. The Leipsic station of the Saxon and Bavarian railroad

PLATE 144.

Figs. 1-6. Illustrating the motive power on railroads
" 7-14. Illustrating the construction of inclined planes

PLATE 145.

Figs. 1-8. Illustrating Stephenson's locomotive with variable expansion
" 9-28. Illustrating the construction of tenders and railroad cars
" 29. Interior of the Duke of Brunswick's car on the Brunswick railroad
" 30. Interior of Queen Victoria's car on the London and Dover railroad

PLATE 146.

Figs. 1-10. Illustrating the construction of atmospheric railroads

PLATE 147.

Figs. 1-23. Illustrating the construction of stone bridges

PLATE 148.

Figs. 1-45. Illustrating the construction of wooden bridges

PLATE 149.

Figs. 1-33. Illustrating the construction of iron bridges

PLATE 150.

Figs. 1-28. Illustrating the construction of canals and dams
Fig. 29. View of a chain of locks on the Rideau canal near Bytown in Canada

PLATE 151.

Figs. 1-5. Details from the Languedoc canal in France
" 6-8. The Cesse Aqueduct
" 9-11. The Croton Aqueduct of New York
" 12-18. Locks and weirs

PLATE 152.

Figs. 1-12. Illustrating the construction of windlasses and cranes

PLATE 153.

Fig. 1. A lift pump
" 2. A forcing pump
Figs. 3-6. Stephenson's double action pump
" 7-9. Pump used in the mine Huelgoet in Normandy
" 10-12. Letestu's pump
" 13, 14. Jordan's hydraulic ram in Clausthal
" 15-21. Reichenbach's hydraulic ram in the saltworks at Illfang in Bavaria

PLATE 154.

Figs. 1, 2. The simplest construction of a fire-engine
" 3, 4. Portable fire-engine
" 5, 6. Pontifex's fire-engine
" 7-10. Repsold's fire-engine
" 11-15. Letestu's fire-engine
" 16-21. Common double-action fire-engine
" 22, 23. Bramah's fire-engine
" 24. Steam fire-engine
" 25-29. Apparatus to save persons and property at fires

PLATE 155.

Figs. 1-18. Illustrating the construction of vertical water-wheels
" 19-29. Illustrating the construction of horizontal water-wheels

PLATE 156.

Figs. 1-15. Illustrating the construction of an American grinding-mill

PLATE 157.

Figs. 1, 2. Cotton gin
" 3-5*a*. Wolf or willow
" 6. Spreading machine
" 7. Lap-machine
" 8-16. Carders and carding machines
" 17-20. Drawing frame
" 21-24. Roving frame

PLATE 158.

Figs. 1-7. Danforth's tube roving-frame
" 8-16. Self-acting mule
" 17-19. Washing-kettle
" 20, 21. Wringing-machine
" 22-24. Scales for weighing yarn
" 25, 26. Starching and steam drying apparatus
" 27-29. Press for packing yarn
" 30-33. Woollen willow

PLATE 159.

Fig. 1. Beaming for handweaving
" 2, 3. Warping-mill
" 4. Simplest loom
" 5-7. Power-loom
" 8, 9. Shuttle
" 10, 11. Jaw-temples
" 12, 13. Singing-oven
" 14, 15. Wash-wheel
" 16-19. Gassing-machine
" 20. Arrangement of spools
" 21. Lever of a power-loom
" 22. Batten of a power-loom

PLATE 160.

Figs. 1, 2. Casting-machine of mints
" 3, 4. Rolling mill of mints
Fig. 5, 6. Circular shears
" 7-9. Flattening-mill
" 10-14. Drawing machine of mints
" 15, 16. Coin-punch
" 17-20. Milling machine
" 21-37. Stamping-machines

PLATE 161.

Fig. 1. Stamping machine of the mint in Rio Janeiro

COINS OF VARIOUS NATIONS:

With their approximate values

	VALUE.
" 2. Persian gold piece of Imam Riza, ...	$7 35
" 3. East India Rupee Zodiac, .	6 20
" 4. Gold piece of the East India Company, ...	4 90
" 5. Gold piece of the Dutch East India Company, . .	4 90
" 6. Double gold sovereign of Brabant of 1800, . . .	13 00
" 7. Gold Sovereign of Brabant of 1796, ...	6 50
" 8. Belgian gold lion of 1790, .	9 00
" 9. Danish Species Ducat, . .	2 35
" 10. Danish double Frederic d'or of 1828, ...	7 90
" 11. Austrian ducat of 1826, . .	2 25
" 12. Bavarian ducat of 1821, . .	2 25
" 13. Hamburg ducat of 1818, . .	2 40
" 14. Ducat of Electoral Saxony of 1797, ...	2 25
" 15. Ducat of Canton of Berne, .	2 15
" 16. Carl d'or of the duchy of Brunswick, 1799, . . .	4 00
" 17. Hanoverian double pistole, 1829,	7 90
" 18. Wilhelm d'or of the Electorate of Hesse, 1829, . . .	4 00
" 19. Royal Prussian double Frederick d'or, 1800, . . .	7 90
" 20. Royal Prussian Frederick d'or, 1822, ...	3 90
" 21. Royal Wirtemberg Frederic d'or, 1810, ...	3 90
" 22. Sixteen franc piece or pistole of the Helvetic Republic, 1800, ...	4 25
" 23. Five guilder piece or imperial ducat of the grand duchy of Baden, 1827, . . .	2 35
" 24. Ten guilder piece or Caroline of the grand duchy of Hesse, 1826, ...	4 50
" 25. Royal guinea of Great Britain, 1801, ...	5 00

PLATE 162.

COINS OF VARIOUS NATIONS:

With their approximate values

Fig. 1. English guinea of George III., 1793, ...	5 00
" 2. English third of a guinea of George III., 1797, . .	1 75
" 3. English sovereign of Queen Victoria, 1845, . . .	4 85
" 4. French Louis d'or of Louis XVI., 1797, ...	3 75
" 5. Napoleon d'or, 1813, . .	3 85
" 6. Italian double Napoleon d'or, 1814, ...	6 50

Heck's
Pictorial Archive of
Nature and Science

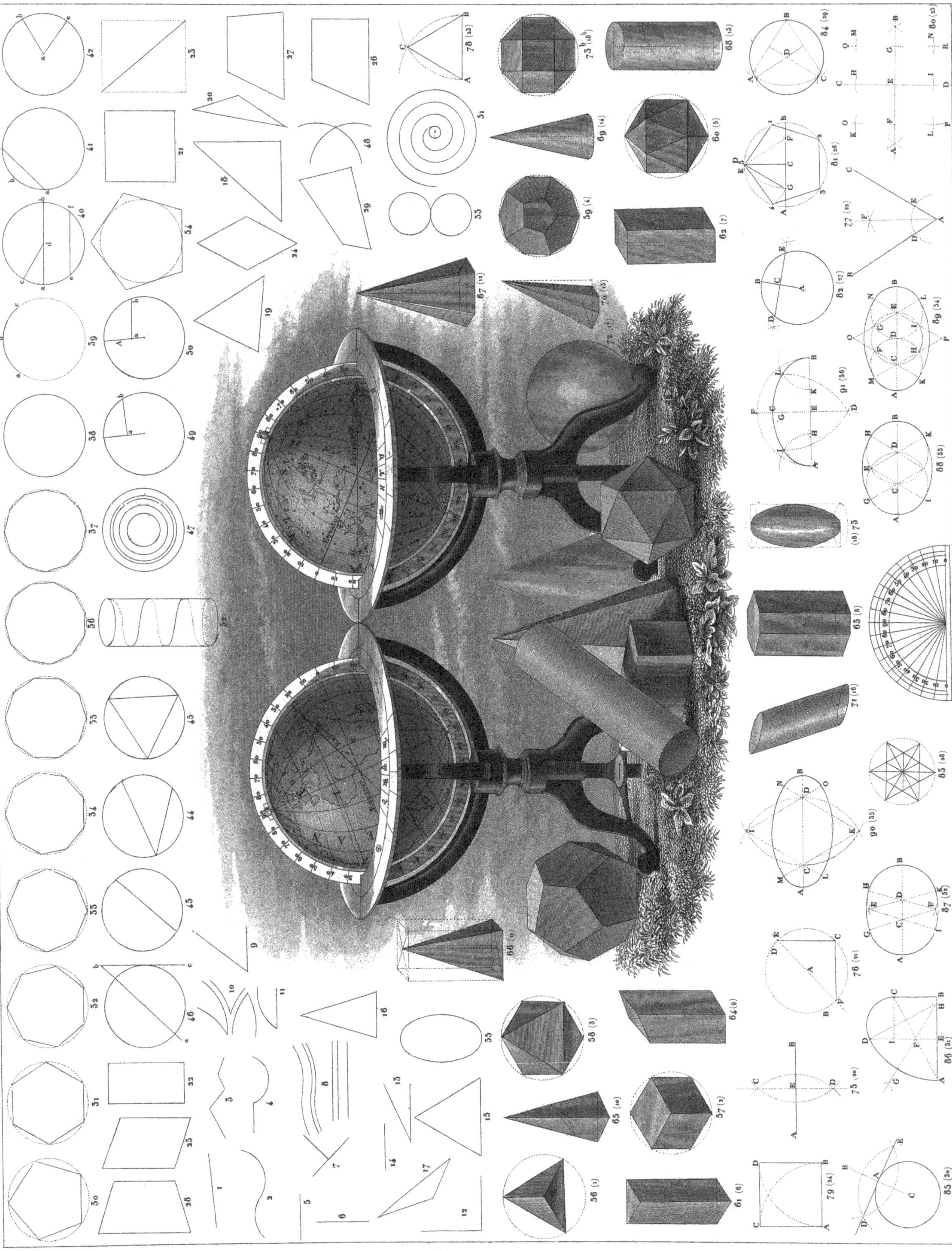

Plates 1 and 2 (Mathematics and Astronomy, Plates 1 and 2)

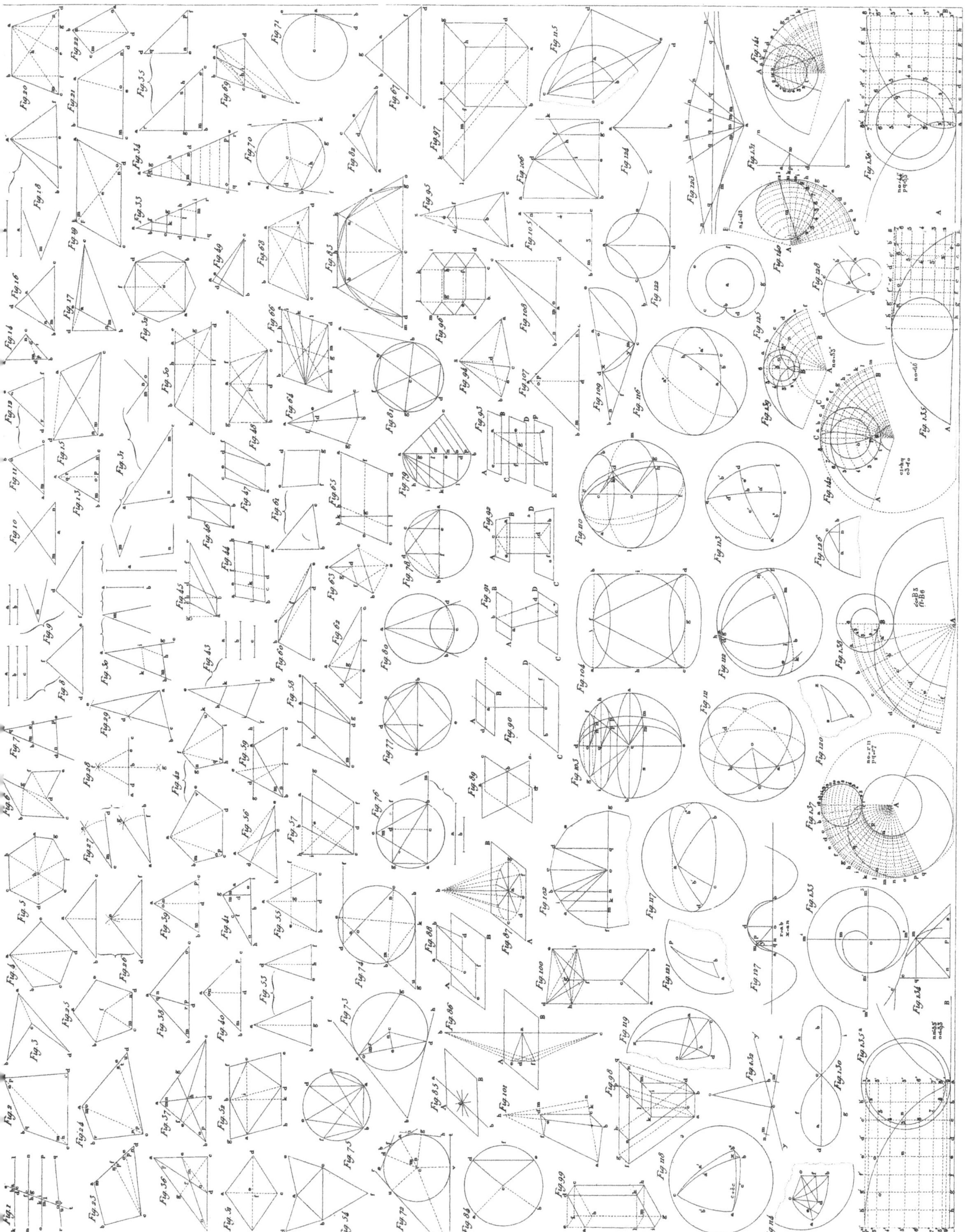

Plate 3 (Mathematics and Astronomy, Plate 3)

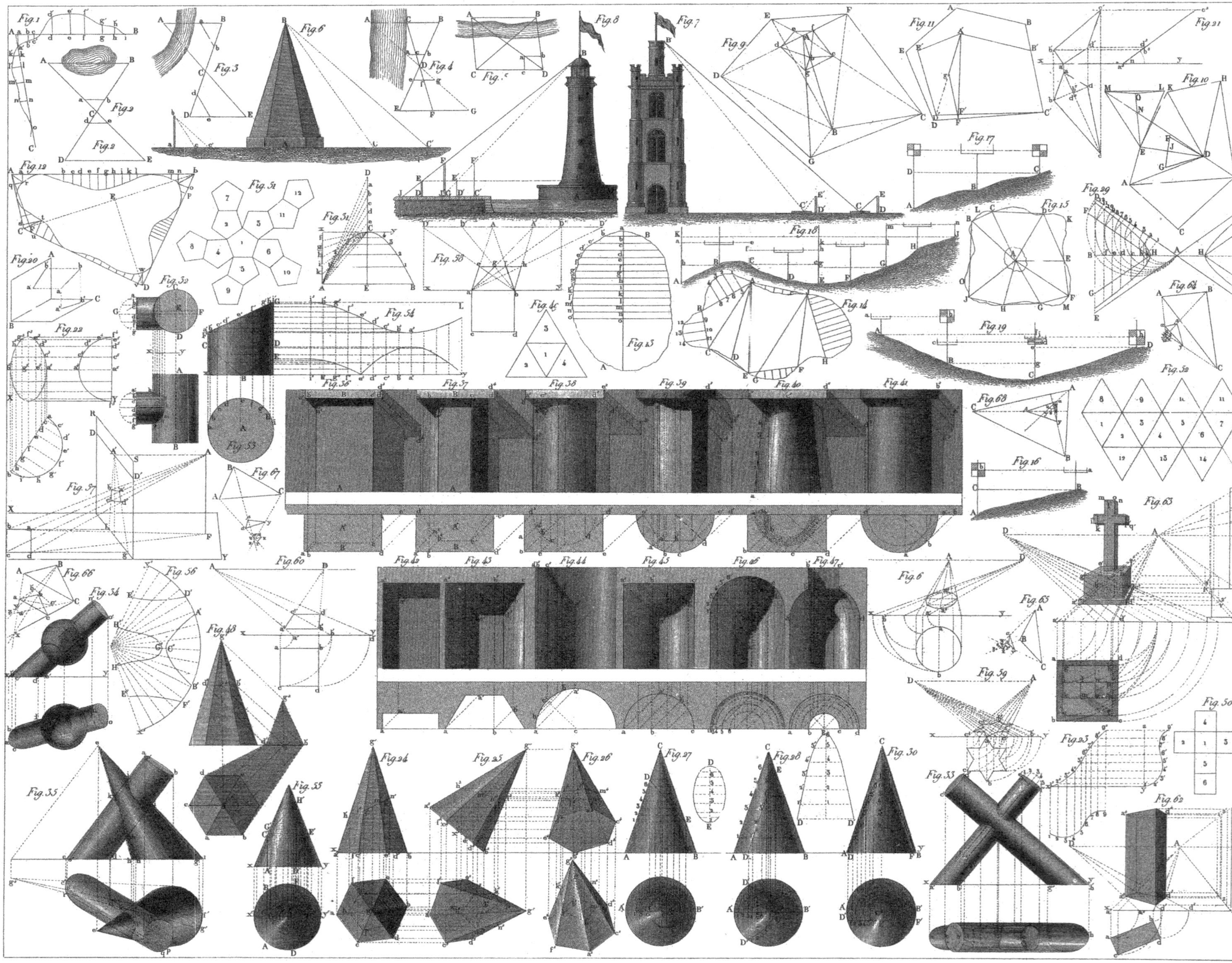

Plate 4 (Mathematics and Astronomy, Plate 4)

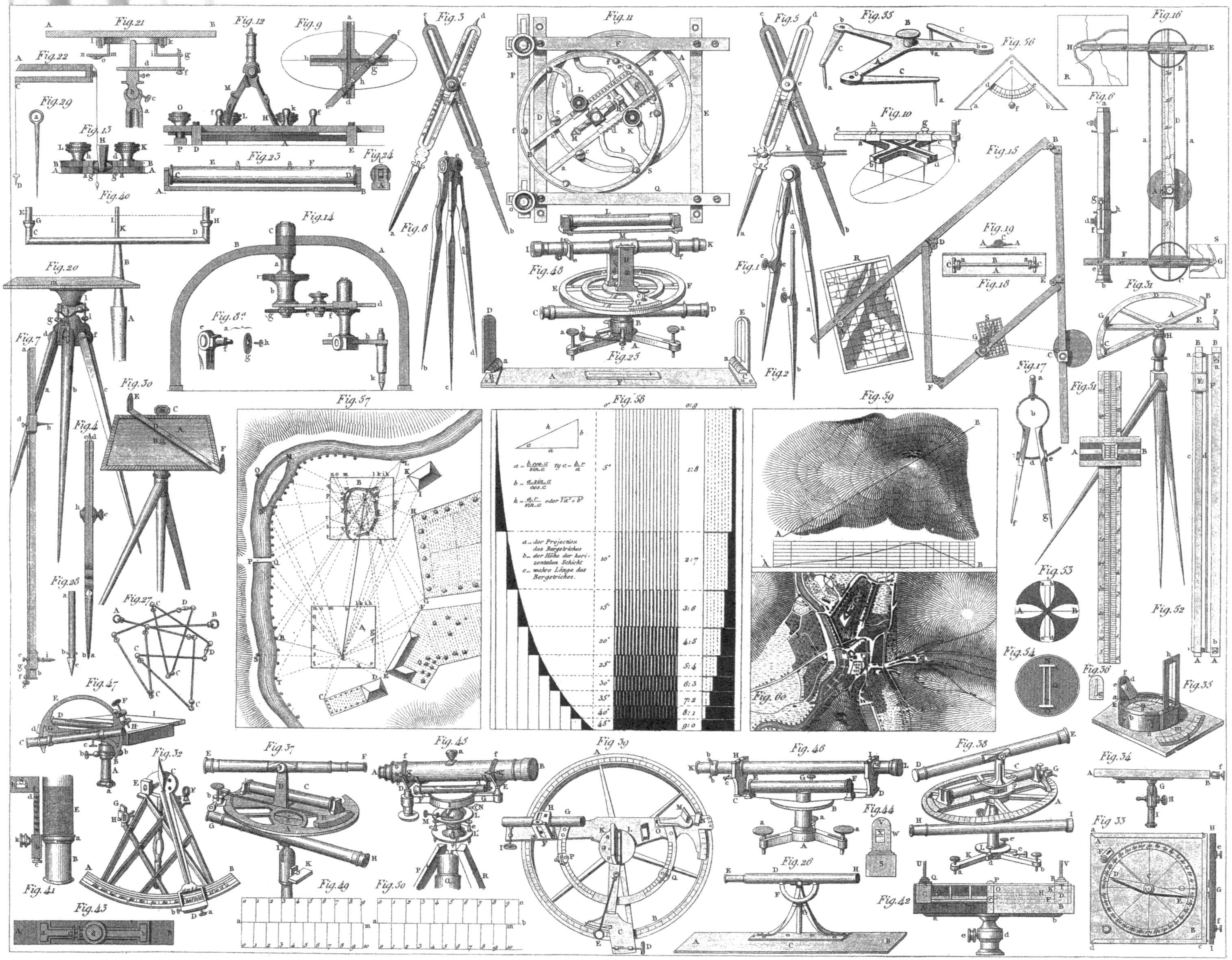

Plate 5 (Mathematics and Astronomy, Plate 5)

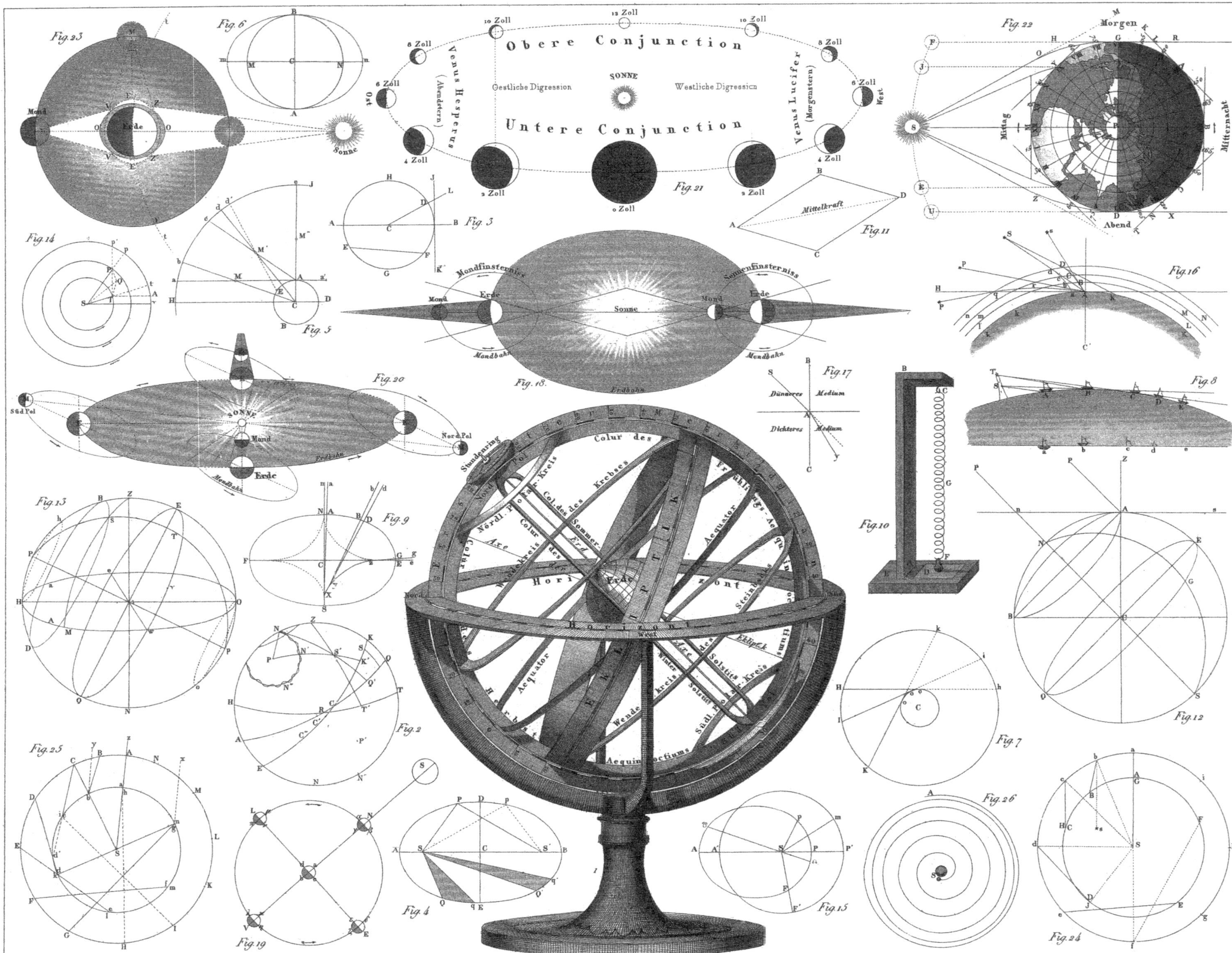

Plate 6 (Mathematics and Astronomy, Plate 6)

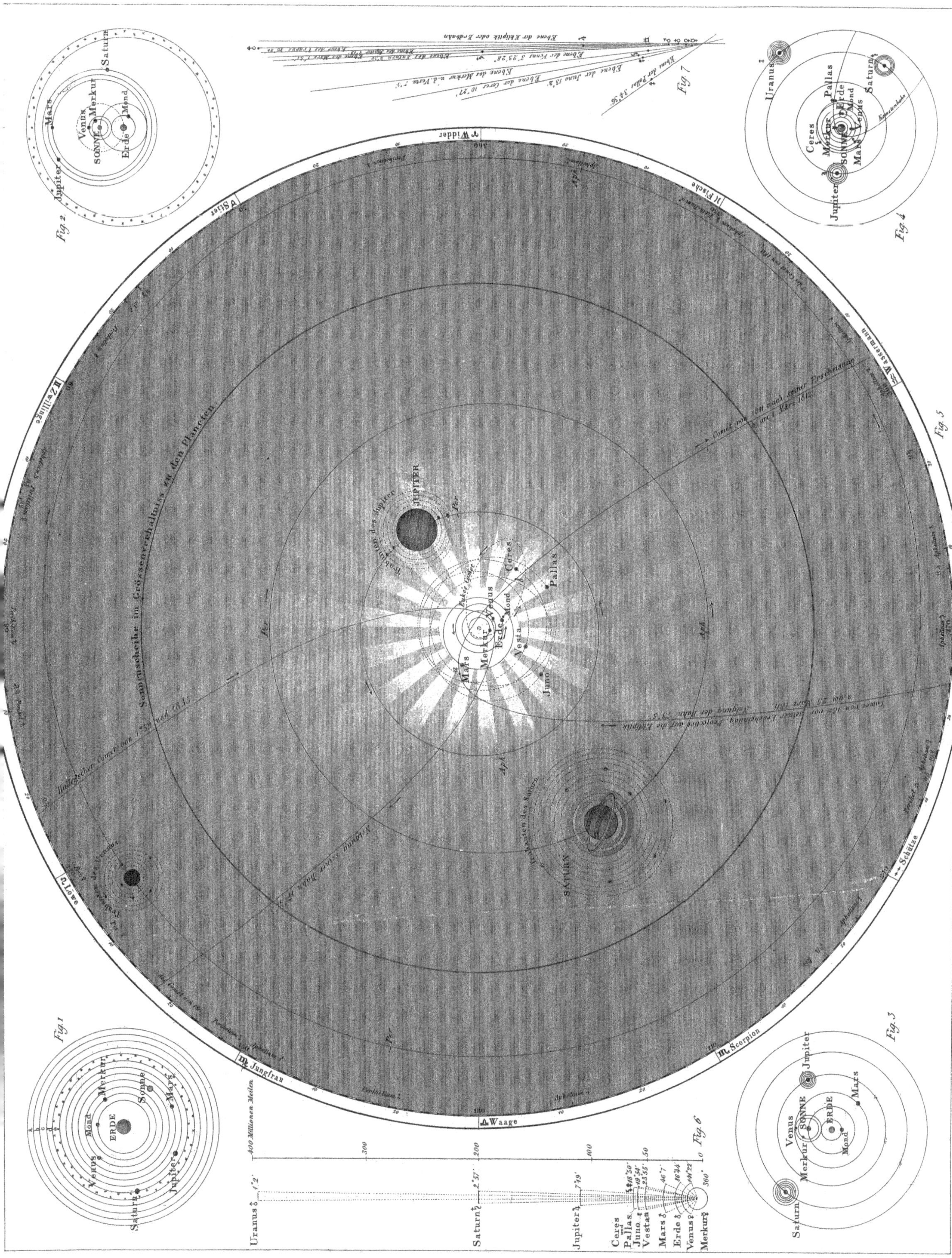

Plate 7 (Mathematics and Astronomy, Plate 7)

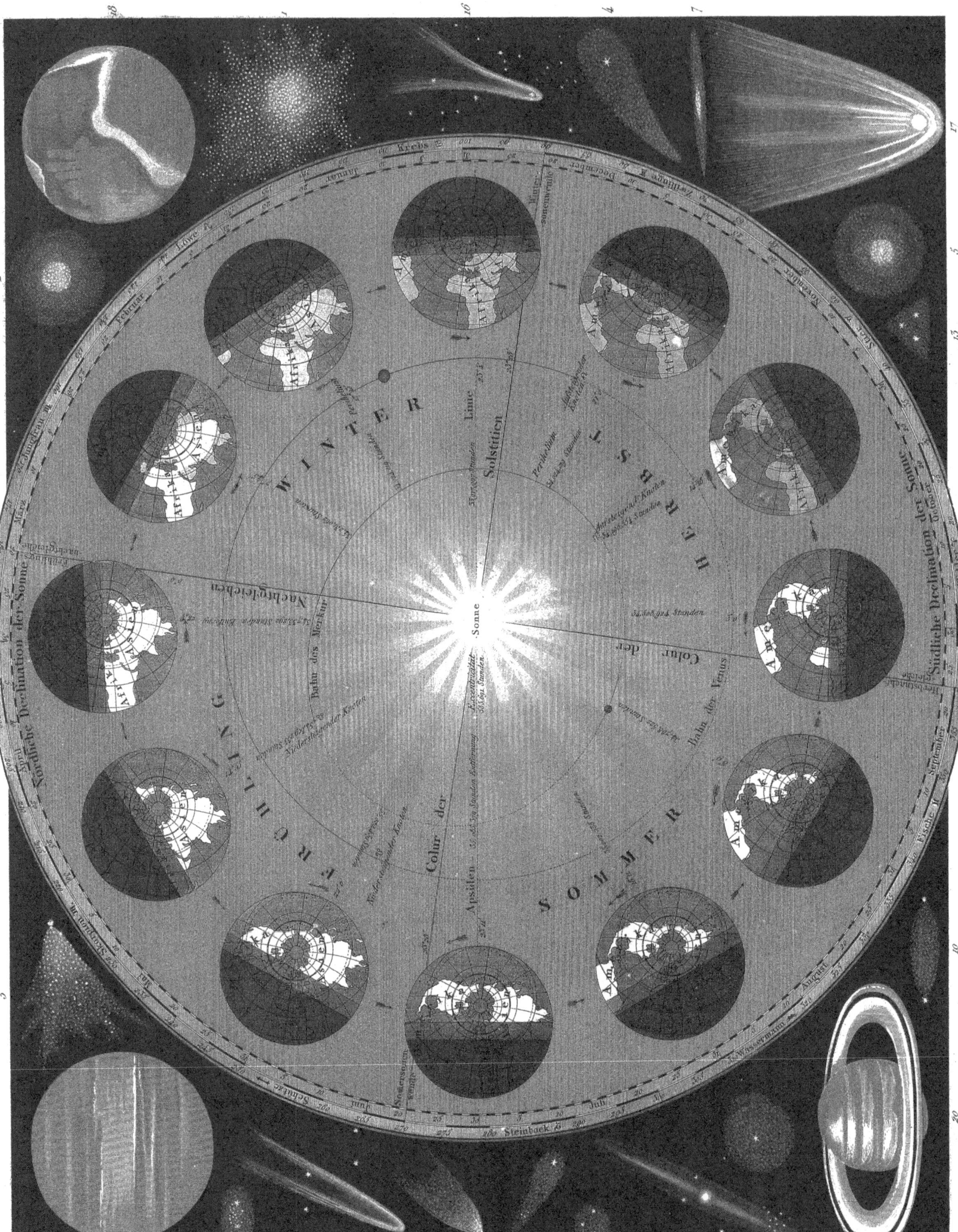

WINTER
HERBST
SOMMER
Sonne
Nachtgleichen
Solstitien
Linie
Colur der
Bahn des Merkur
Bahn der Venus
Nördliche Declination der Sonne
Südliche Declination der Sonne

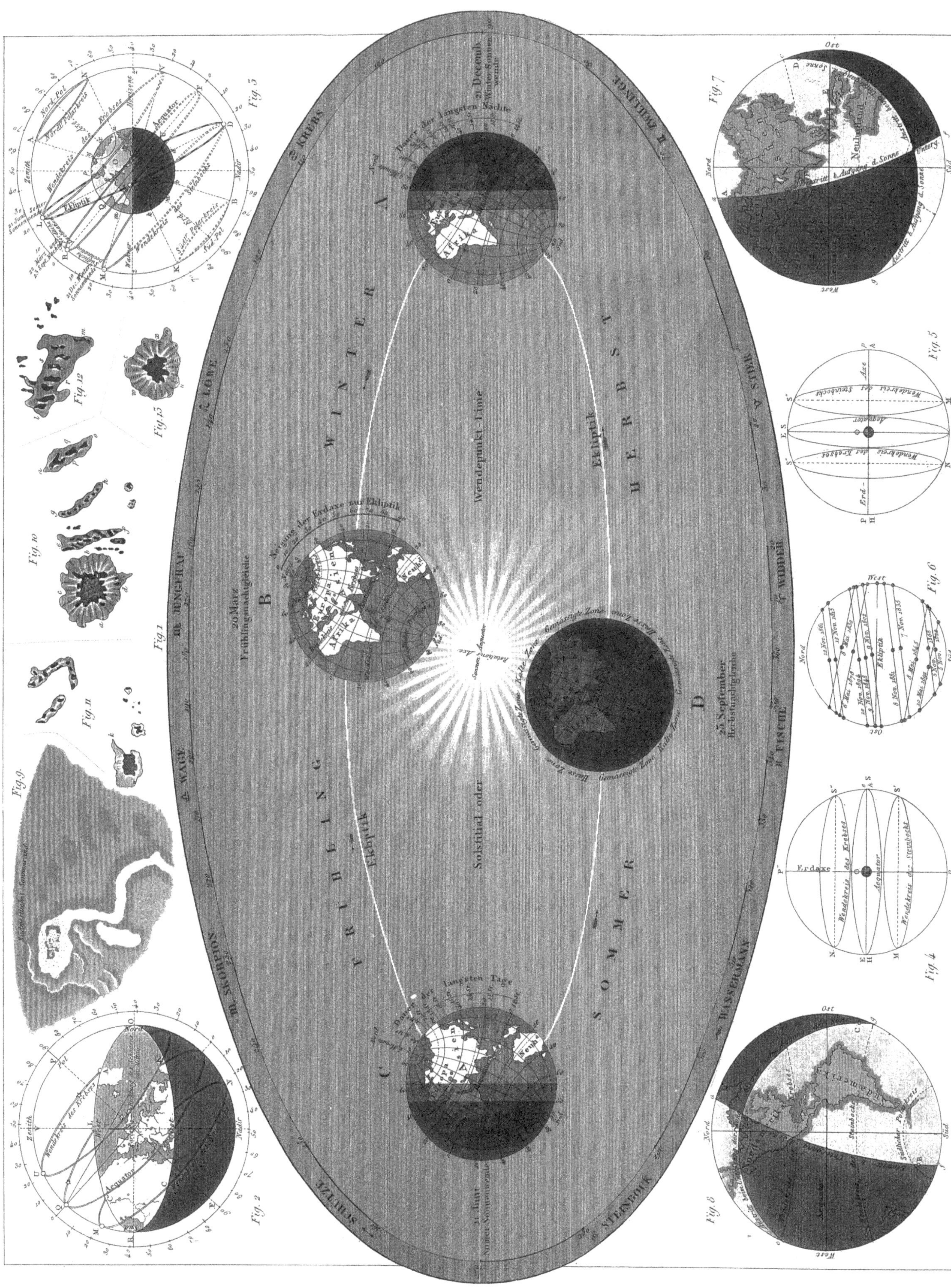

Plate 9 (Mathematics and Astronomy, Plate 9)

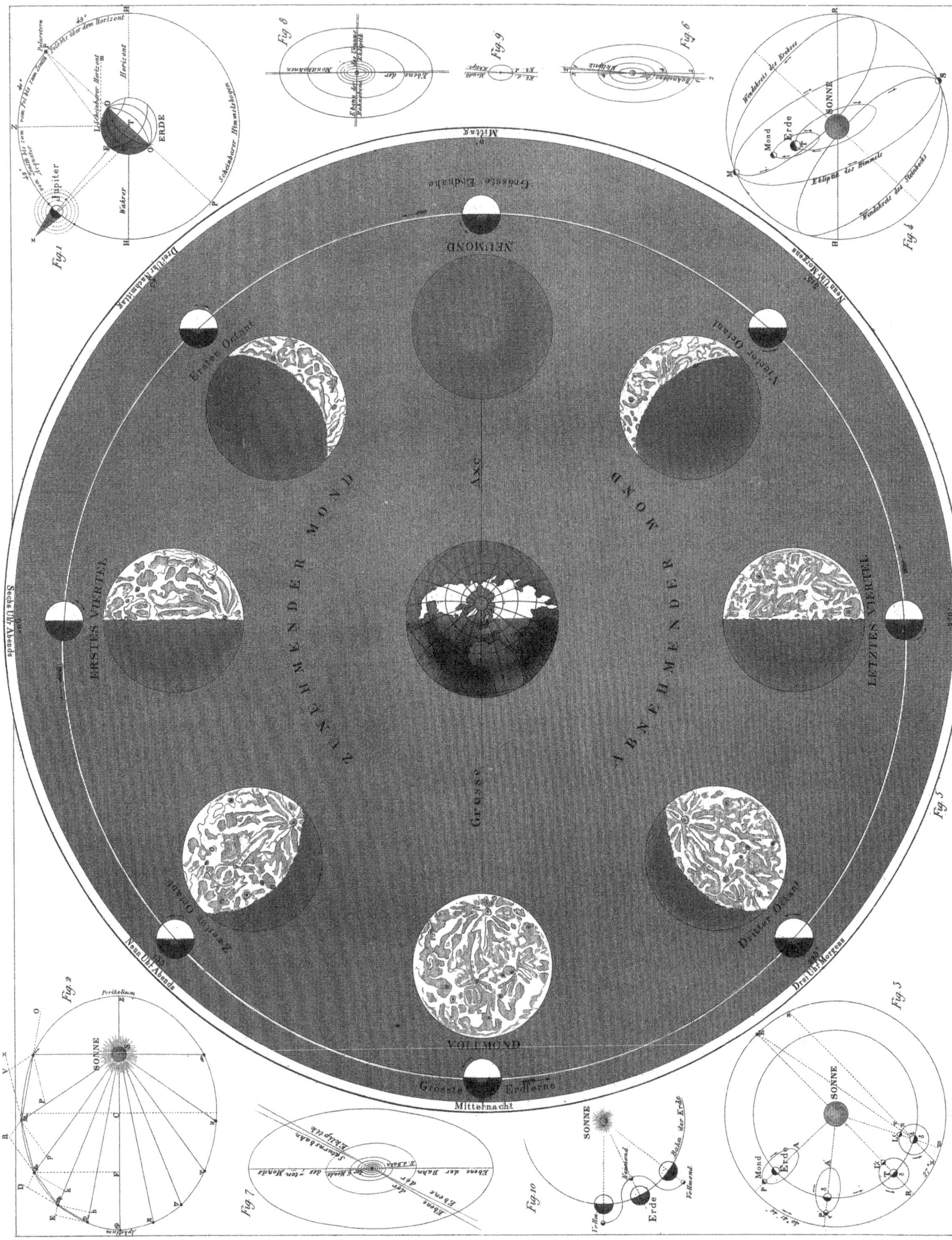
Fig. 1
Fig. 2
Fig. 3
Fig. 4
Fig. 5
Fig. 6
Fig. 7
Fig. 8
Fig. 9
Fig. 10
NEUMOND
VOLLMOND
ERSTES VIERTEL
LETZTES VIERTEL
Erster Octant
Zweiter Octant
Dritter Octant
Vierter Octant
ZUNEHMENDER MOND
ABNEHMENDER MOND
Grosse Axe
Grösste Erdnähe
Grösste Erdferne
Mittag
Mitternacht
Sechs Uhr Abends
Sechs Uhr Morgens
Neun Uhr Abends
Neun Uhr Morgens
Drei Uhr Nachmittags
Drei Uhr Morgens
SONNE
ERDE
Jupiter
Mond
Erde
Ekliptik des Himmels
Wendekreis des Krebses
Wendekreis des Steinbocks
Scheinbarer Himmelsbogen
Perihelium
Aphelium
Bahn der Erde
Vollmond

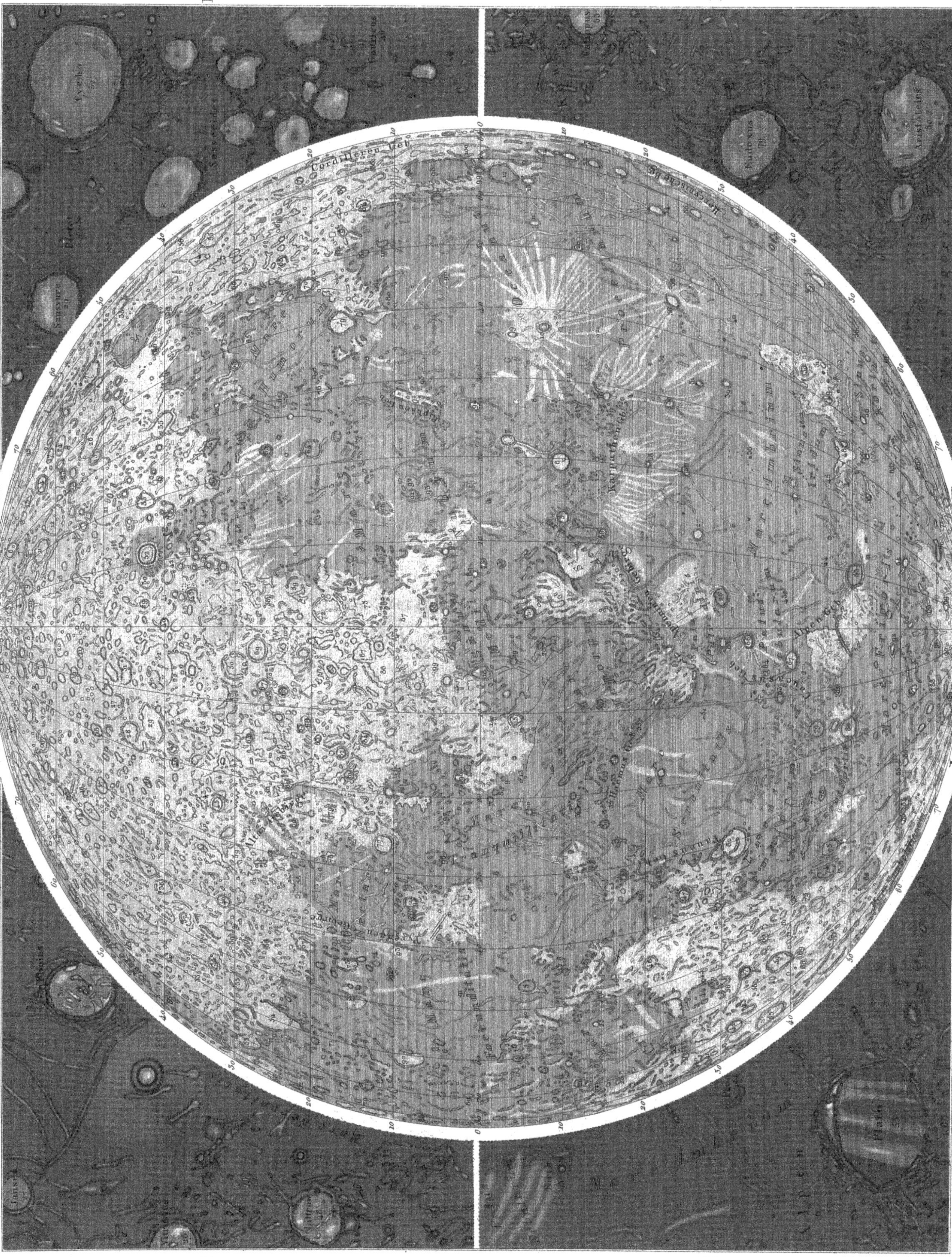

Plate 11 (Mathematics and Astronomy, Plate 11)

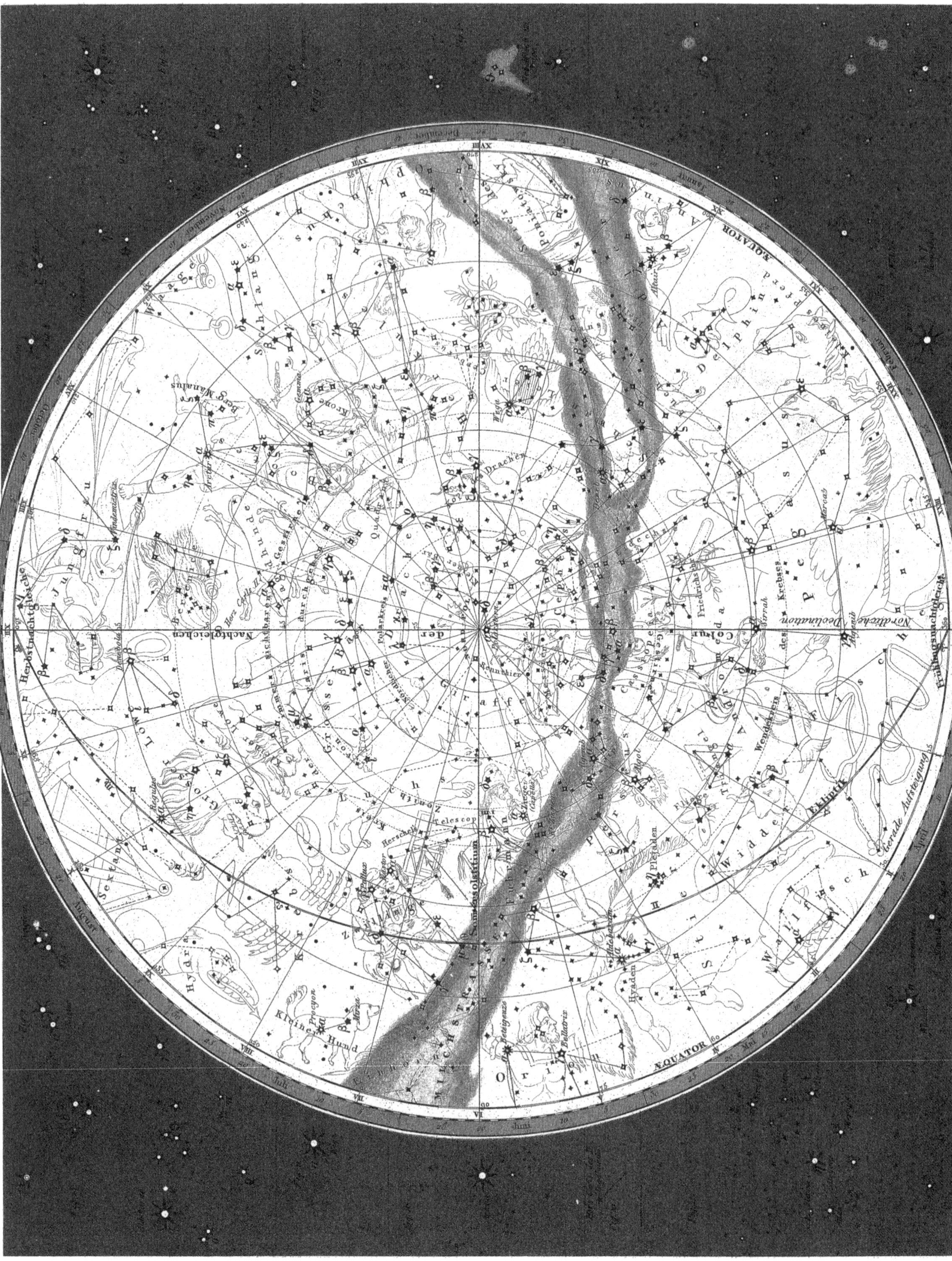
ÆQUATOR
Frühlingsnachtgleiche
Herbstnachtgleiche
Nördliche Declination
Gerade Aufsteigung
Ekliptik
Colur
Nachtgleichen

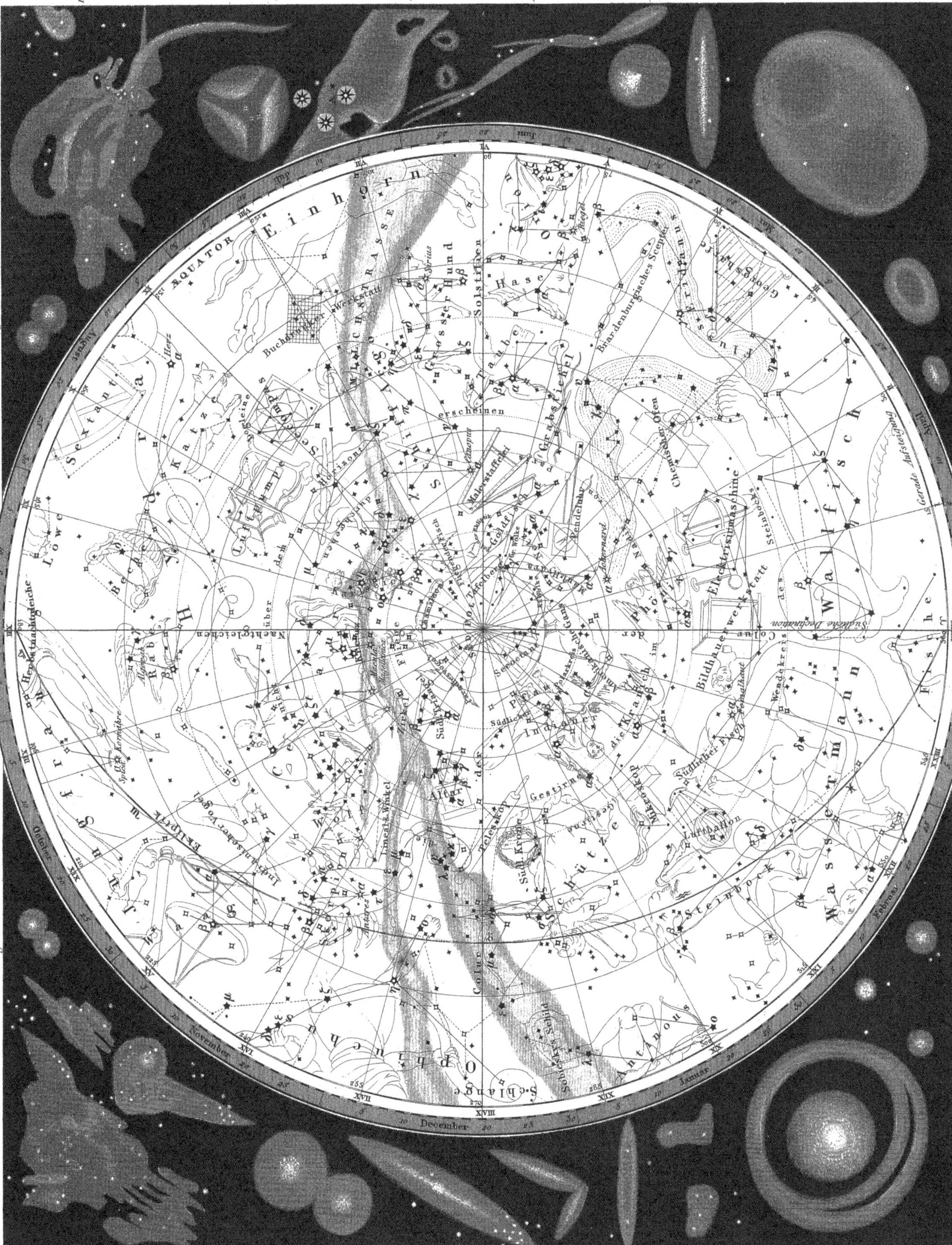

Plate 13 (Mathematics and Astronomy, Plate 13)

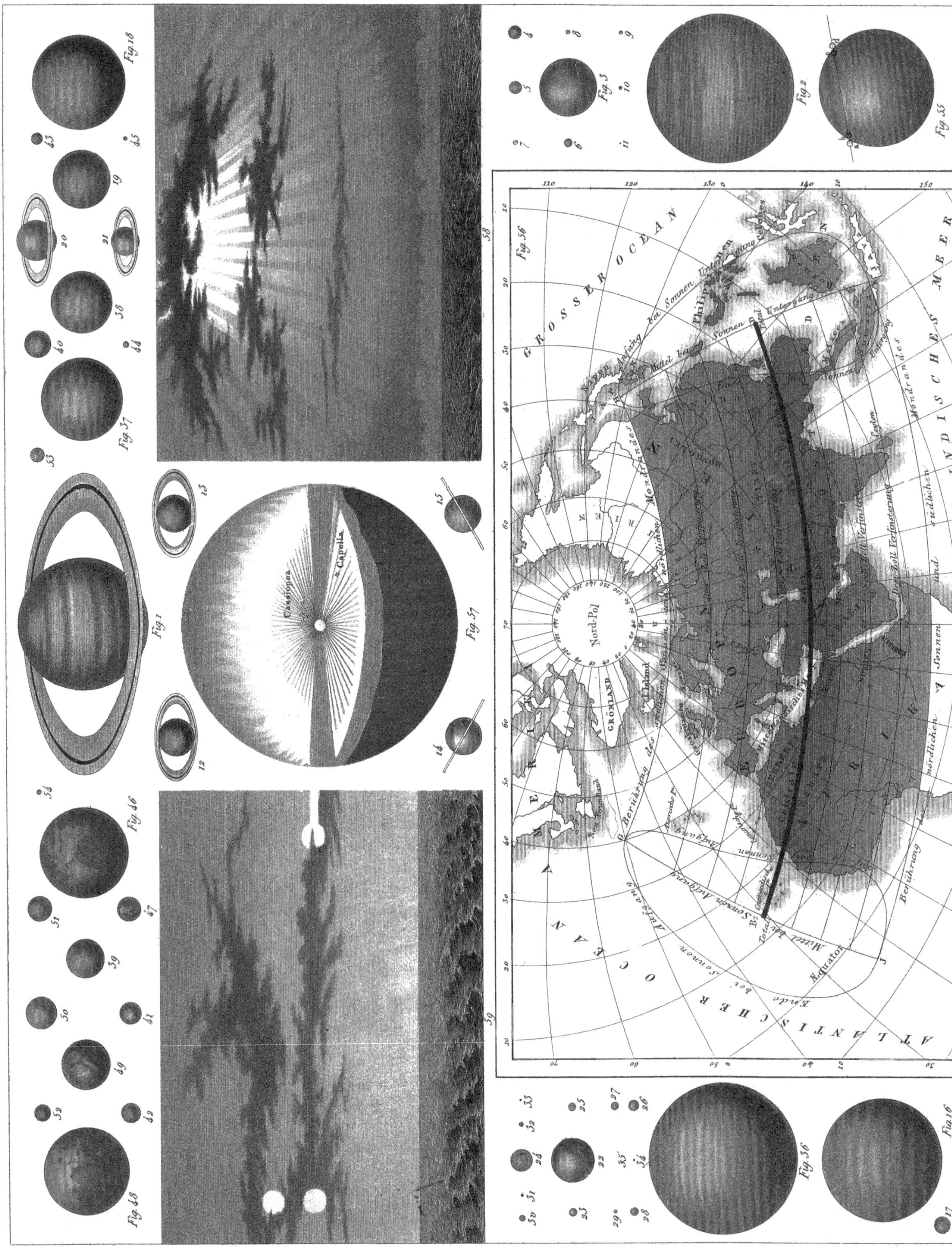

Plate 14 (Mathematics and Astronomy, Plate 14)

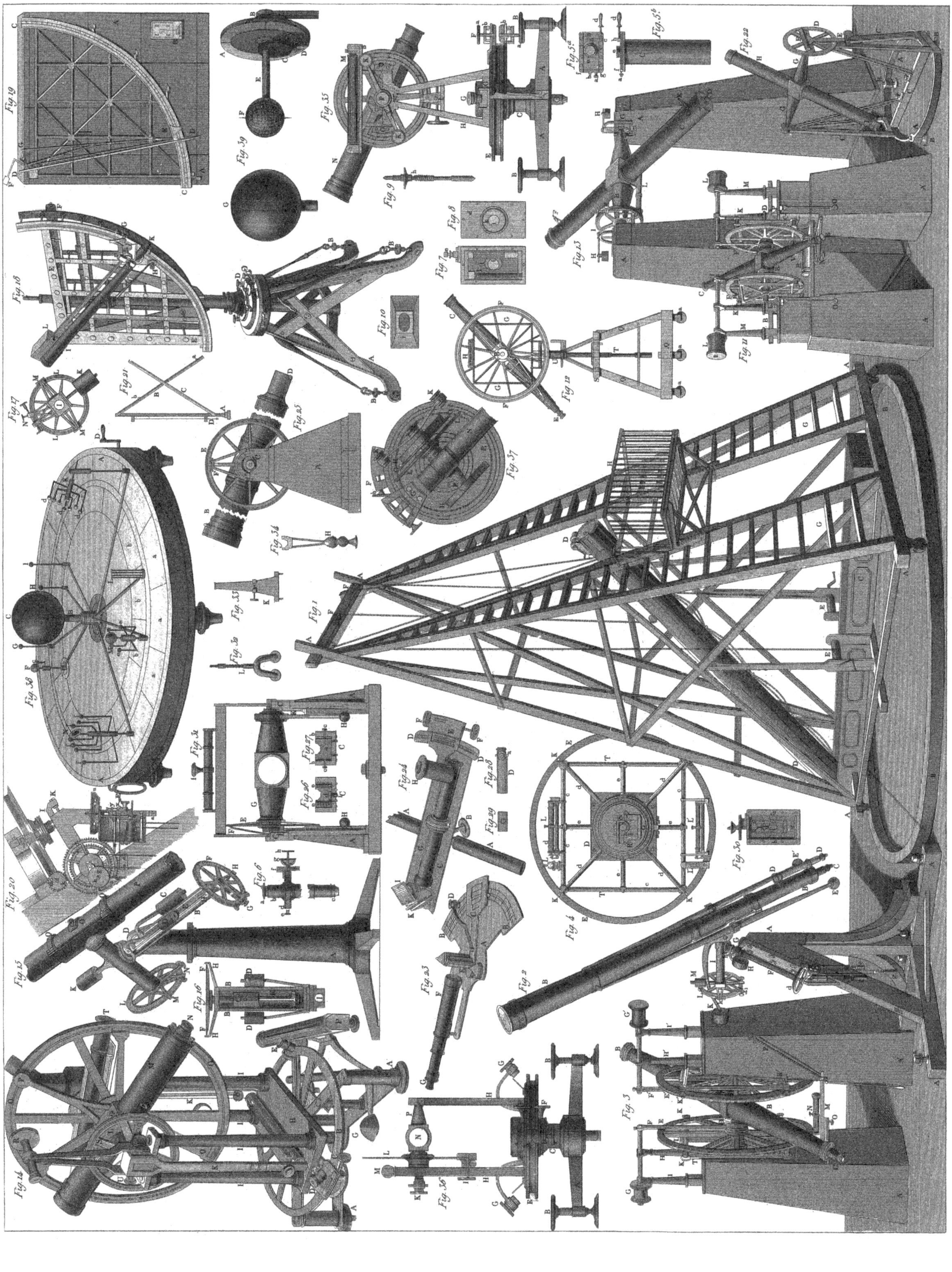

Plate 15 (Mathematics and Astronomy, Plate 15)

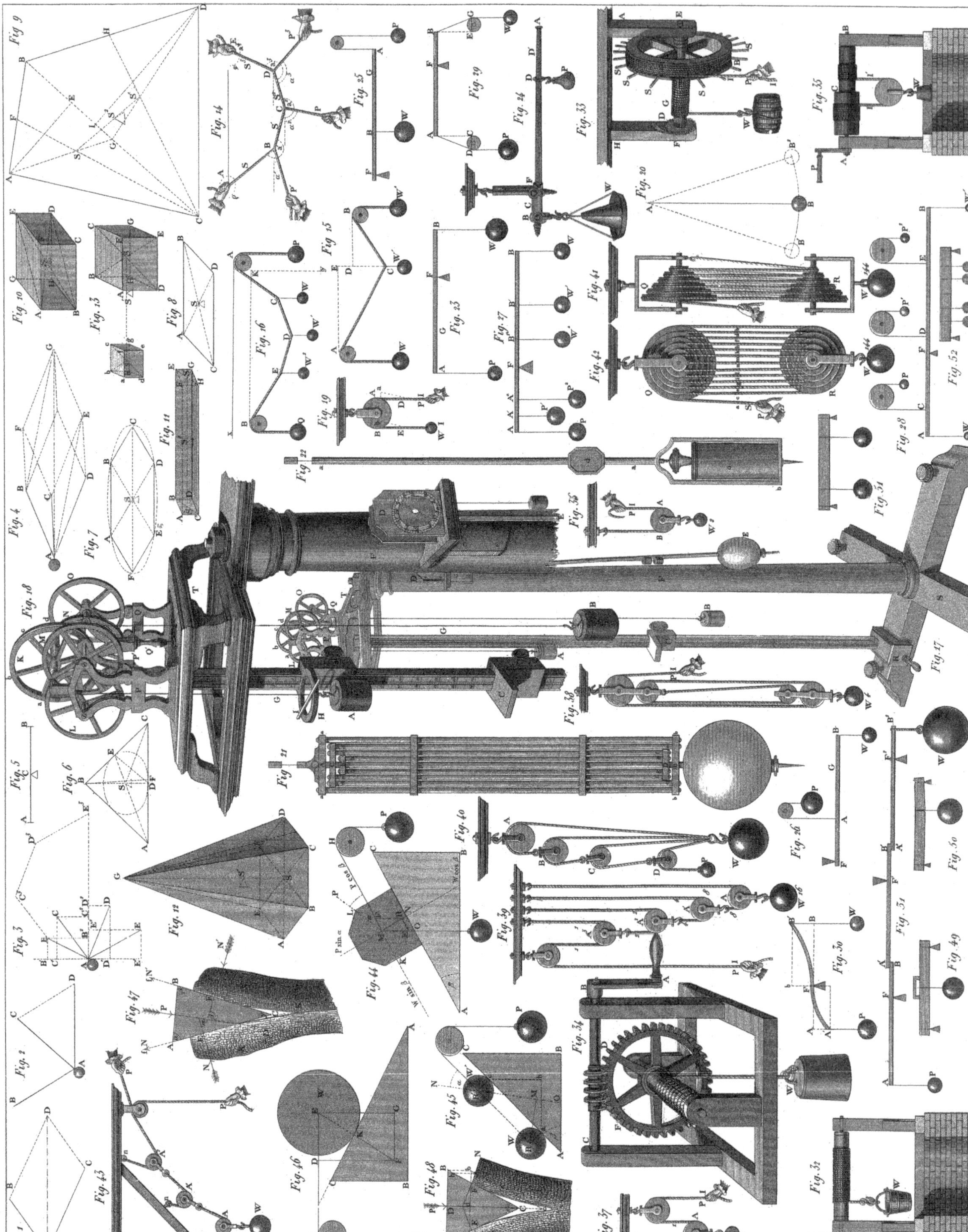

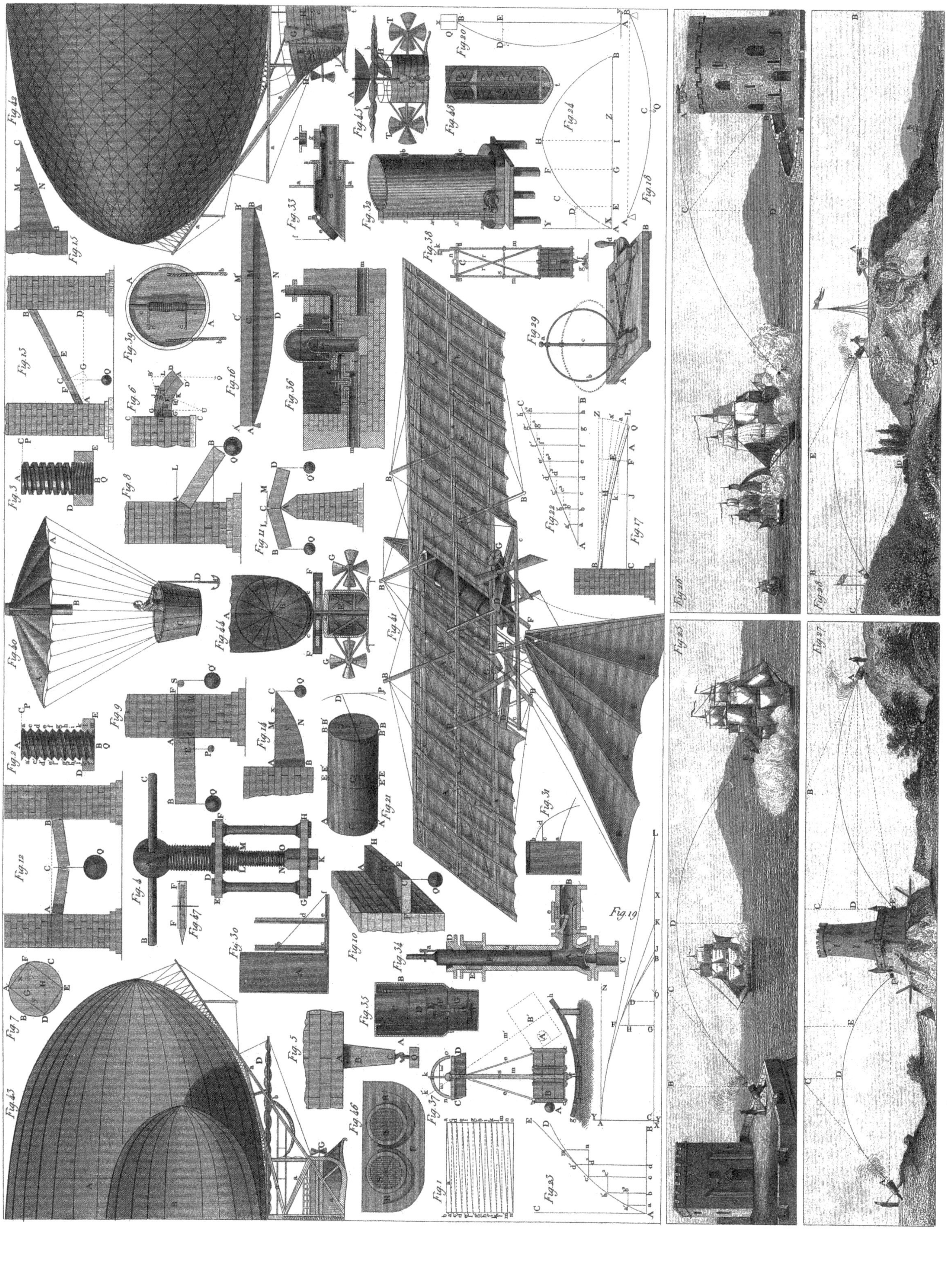

Plate 17 (Physics and Meteorology, Plate 2)

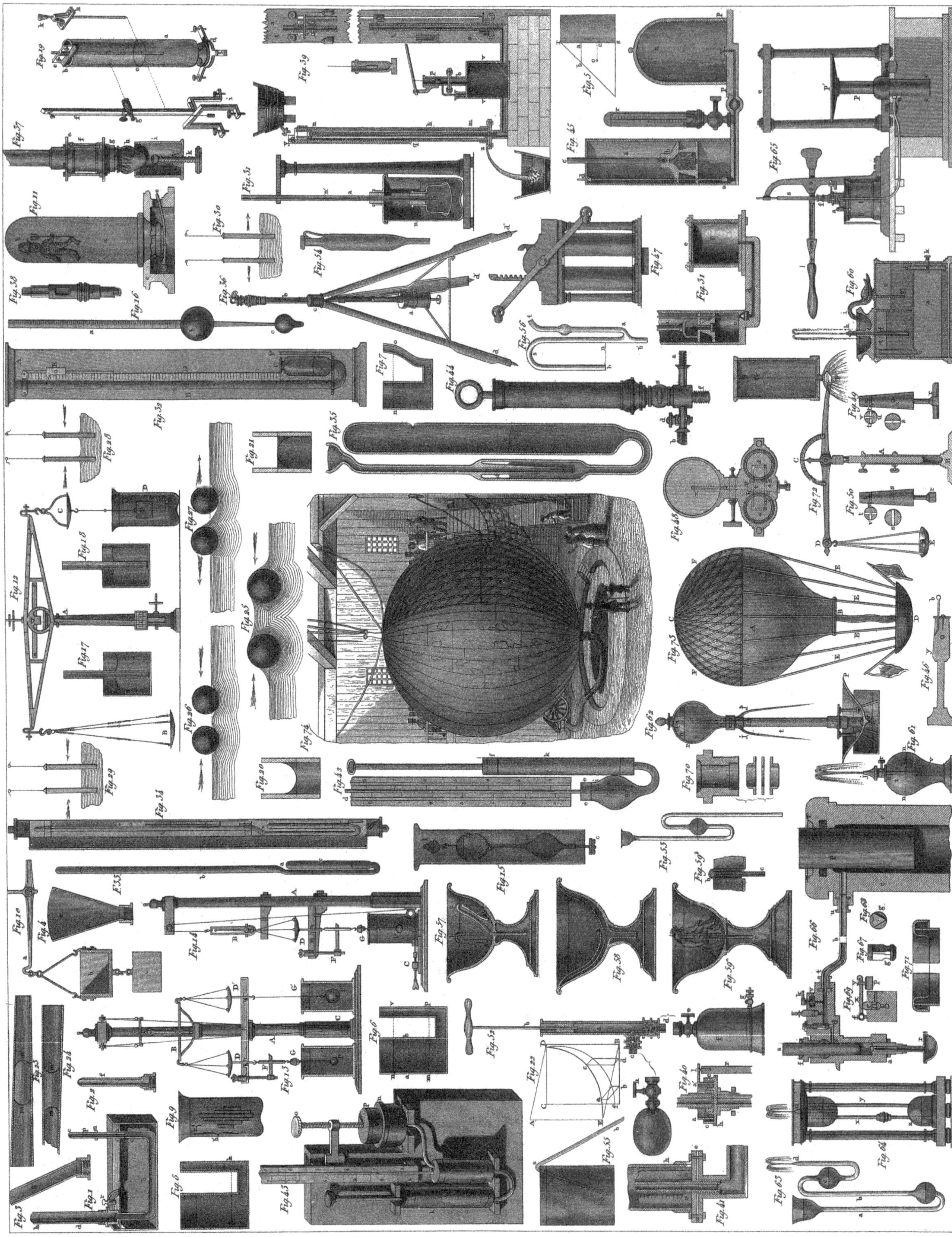

Plate 18 (Physics and Meteorology, Plate 8)

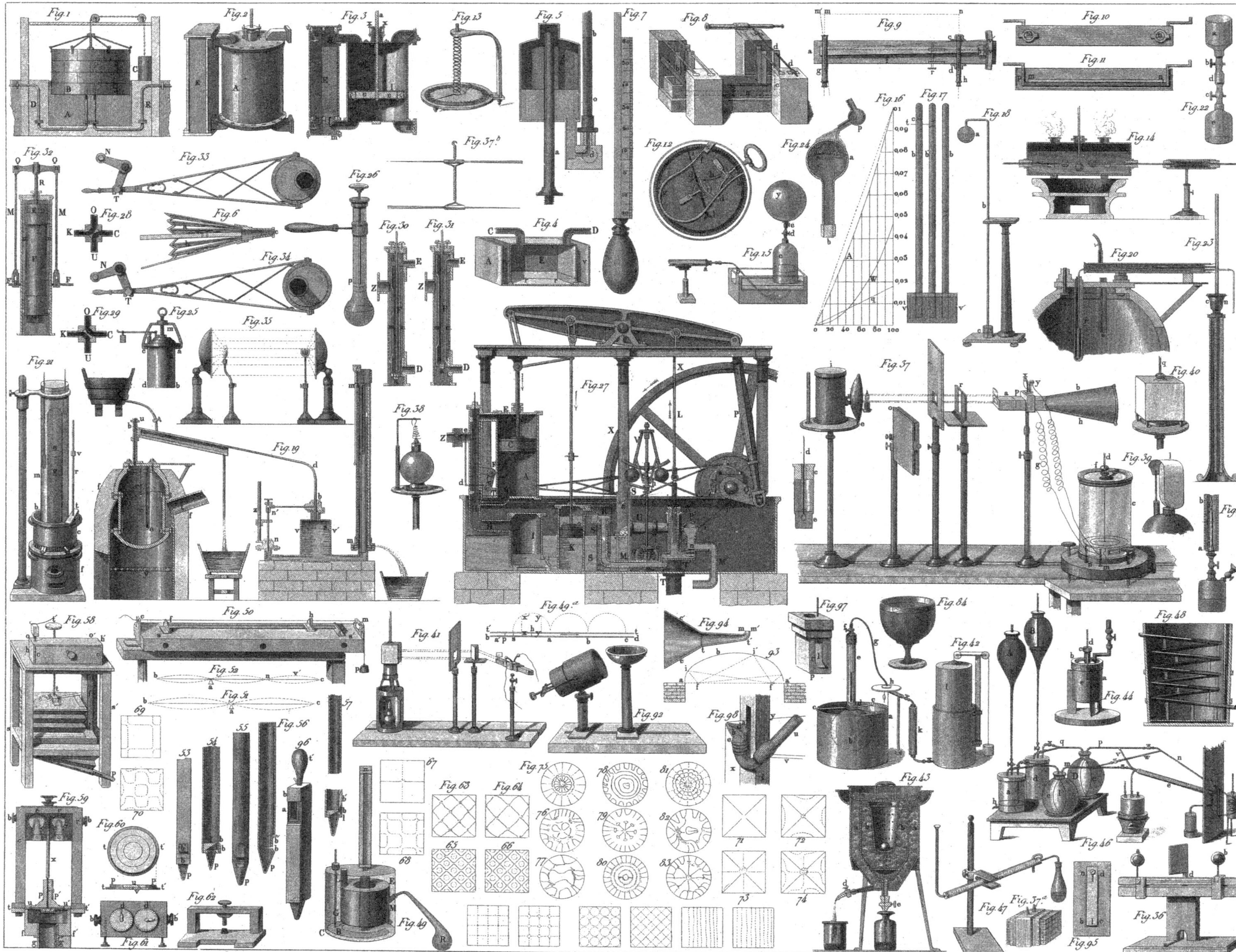

Plate 19 (Physics and Meteorology, Plate 4)

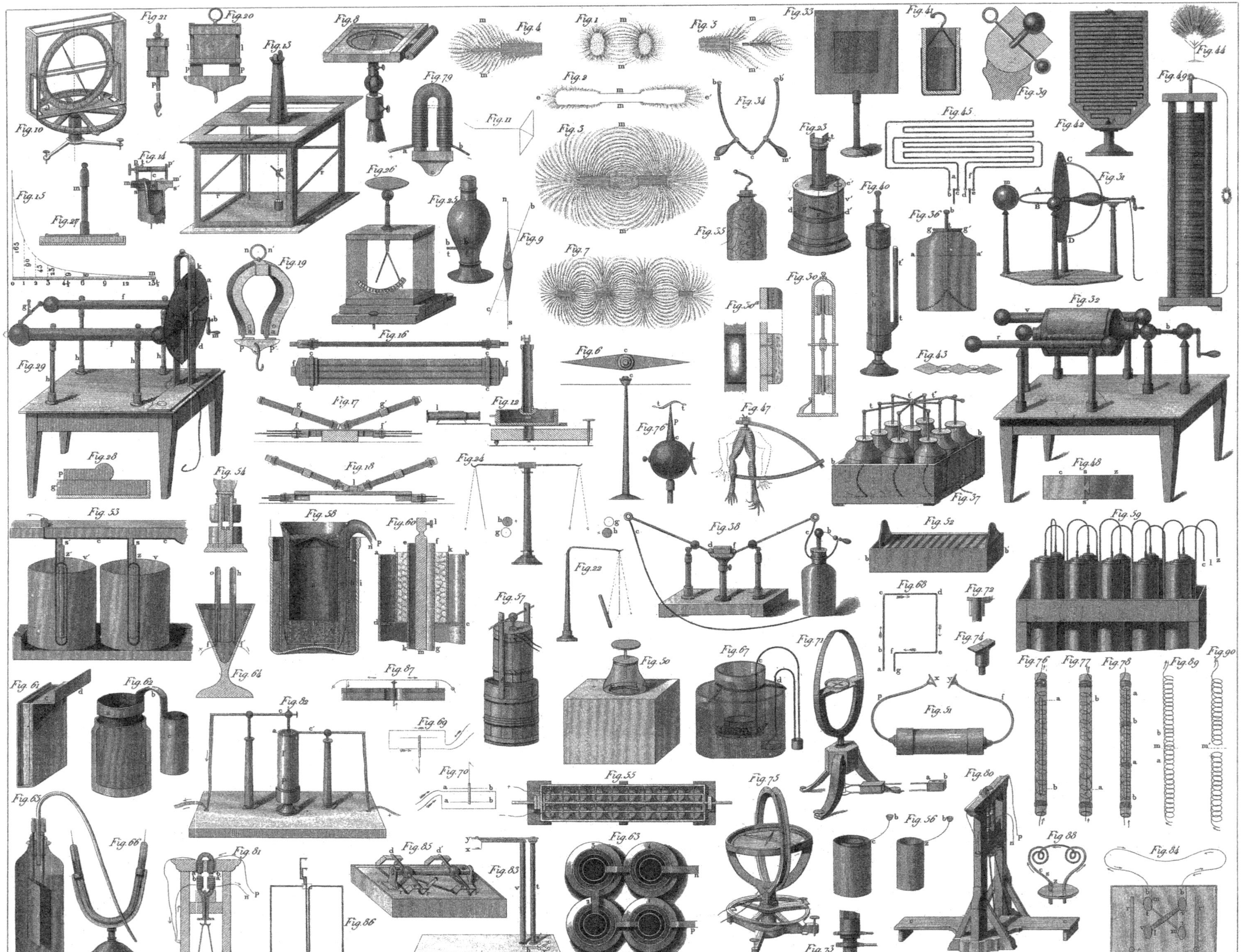

Plate 20 (Physics and Meteorology, Plate 5)

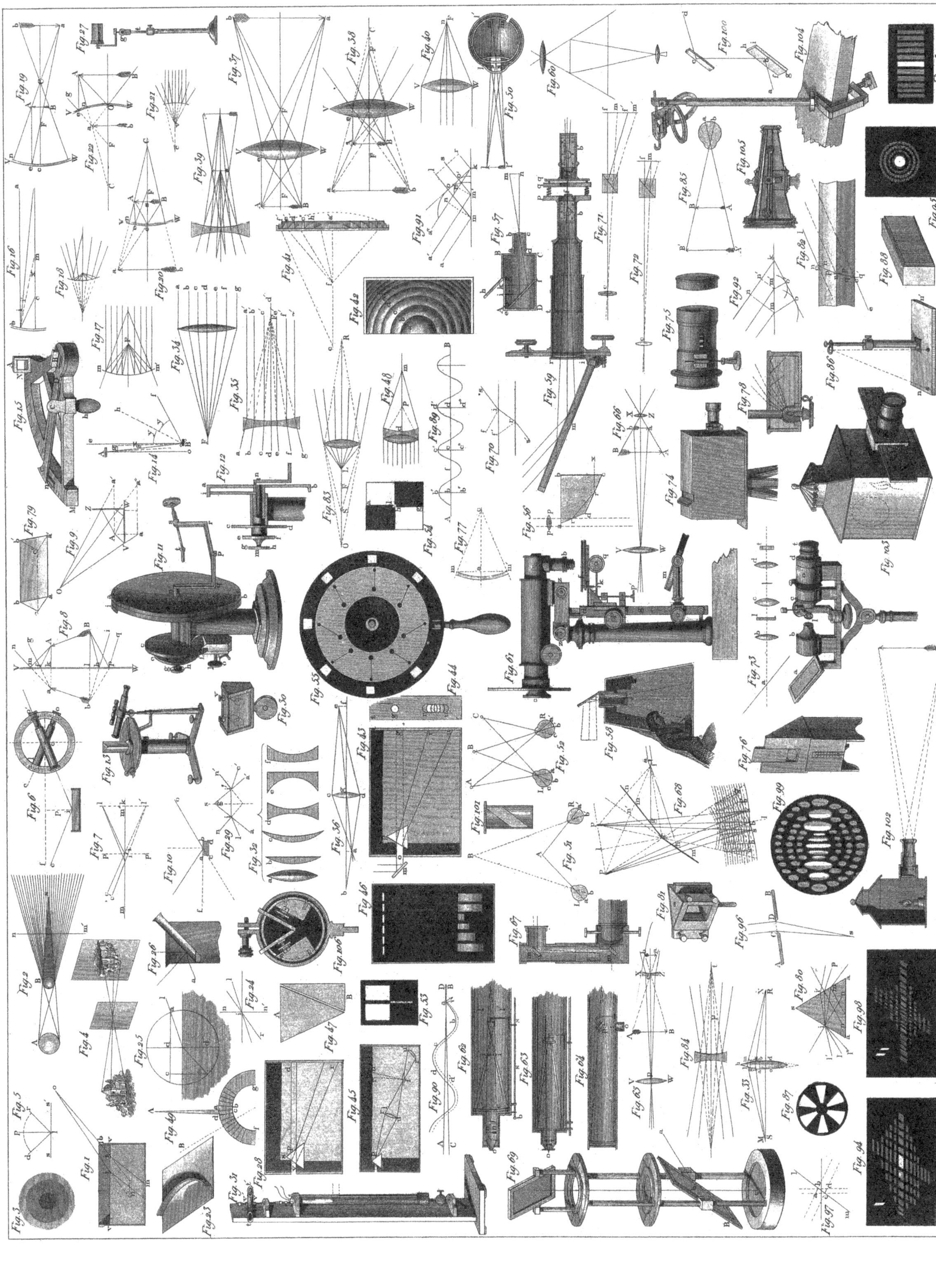

Plate 21 (Physics and Meteorology, Plate 6)

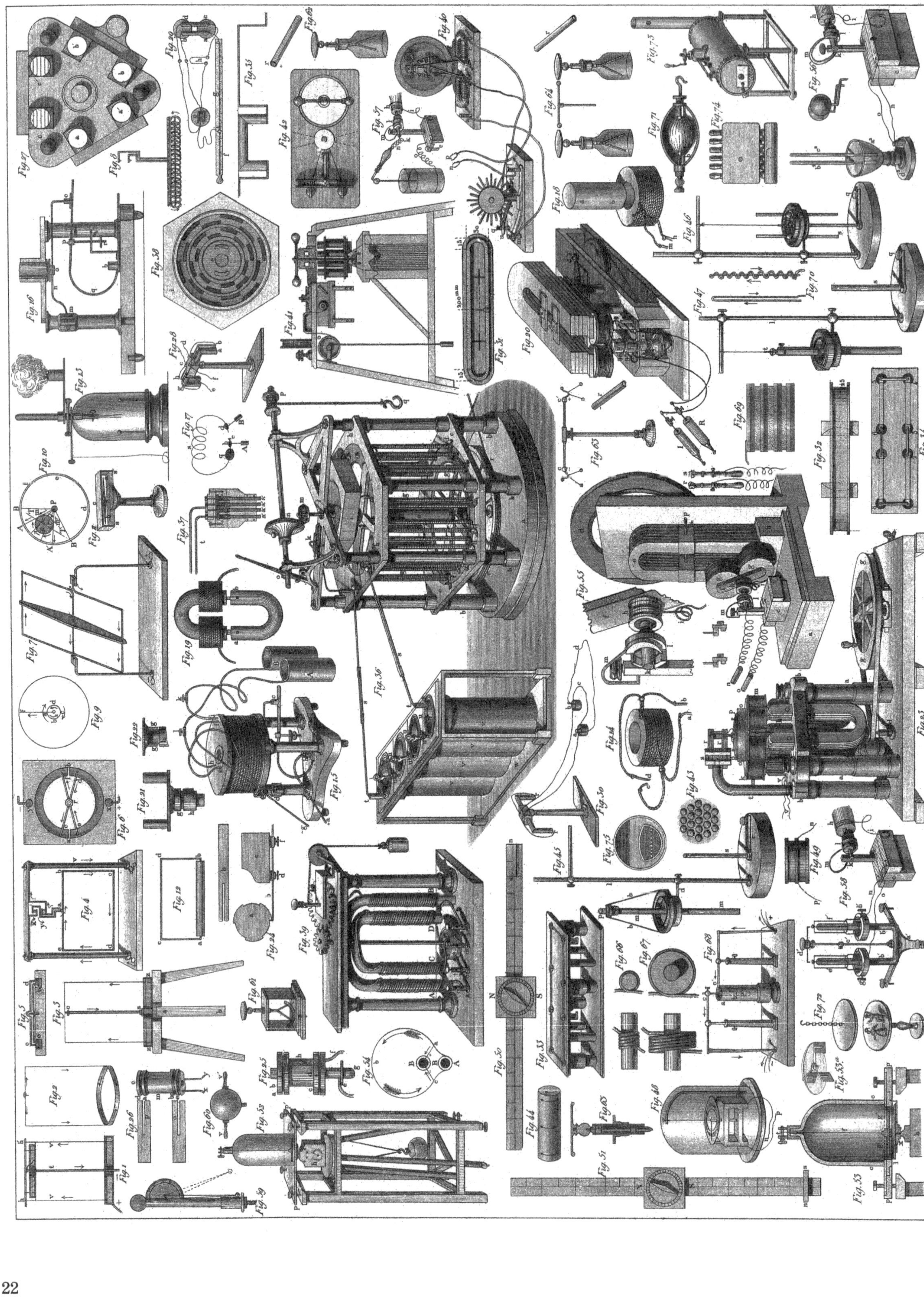

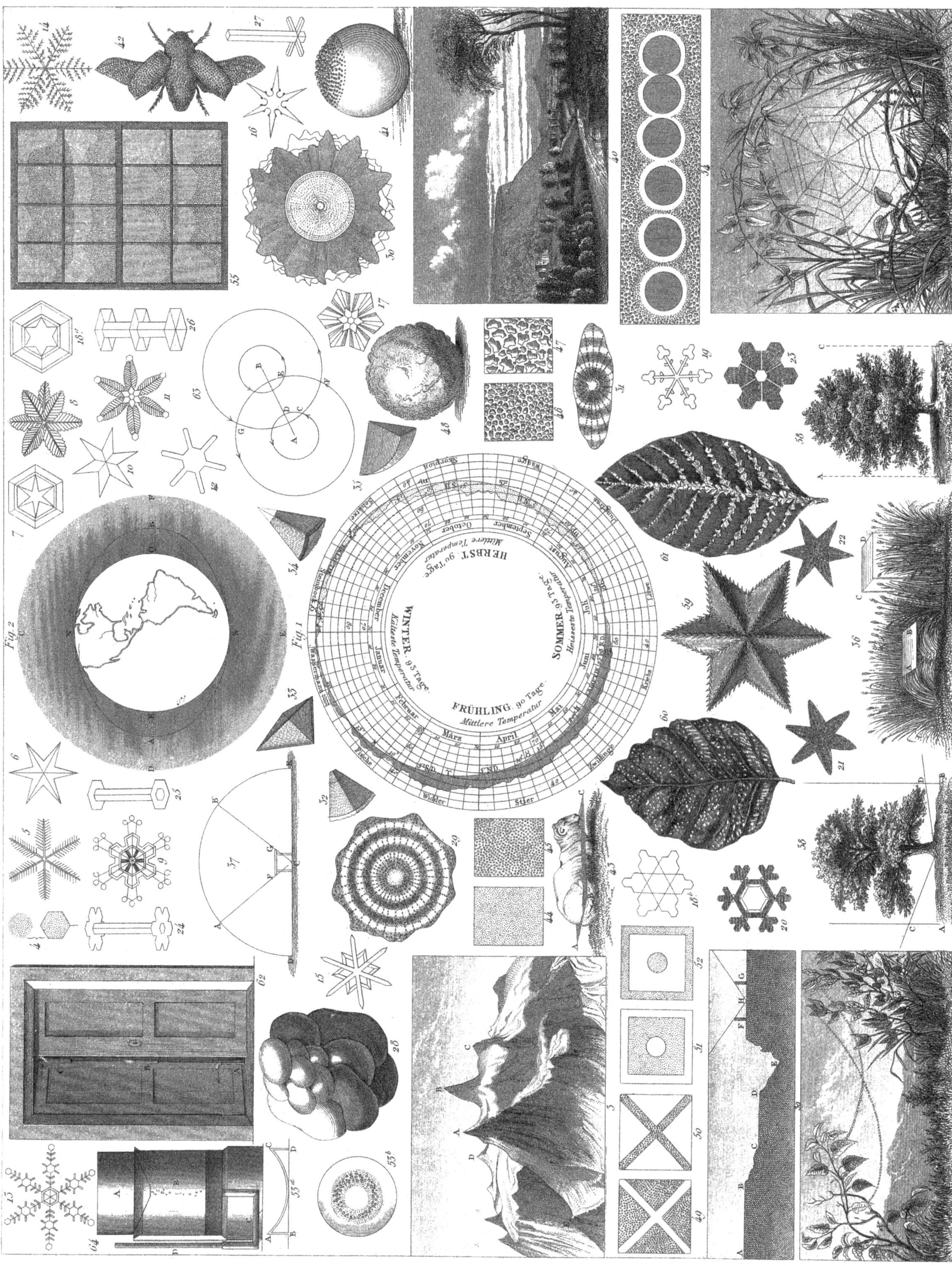

Plate 23 (Physics and Meteorology, Plate 8)

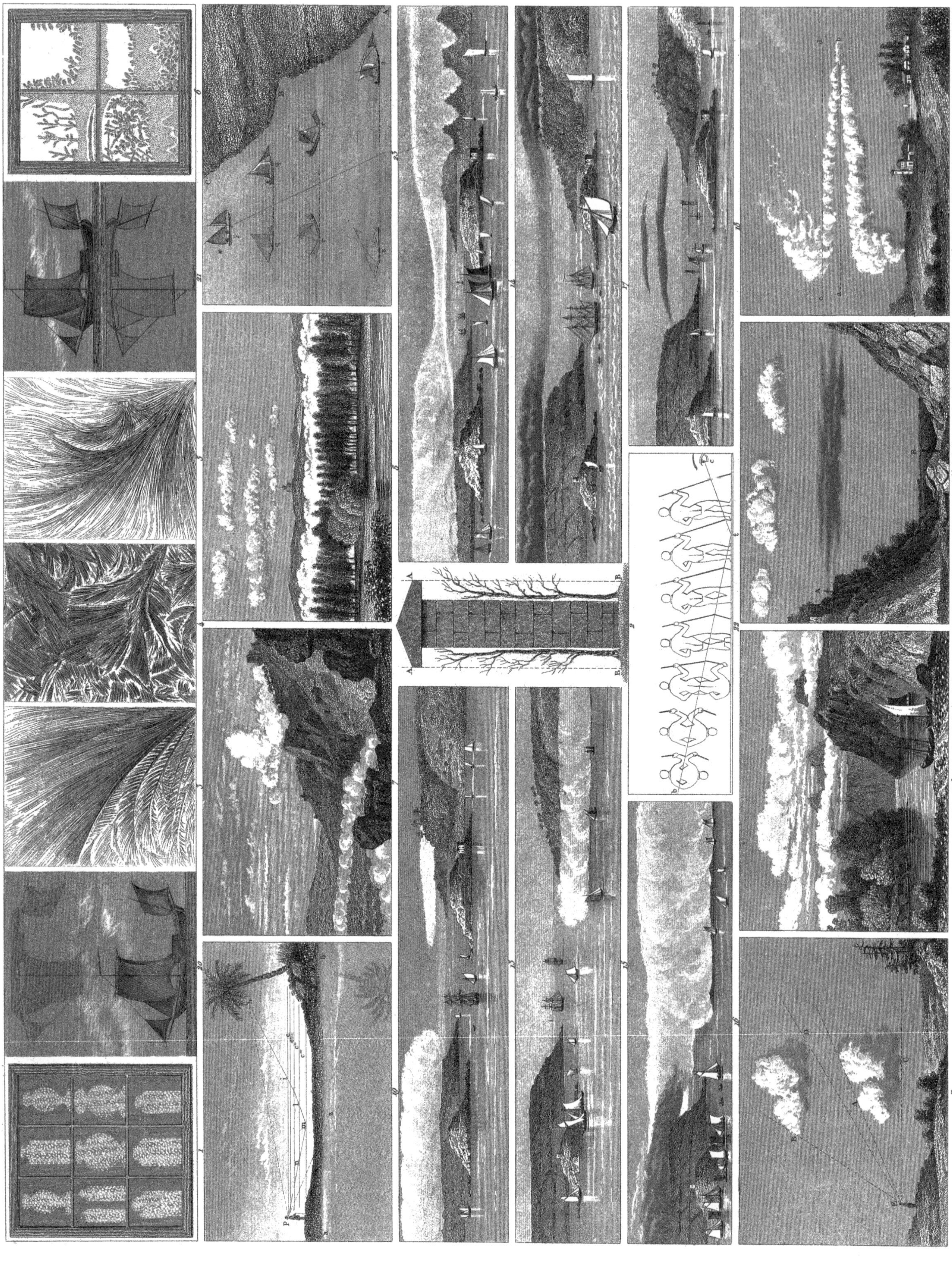

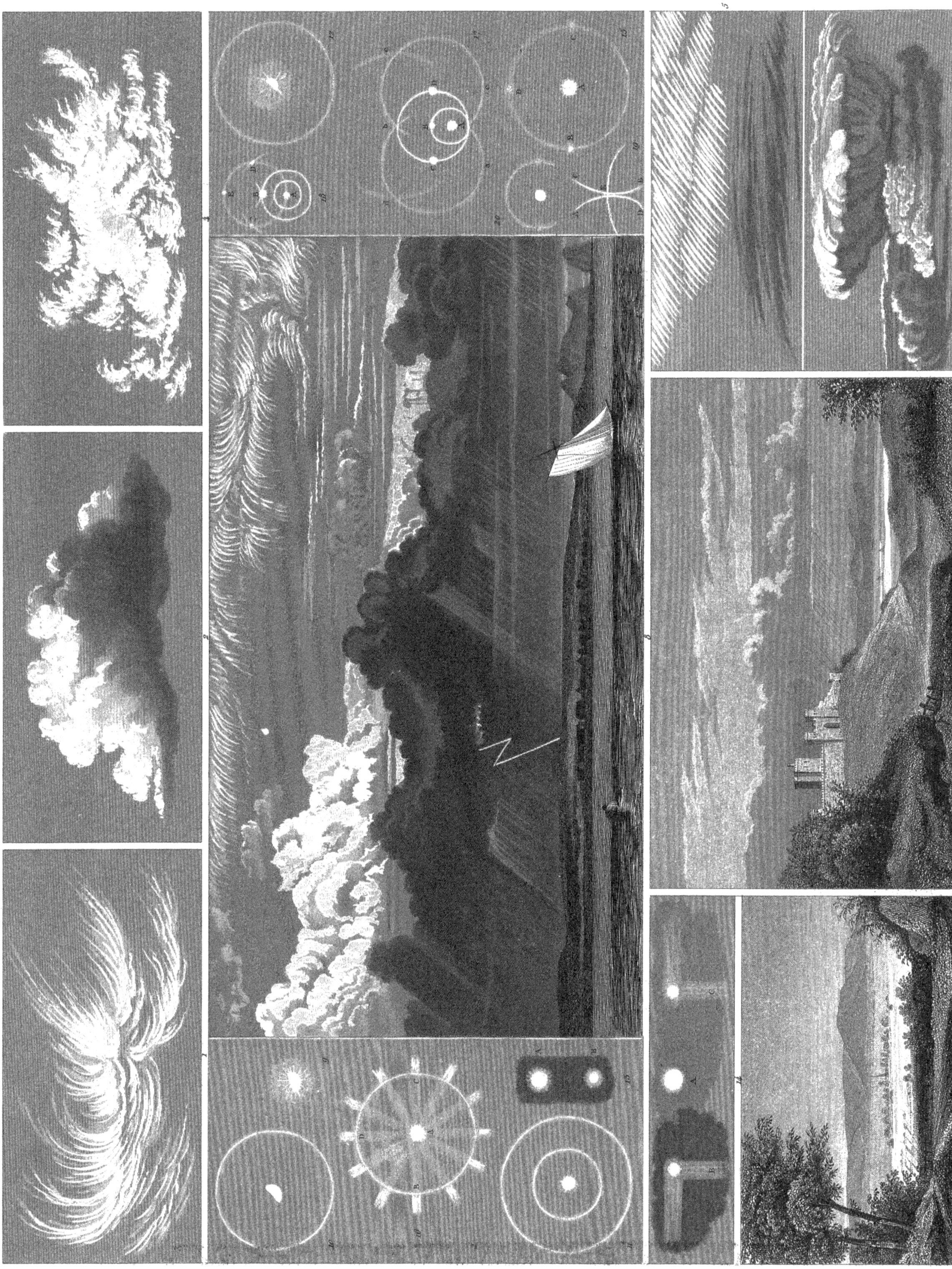

Plate 25 (Physics and Meteorology, Plate 10)

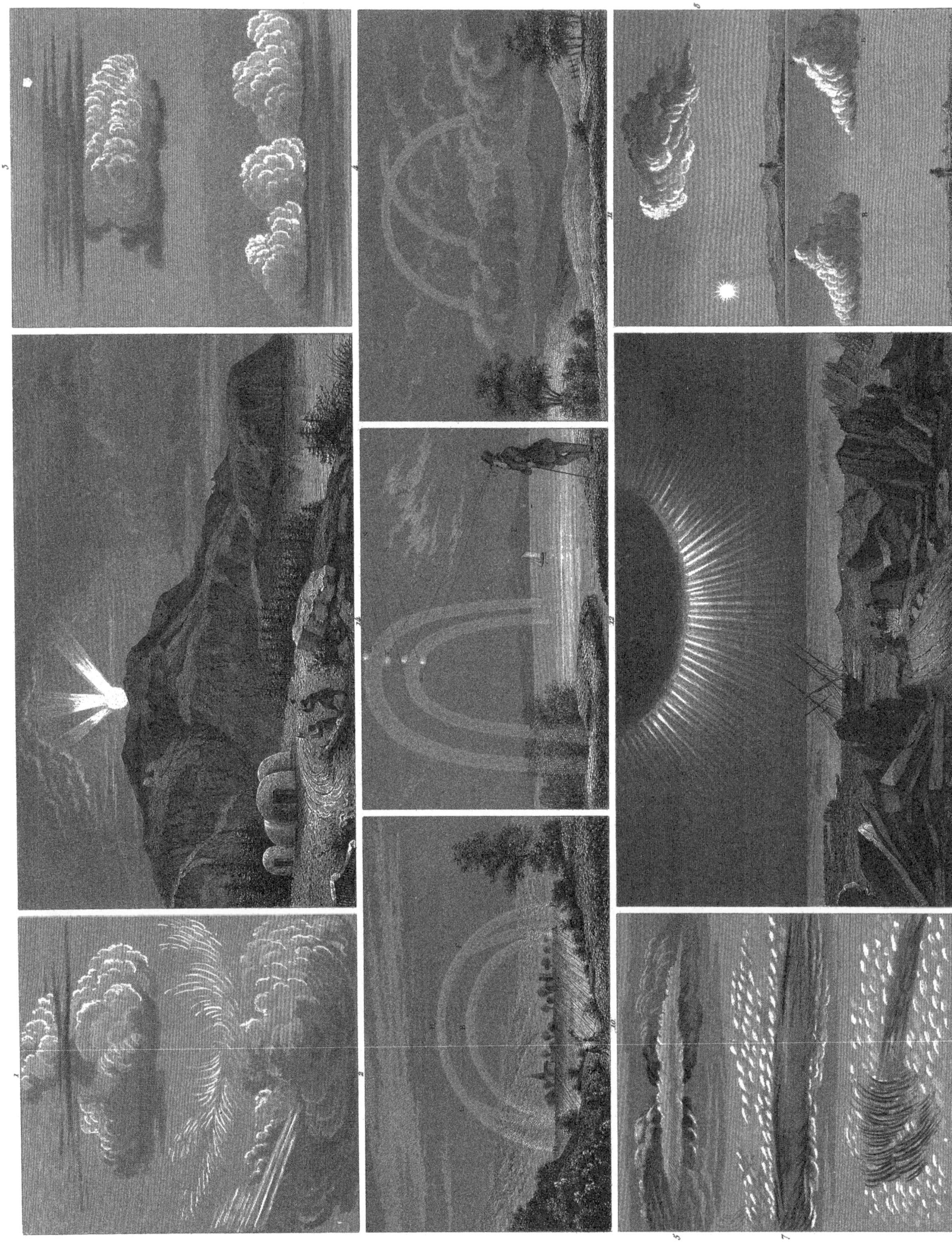

Plate 26 (Physics and Meteorology. Plate 11)

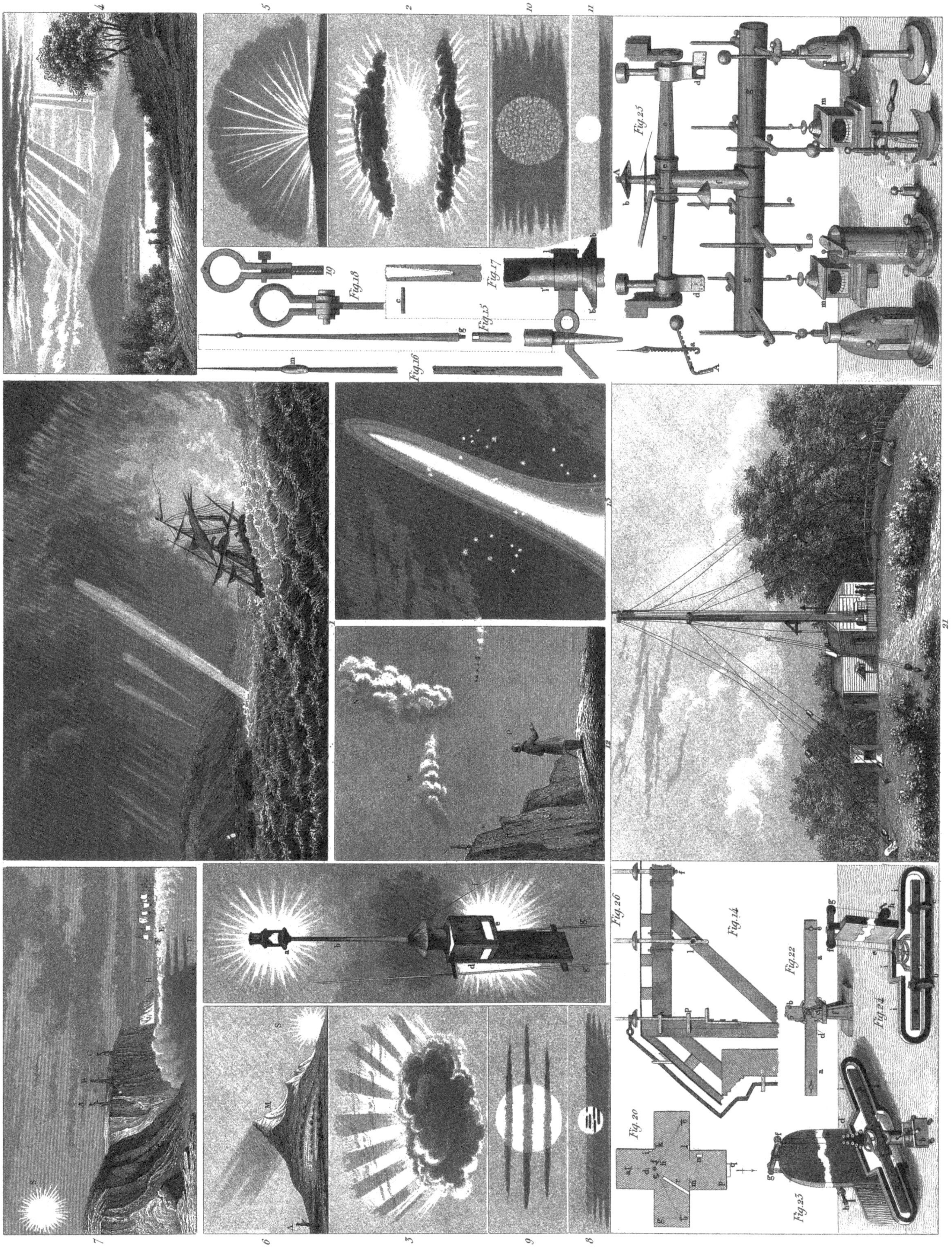

Plate 27 (Physics and Meteorology, Plate 12)

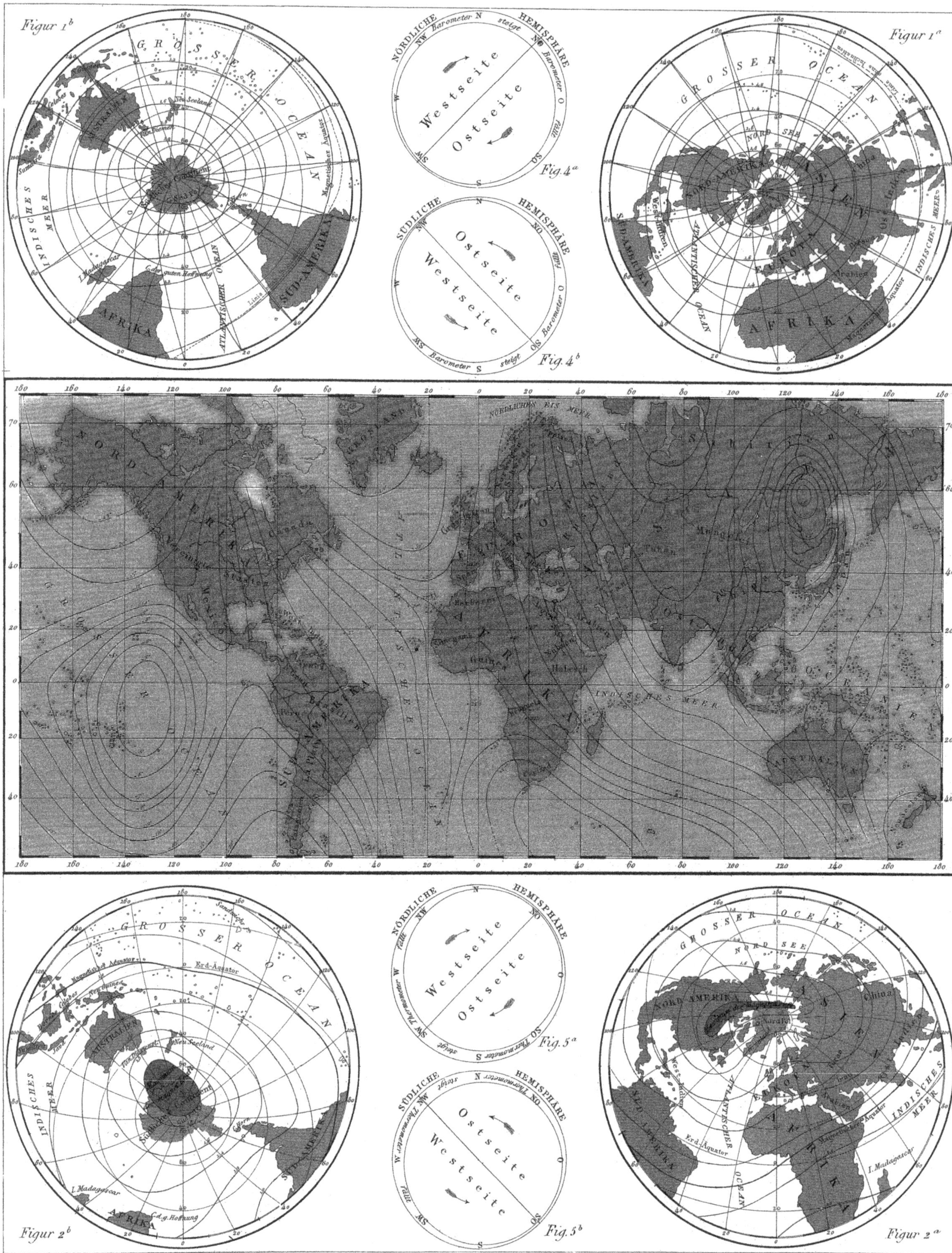

Plate 28 (Physics and Meteorology, Plate 13)

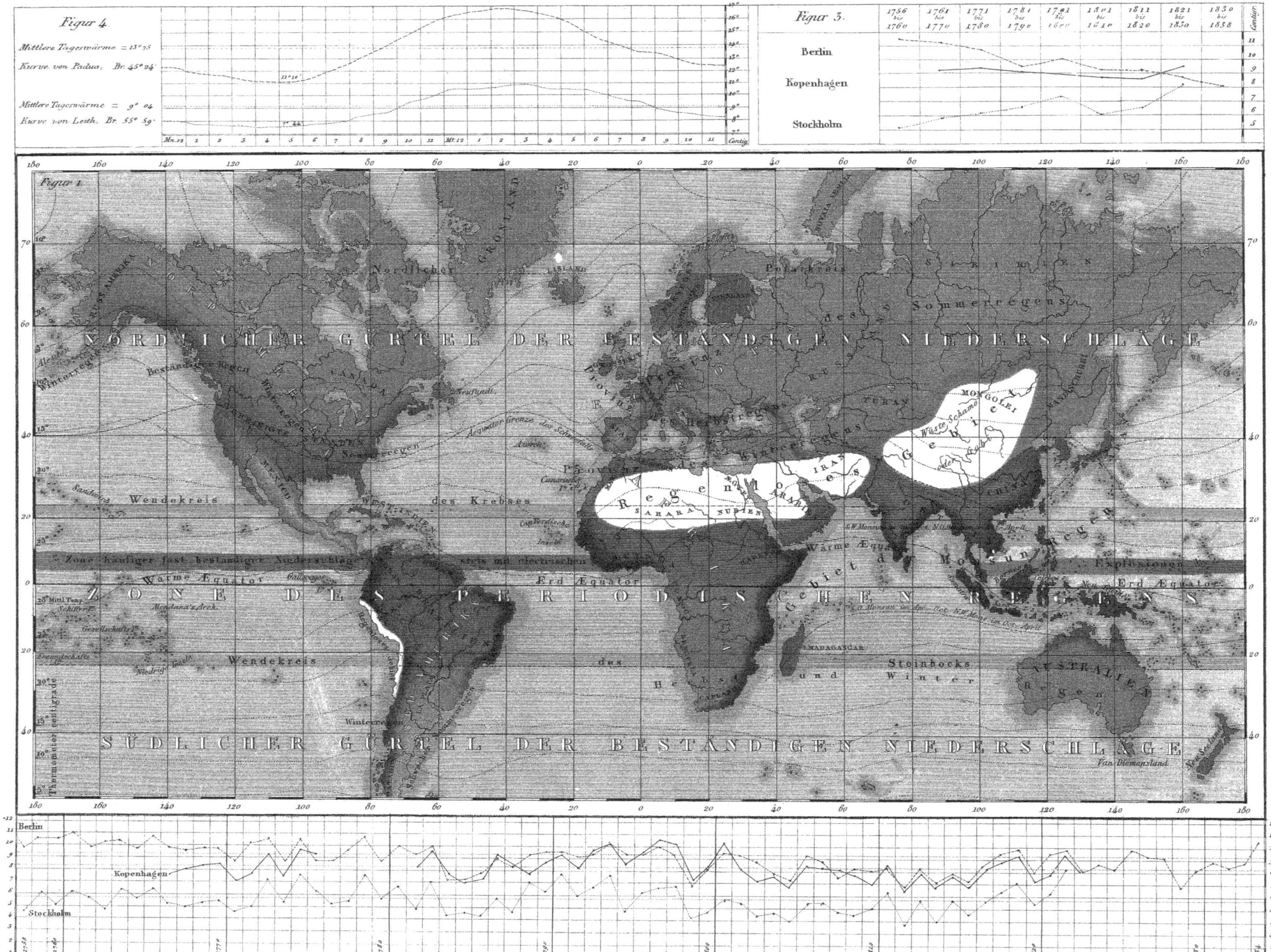

Figur 2.

Plate 29 (Physics and Meteorology, Plate 14)

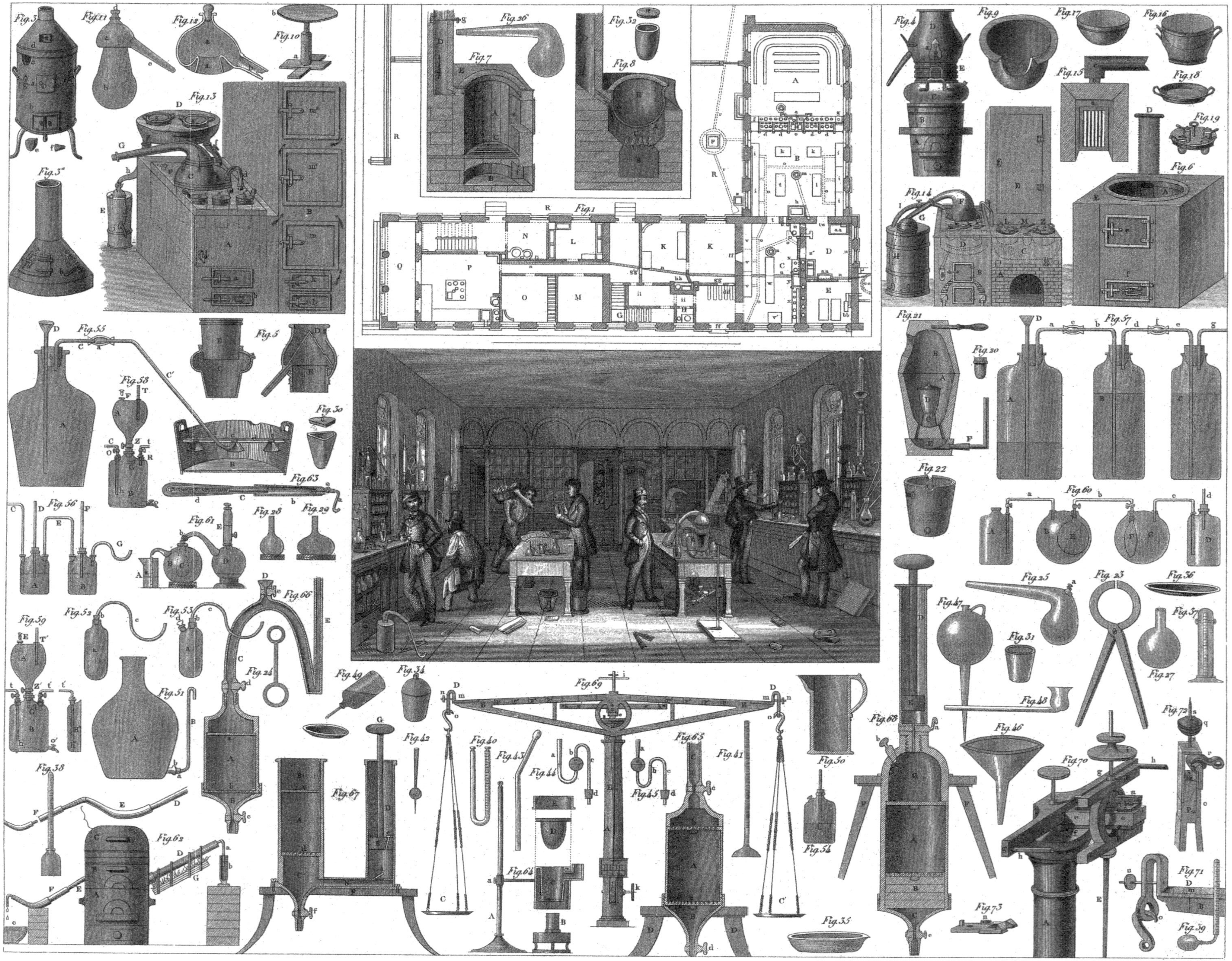

Plate 30 (Chemistry, Mineralogy, and Geology, Plate 1)

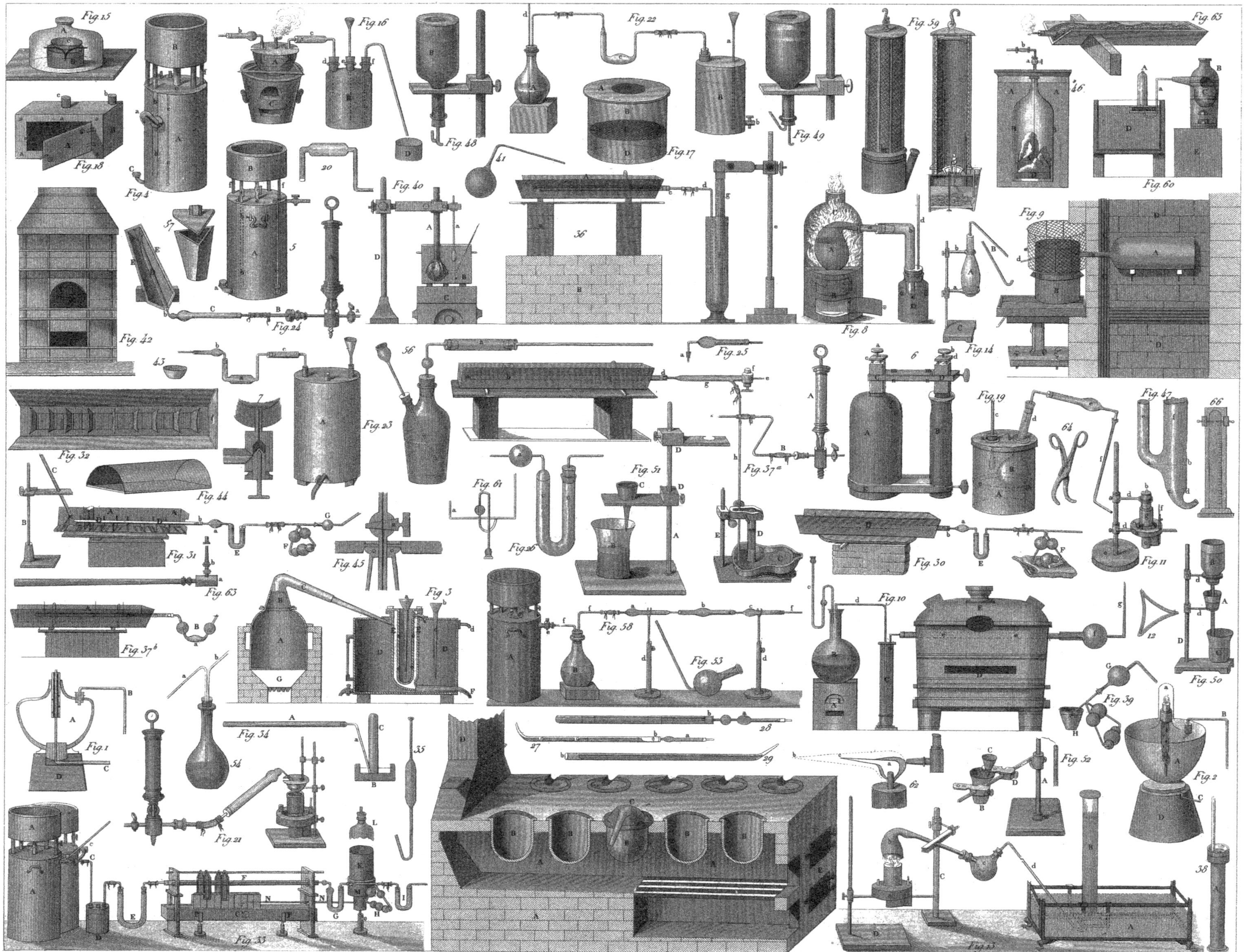

Plate 31 (Chemistry, Mineralogy, and Geology, Plate 2)

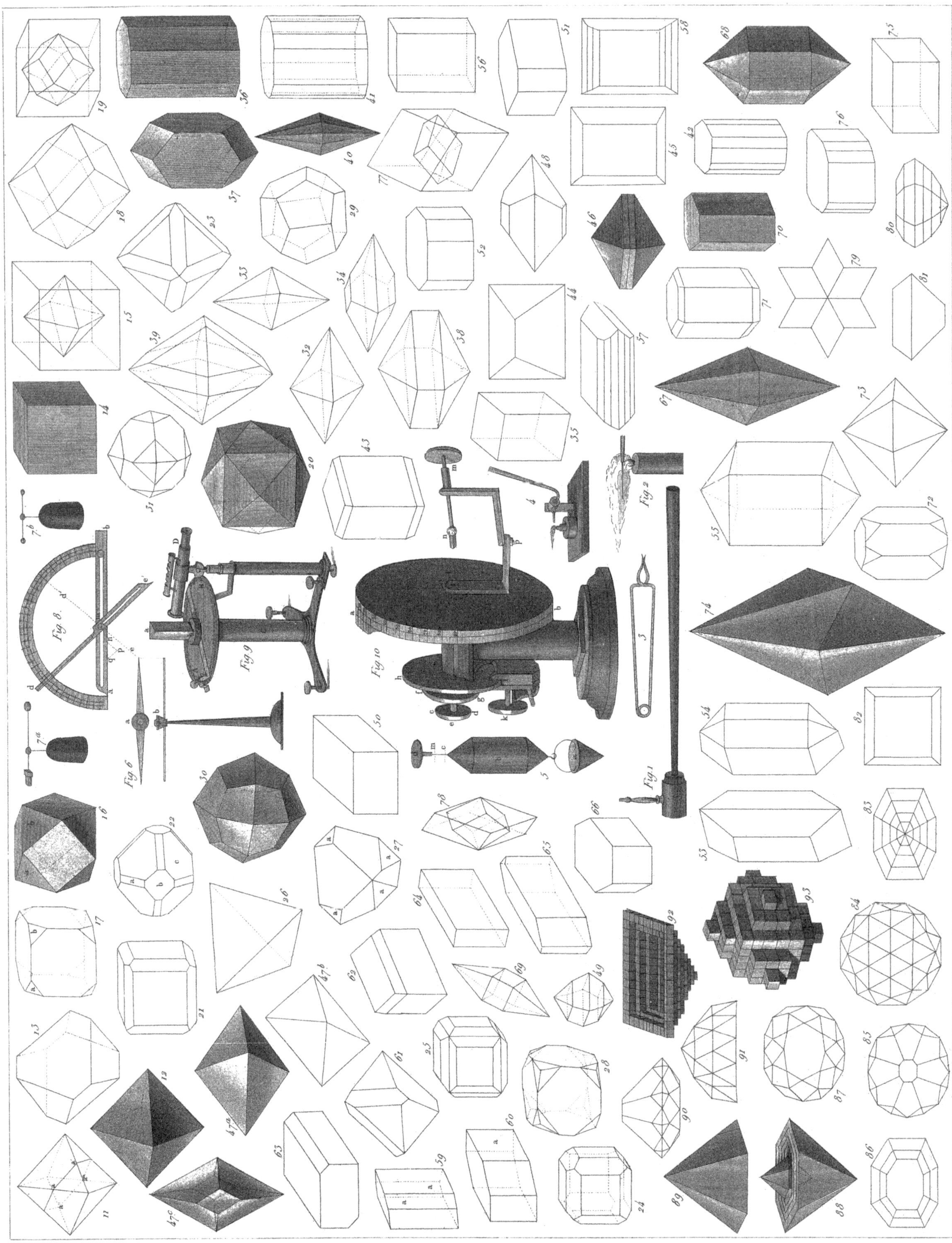
Fig 8
Fig 9
Fig 10
Fig 6
Fig 1
Fig 2

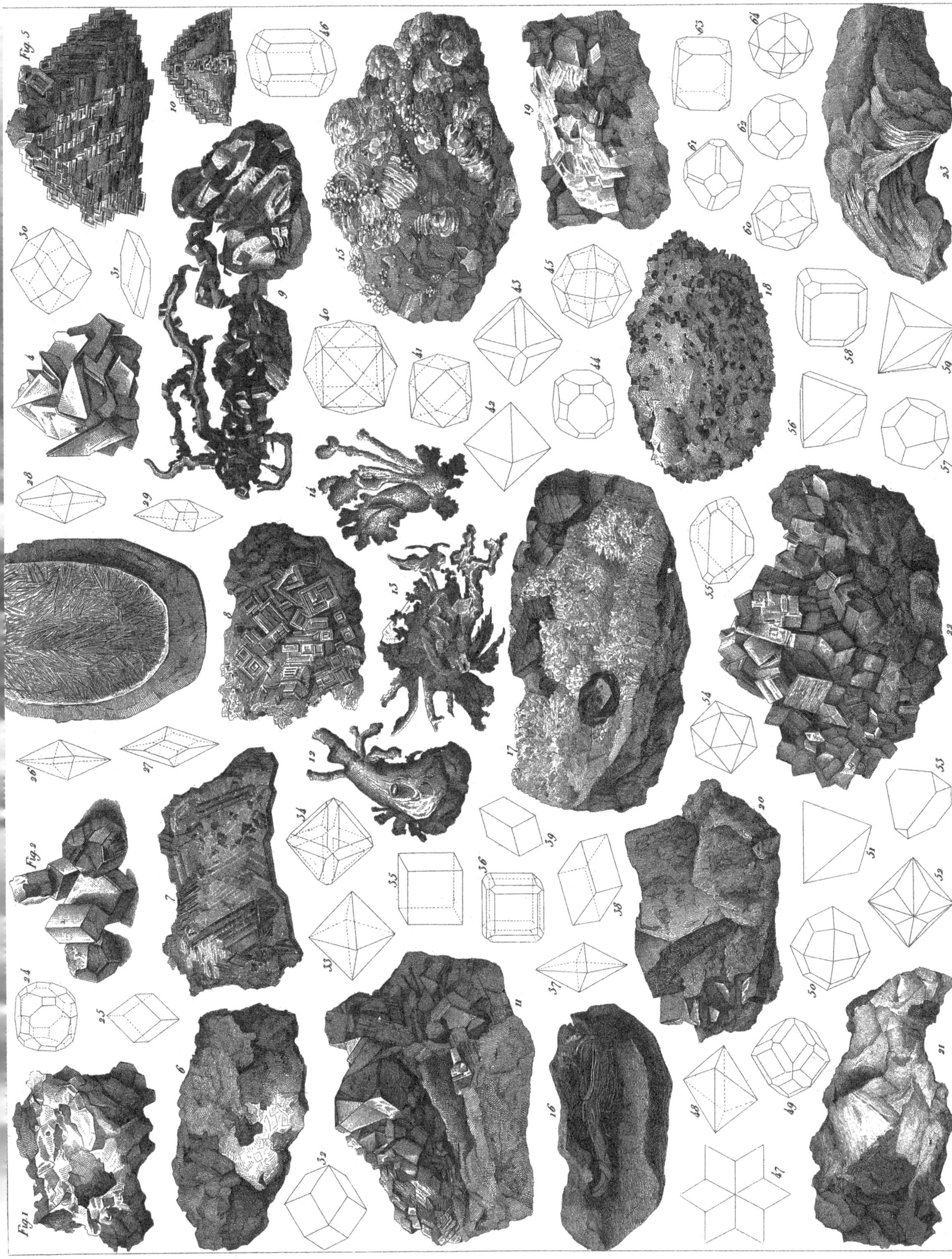

Plate 33 (Chemistry, Mineralogy, and Geology, Plate 4)

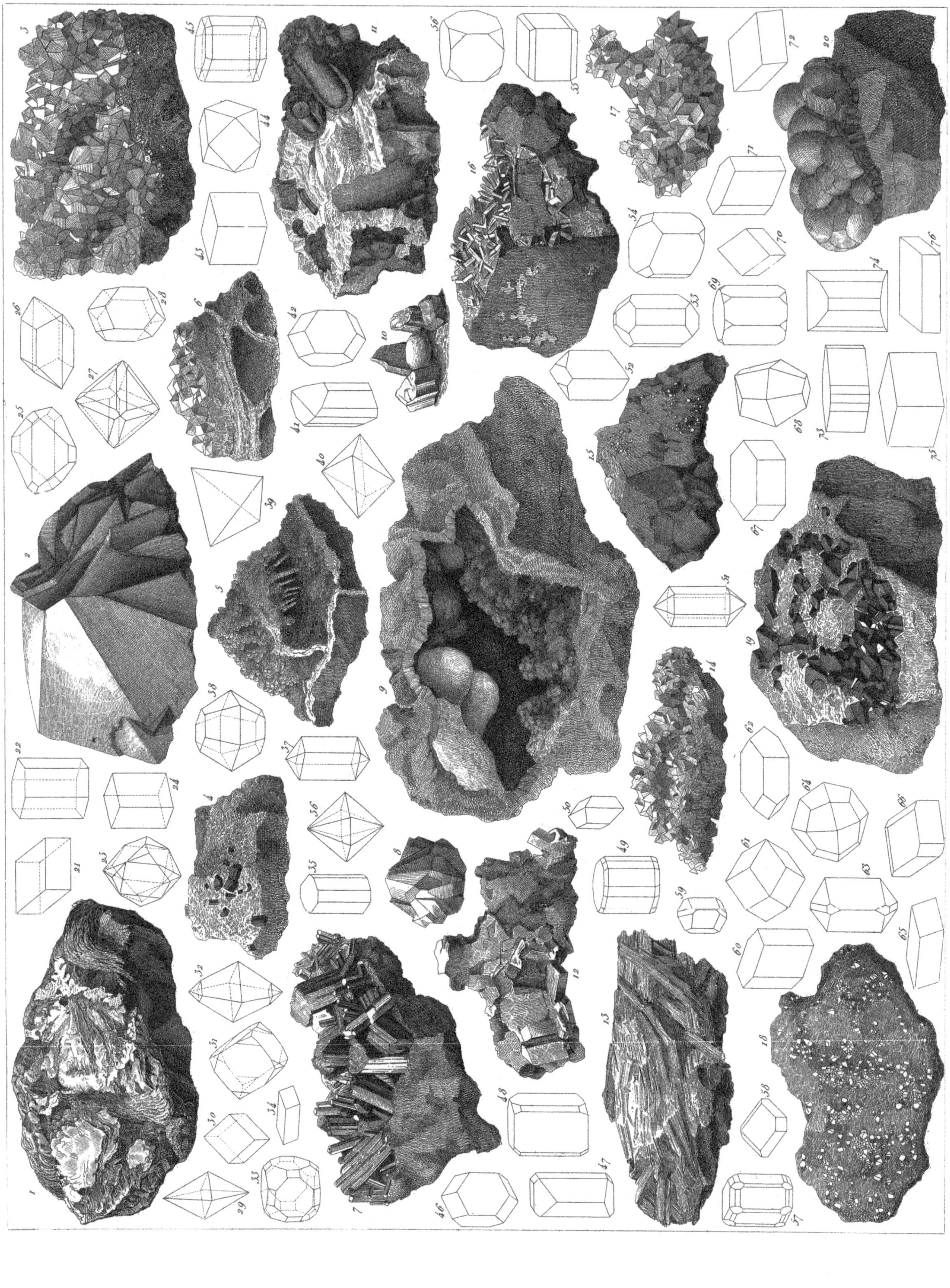

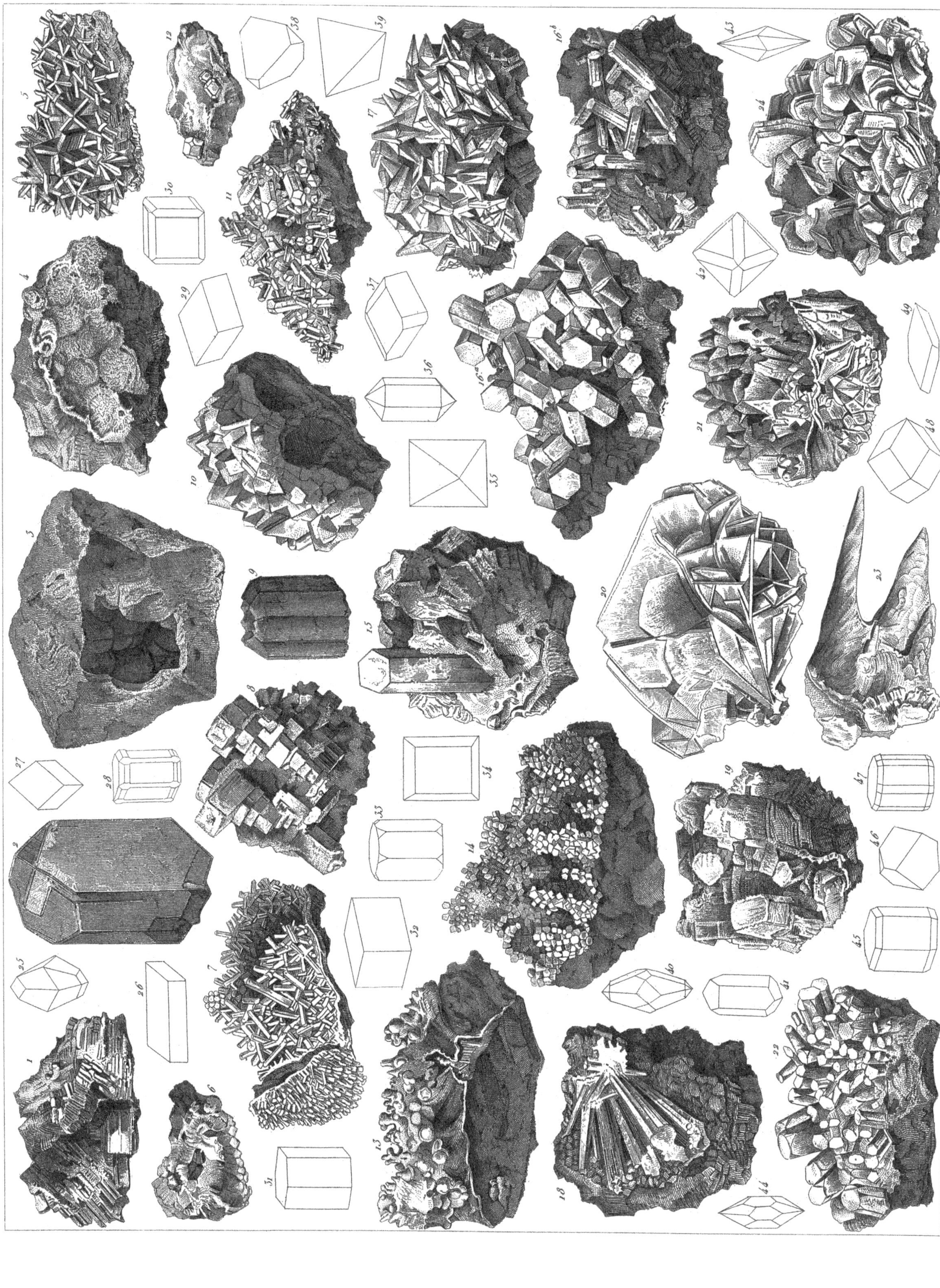

Plate 35 (Chemistry, Mineralogy, and Geology, Plate 6)

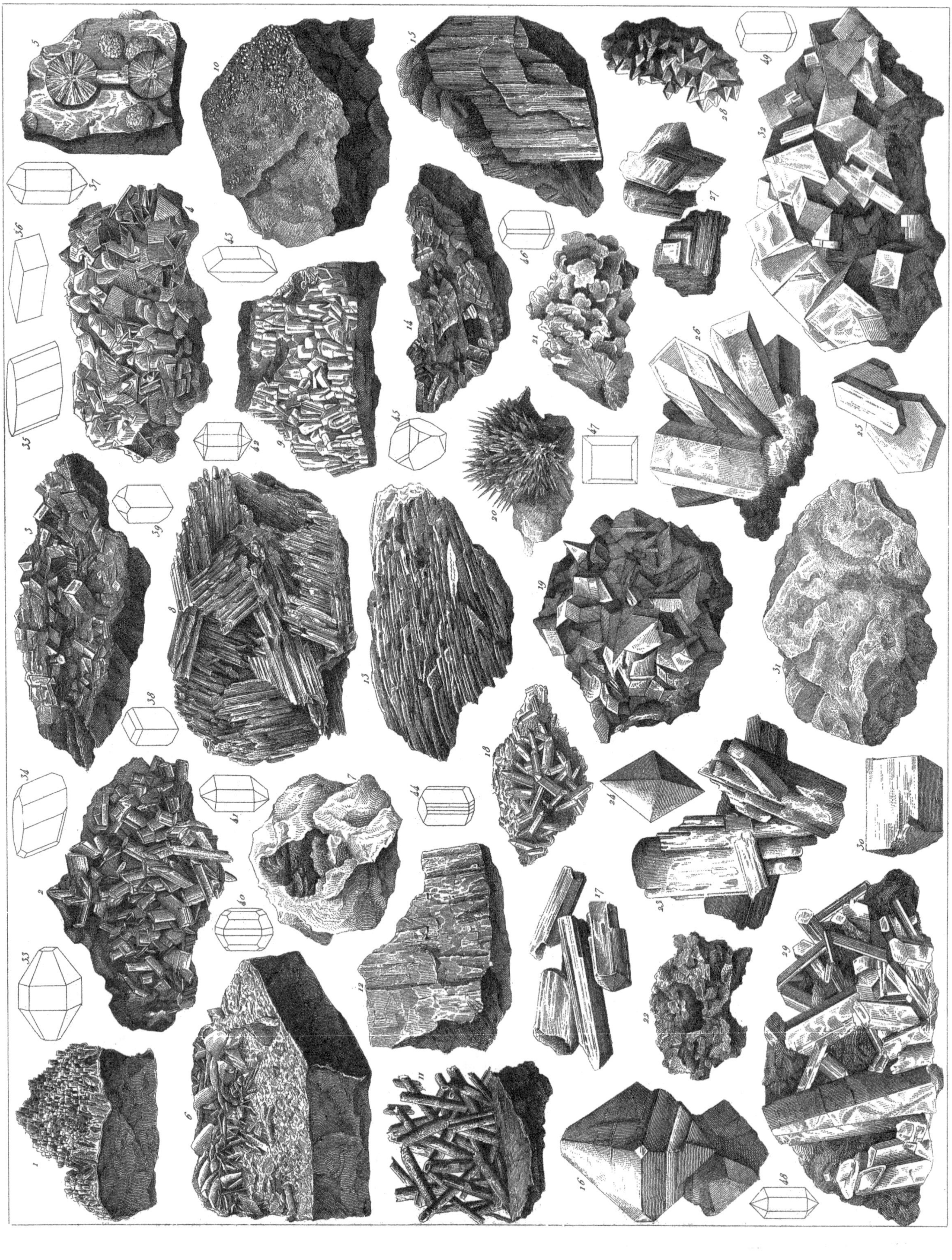

Plate 36 (Chemistry, Mineralogy, and Geology, Plate 7)

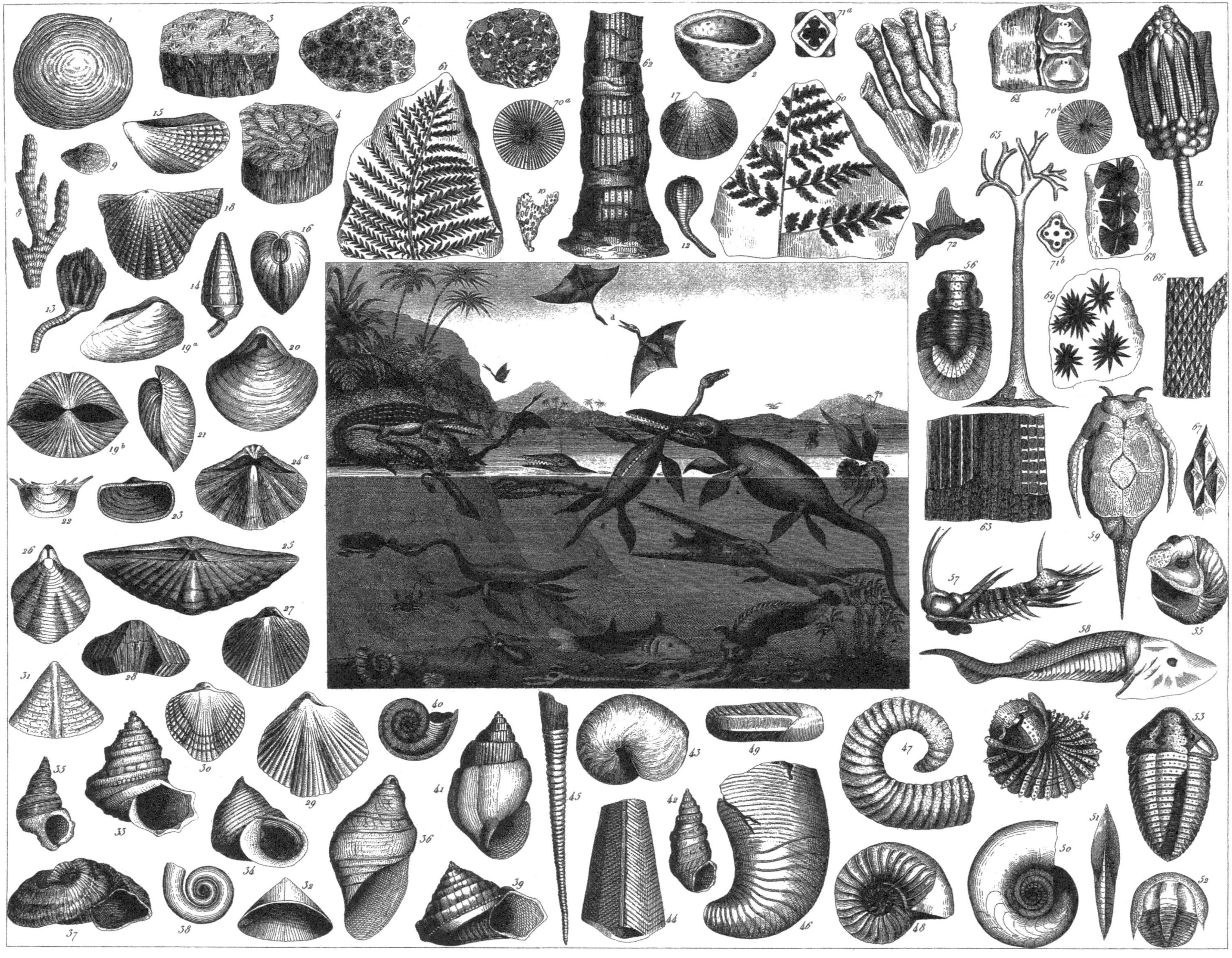

Plate 37 (Chemistry, Mineralogy, and Geology, Plate 8)

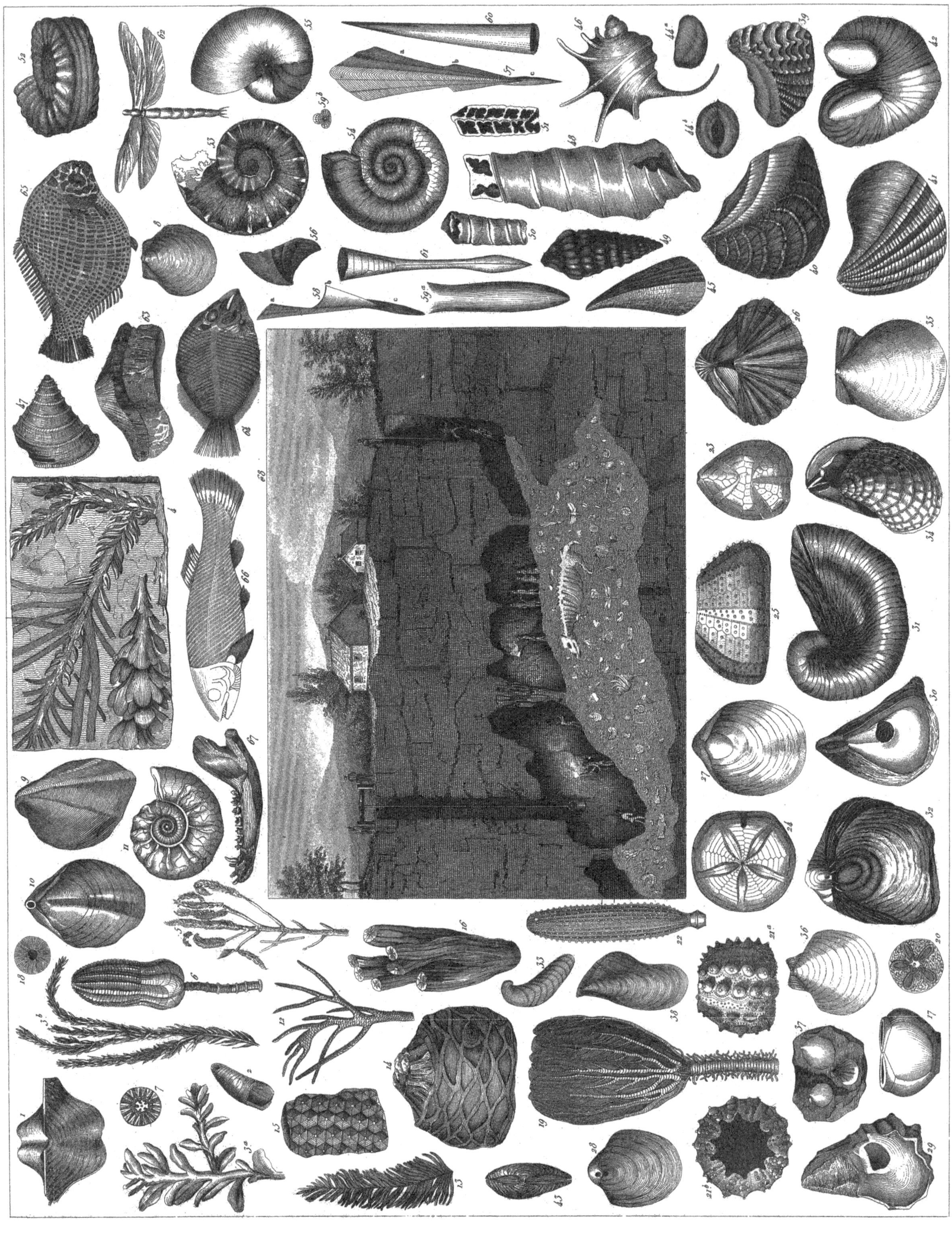

Plate 38 (Chemistry, Mineralogy, and Geology, Plate 9)

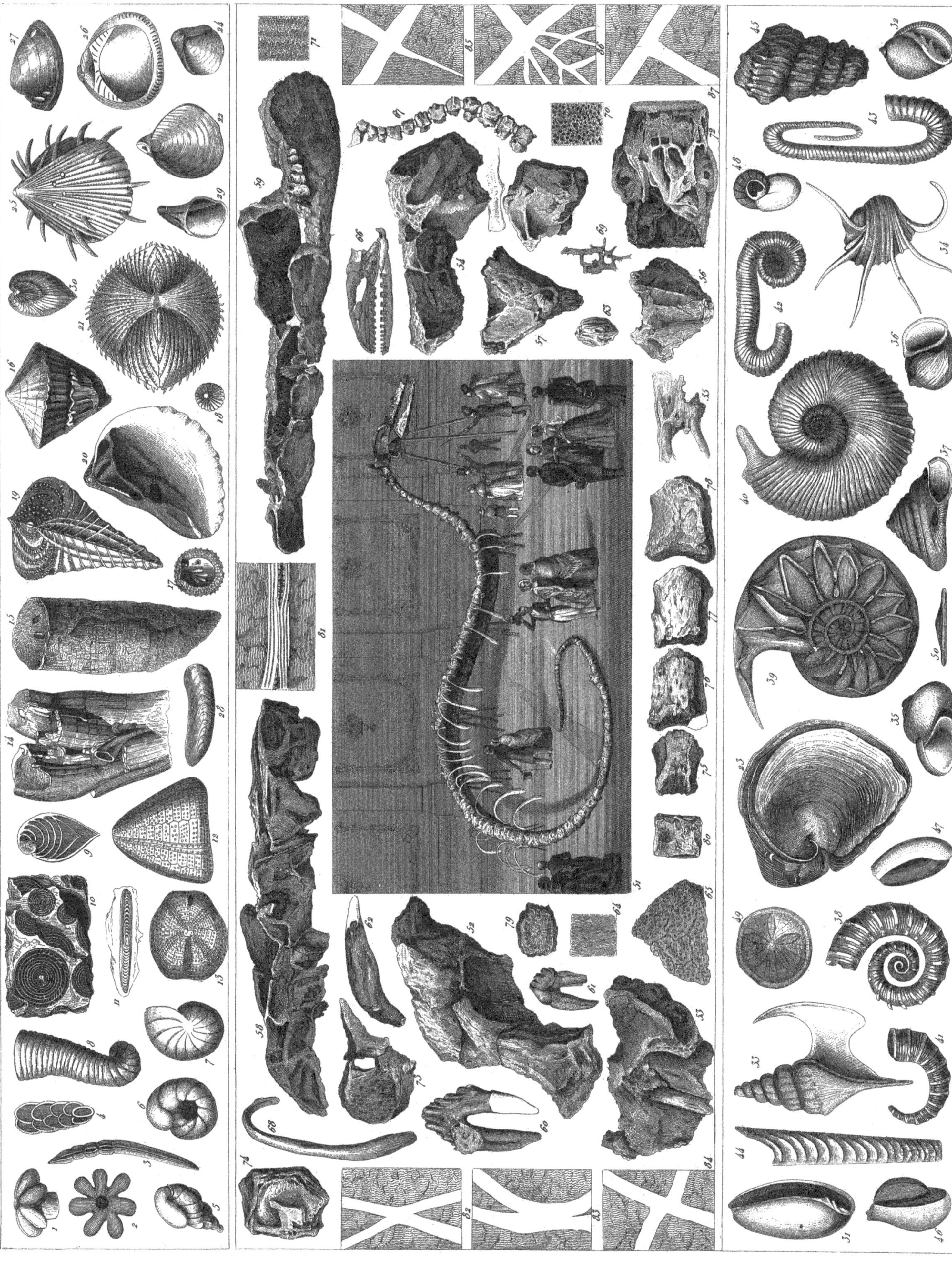

Plate 39 (Chemistry, Mineralogy, and Geology, Plate 10)

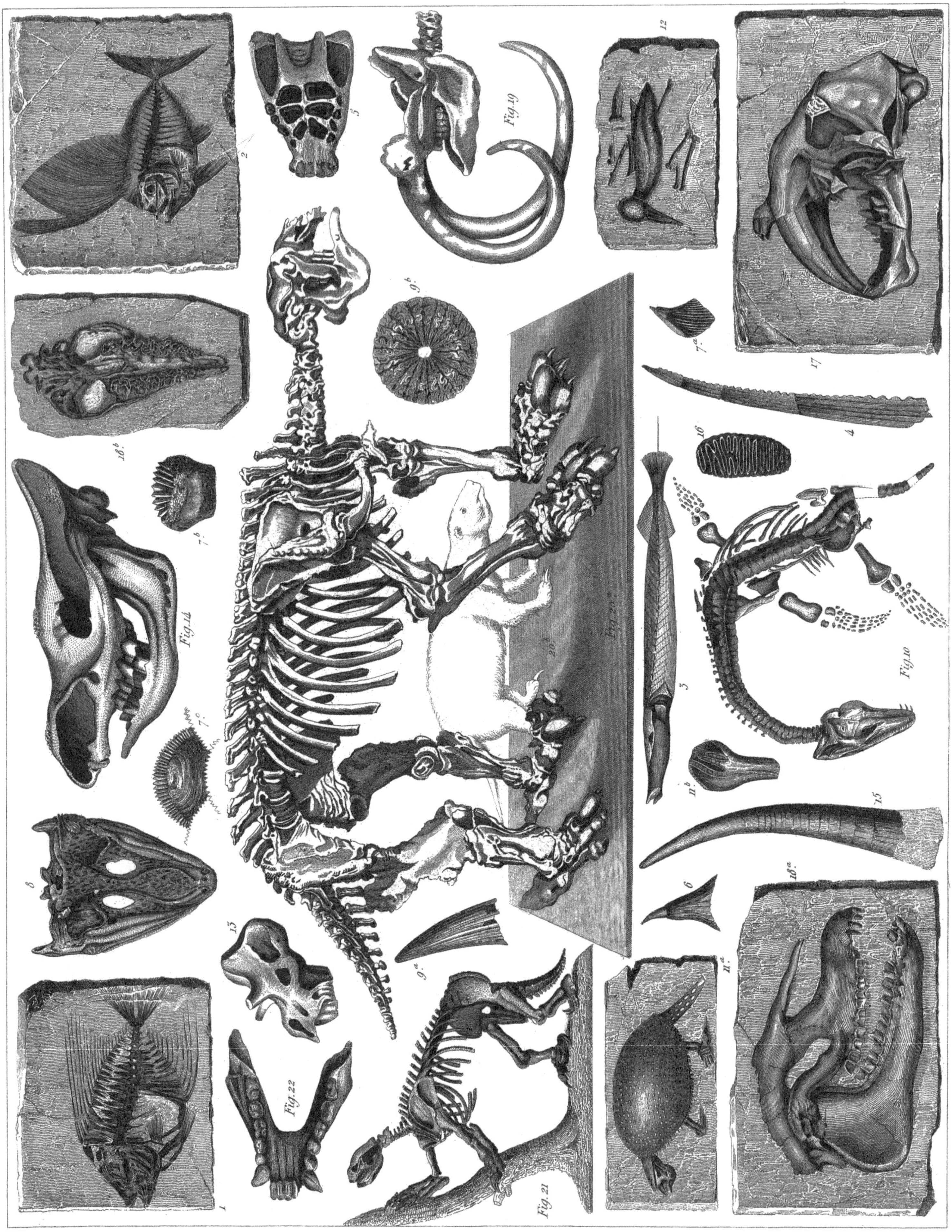
Fig.19
Fig.14
Fig.10
Fig.21
Fig.22
Fig.20ᵃ

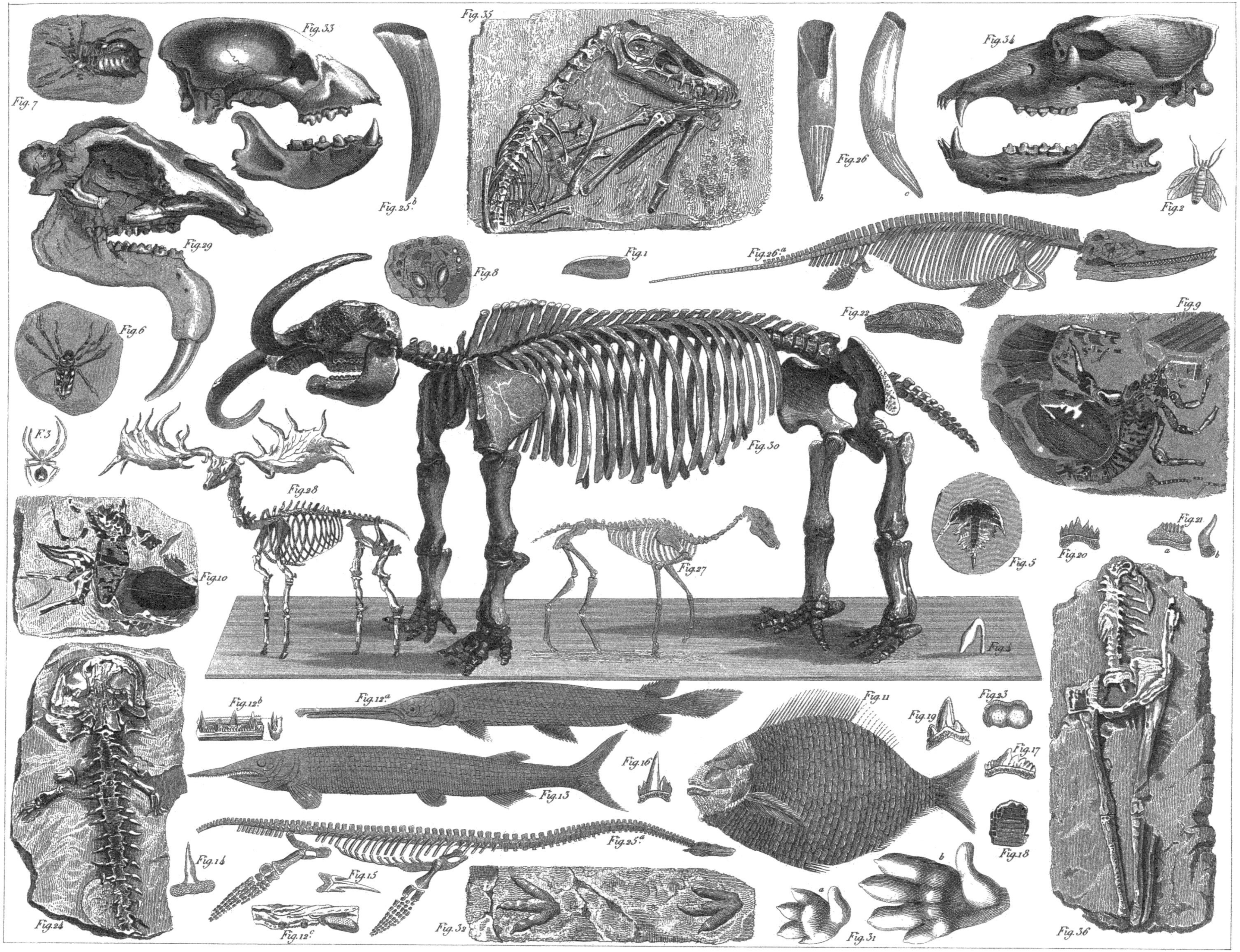

Plate 41 (Chemistry, Mineralogy, and Geology, Plate 12)

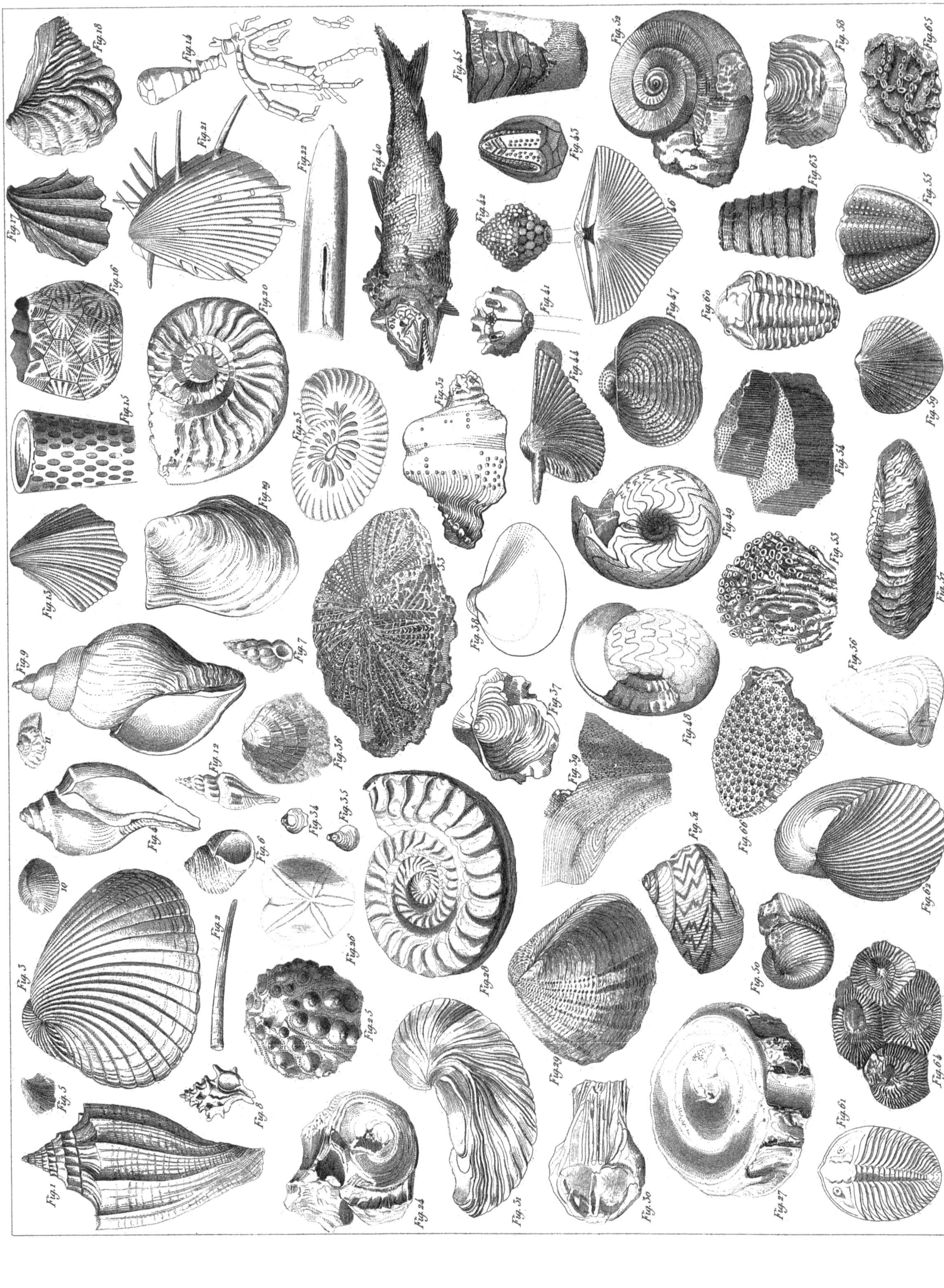

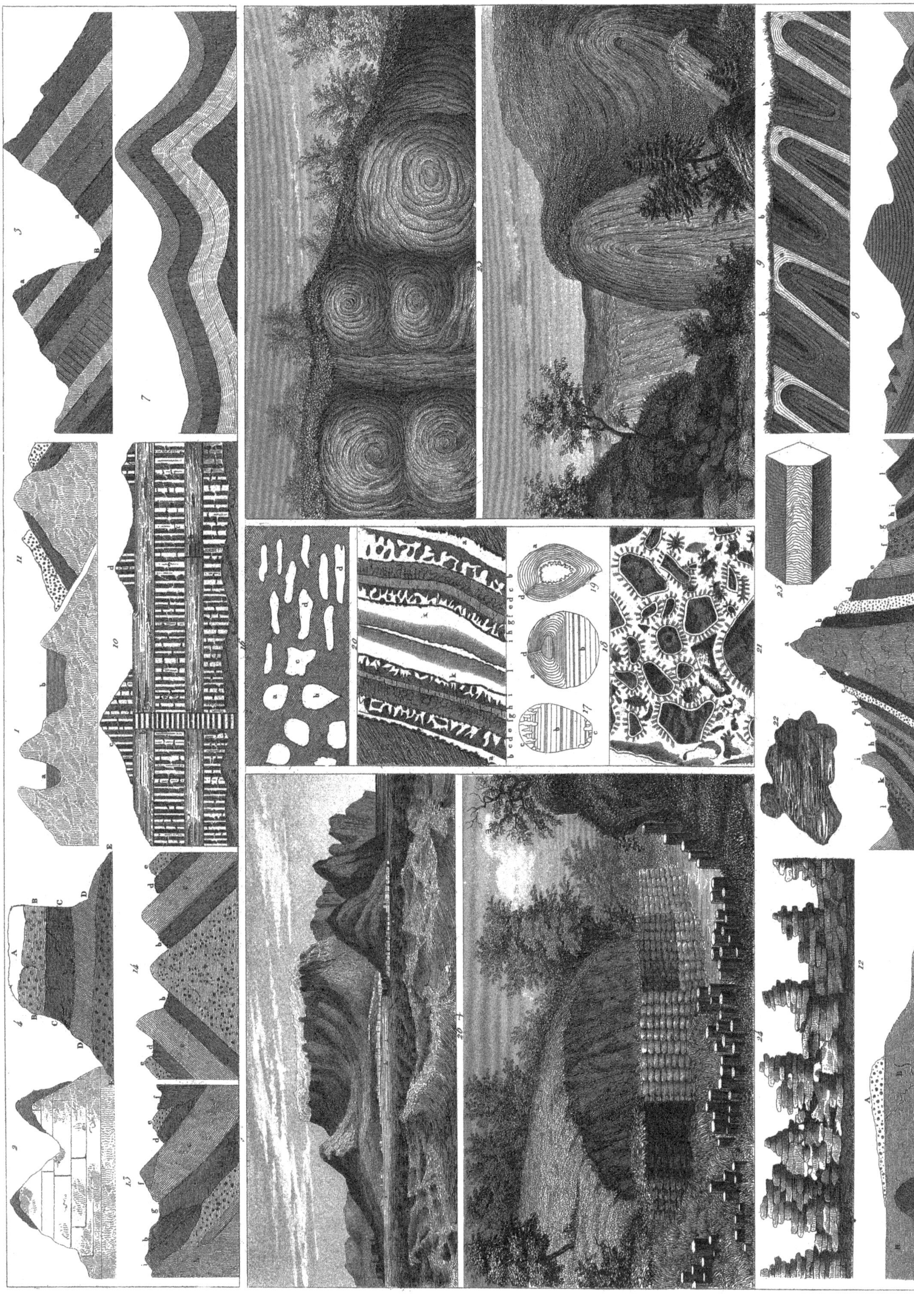

Plate 43 (Chemistry, Mineralogy, and Geology, Plate 14)

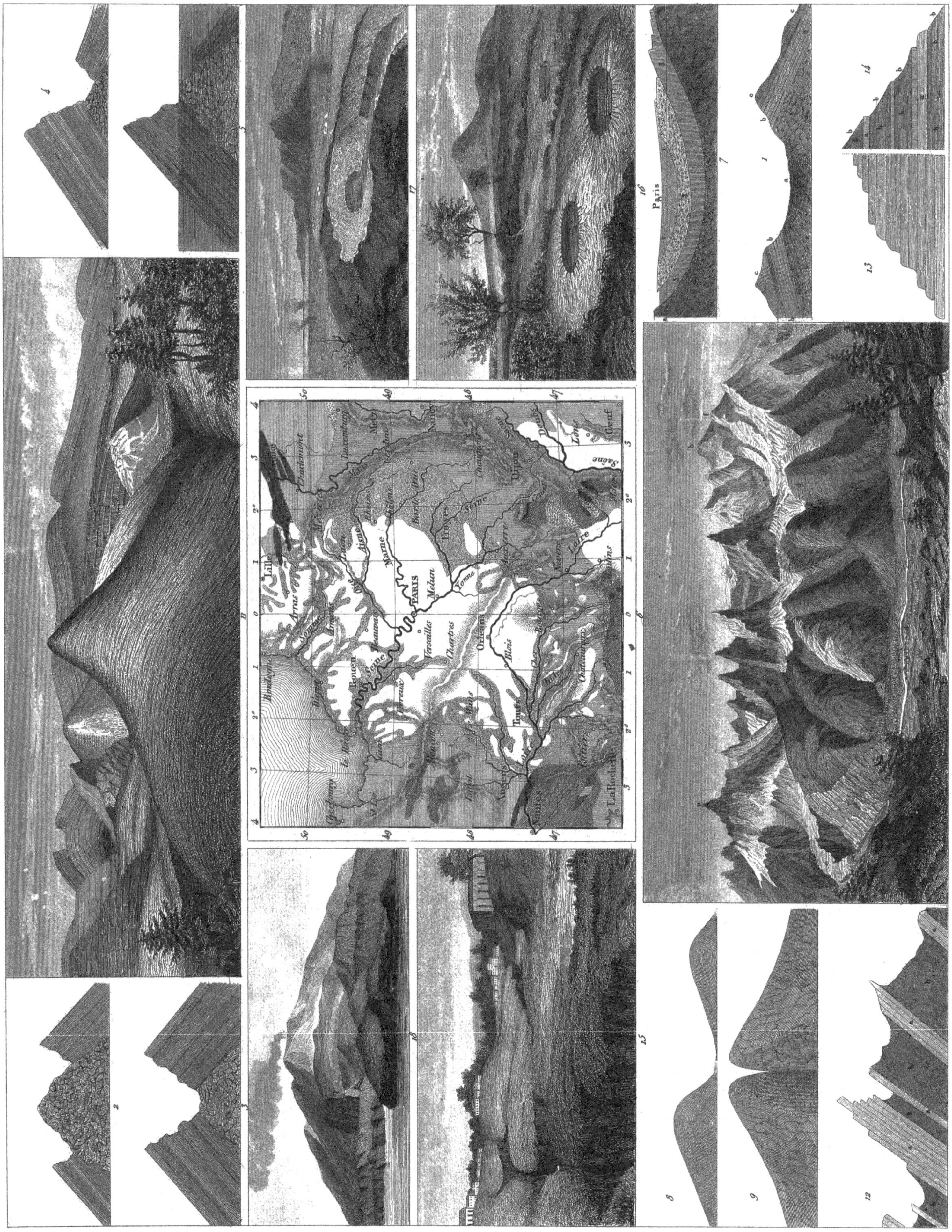

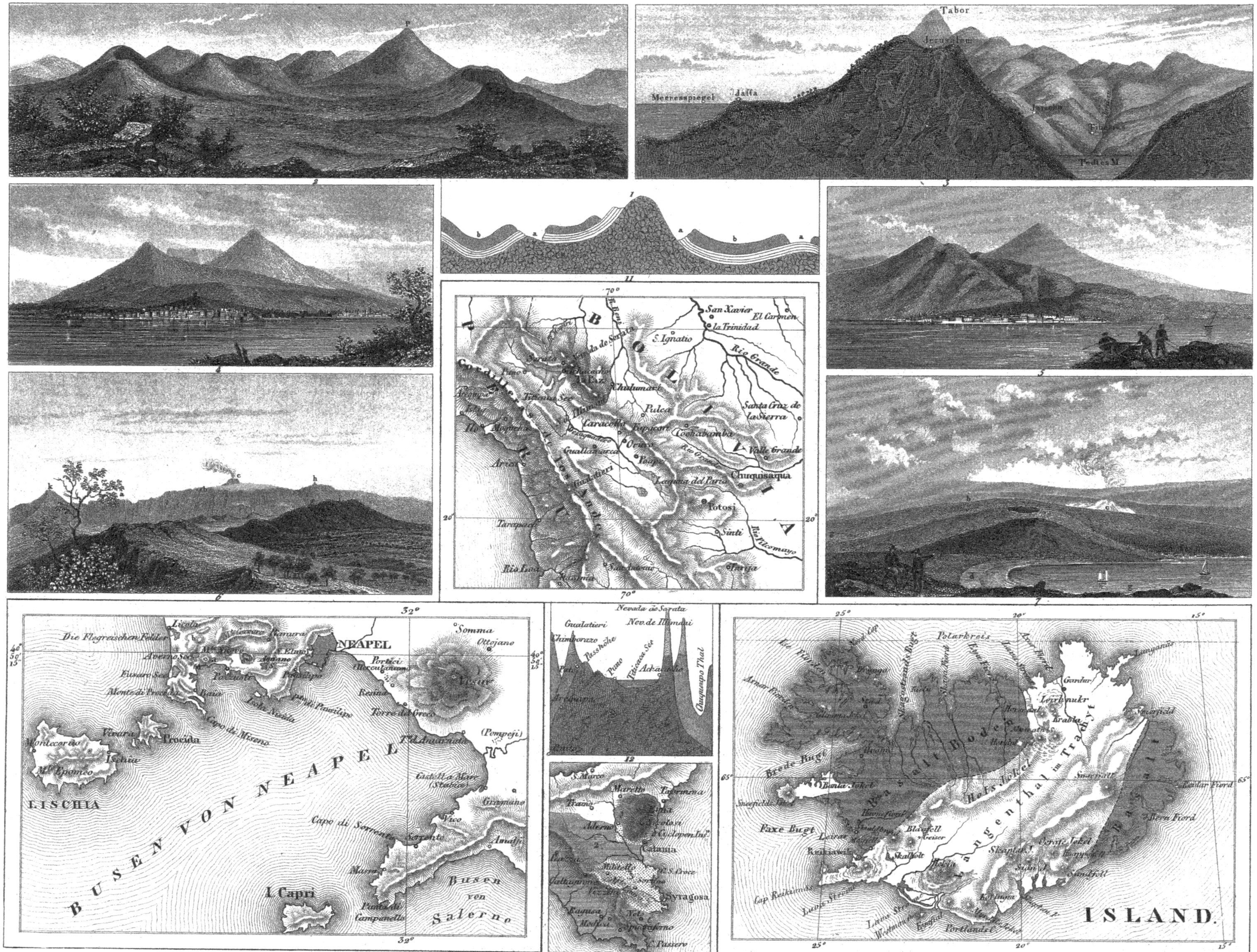

Plate 45 (Chemistry, Mineralogy, and Geology, Plate 16)

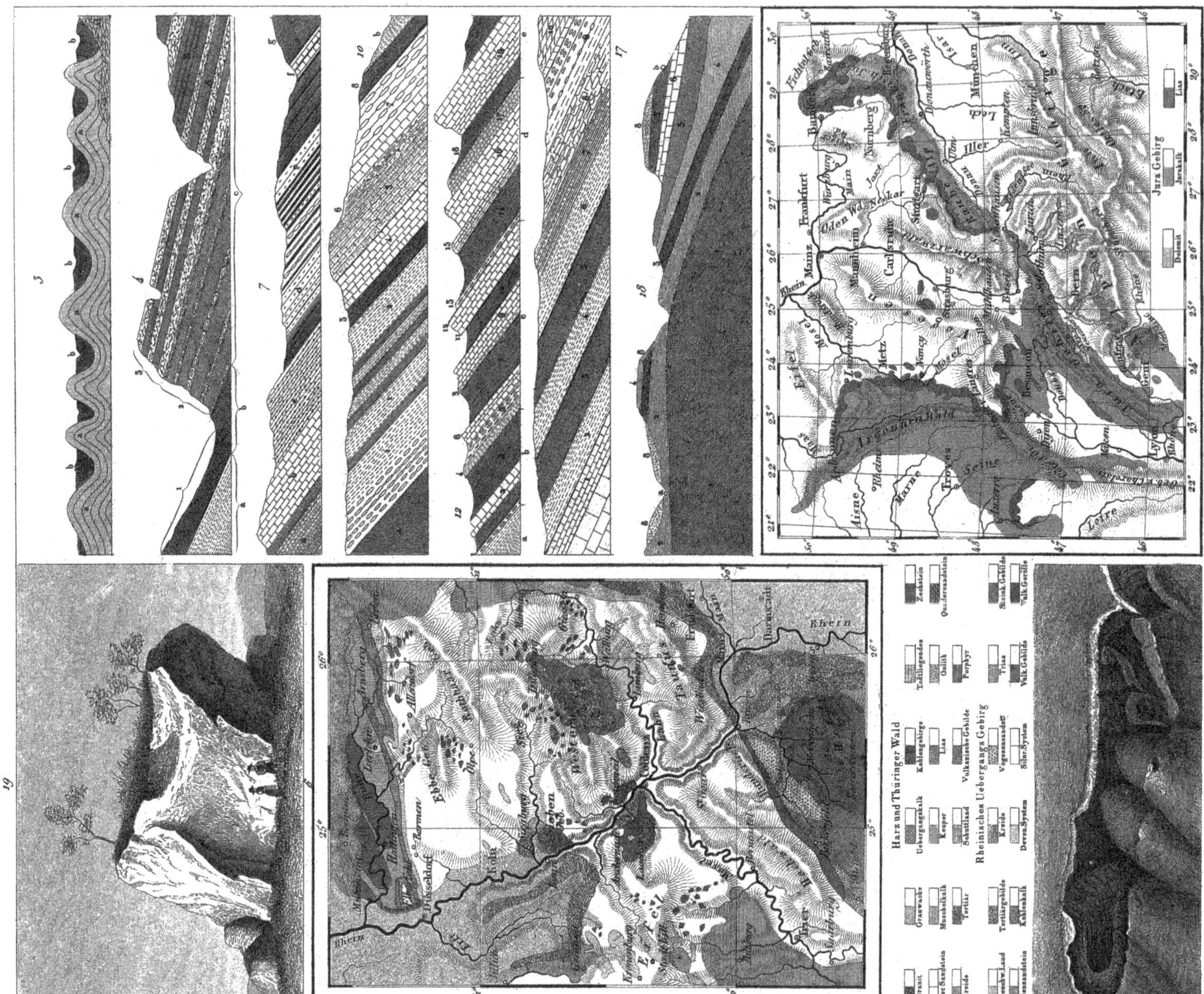
Harz und Thüringer Wald
Rheinisches Uebergangs Gebirg
Jura Gebirg

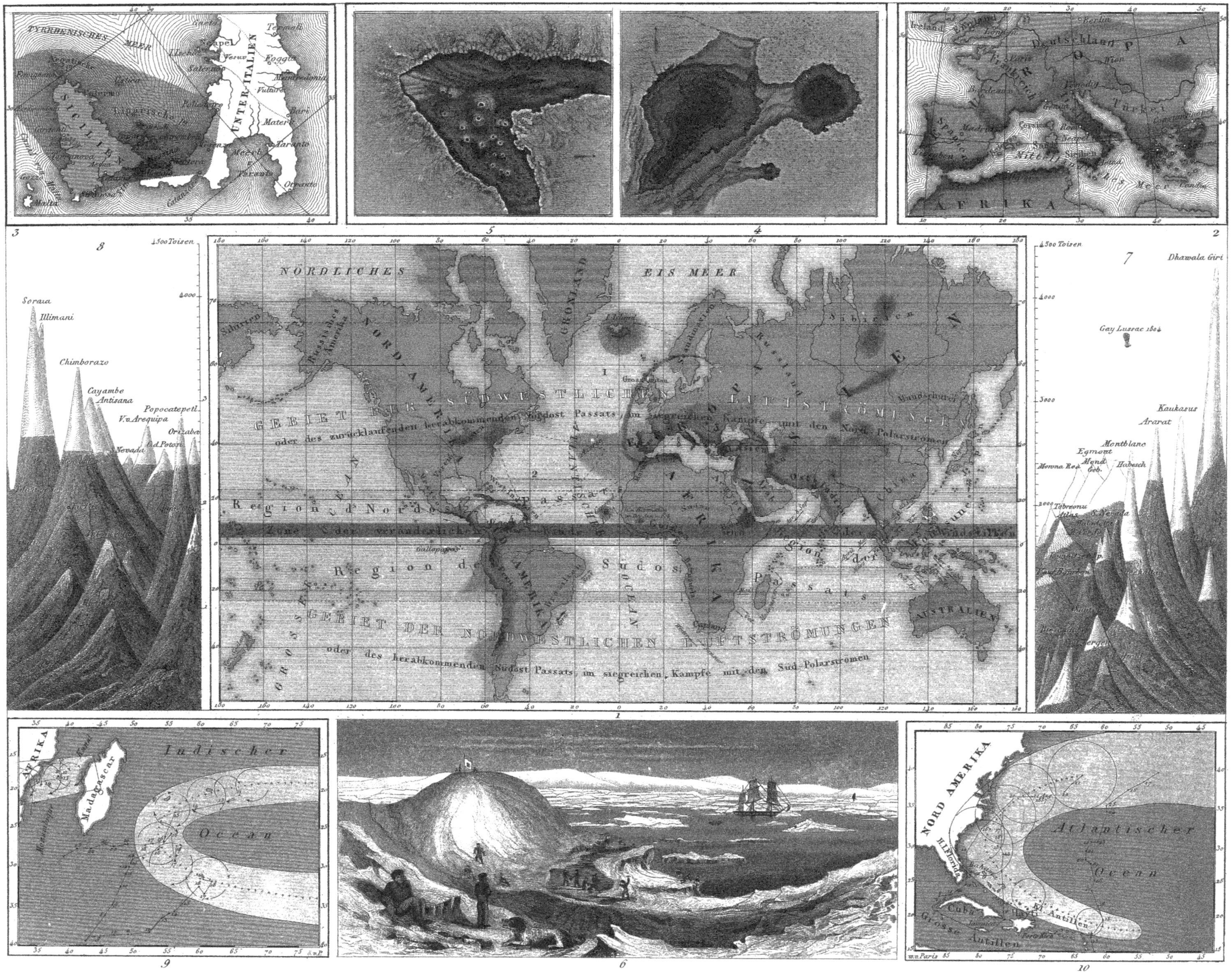

Plate 47 (Chemistry, Mineralogy, and Geology, Plate 18)

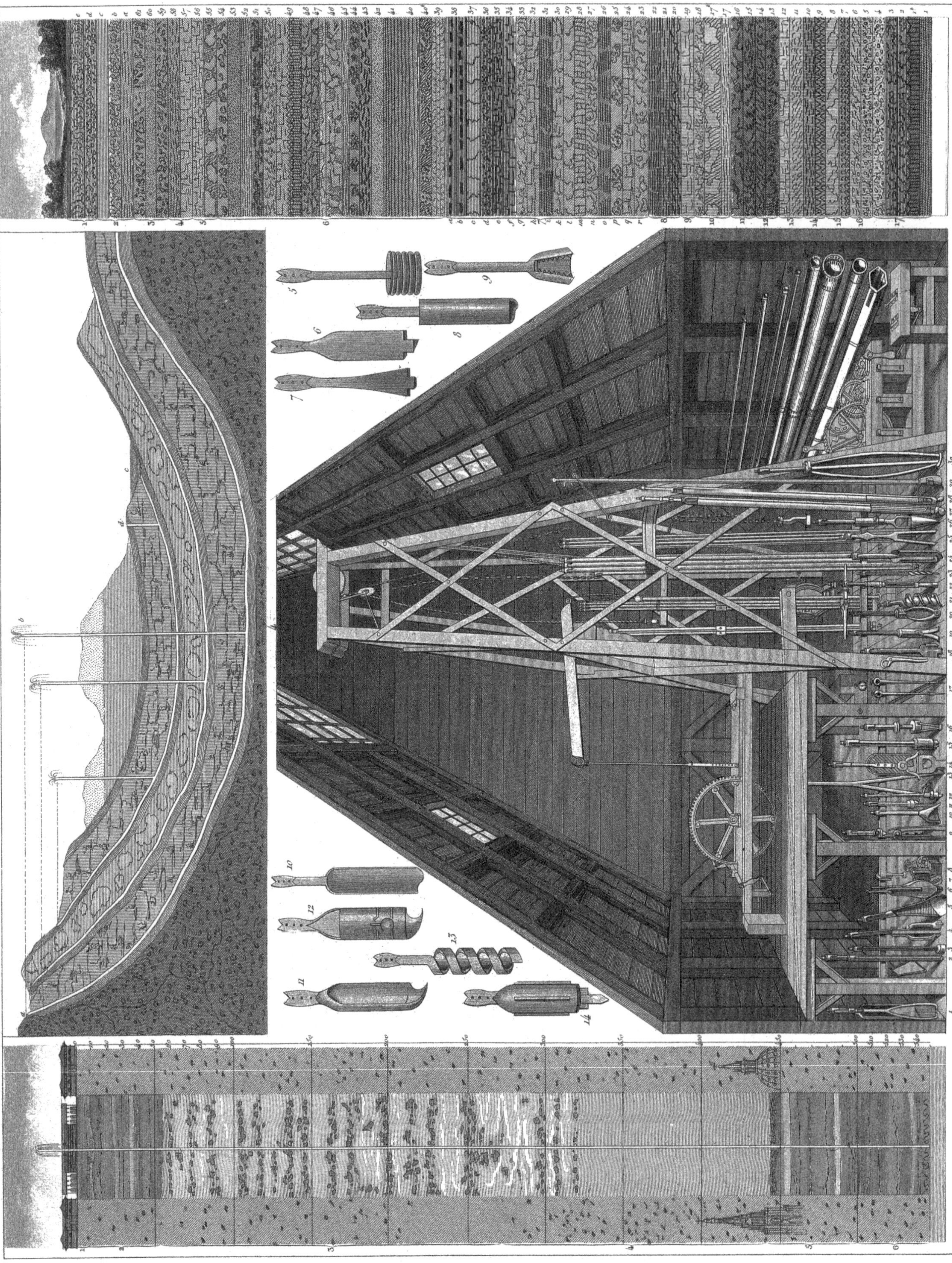

Plate 48 (Chemistry, Mineralogy and Geology, Pl. 4, 10)

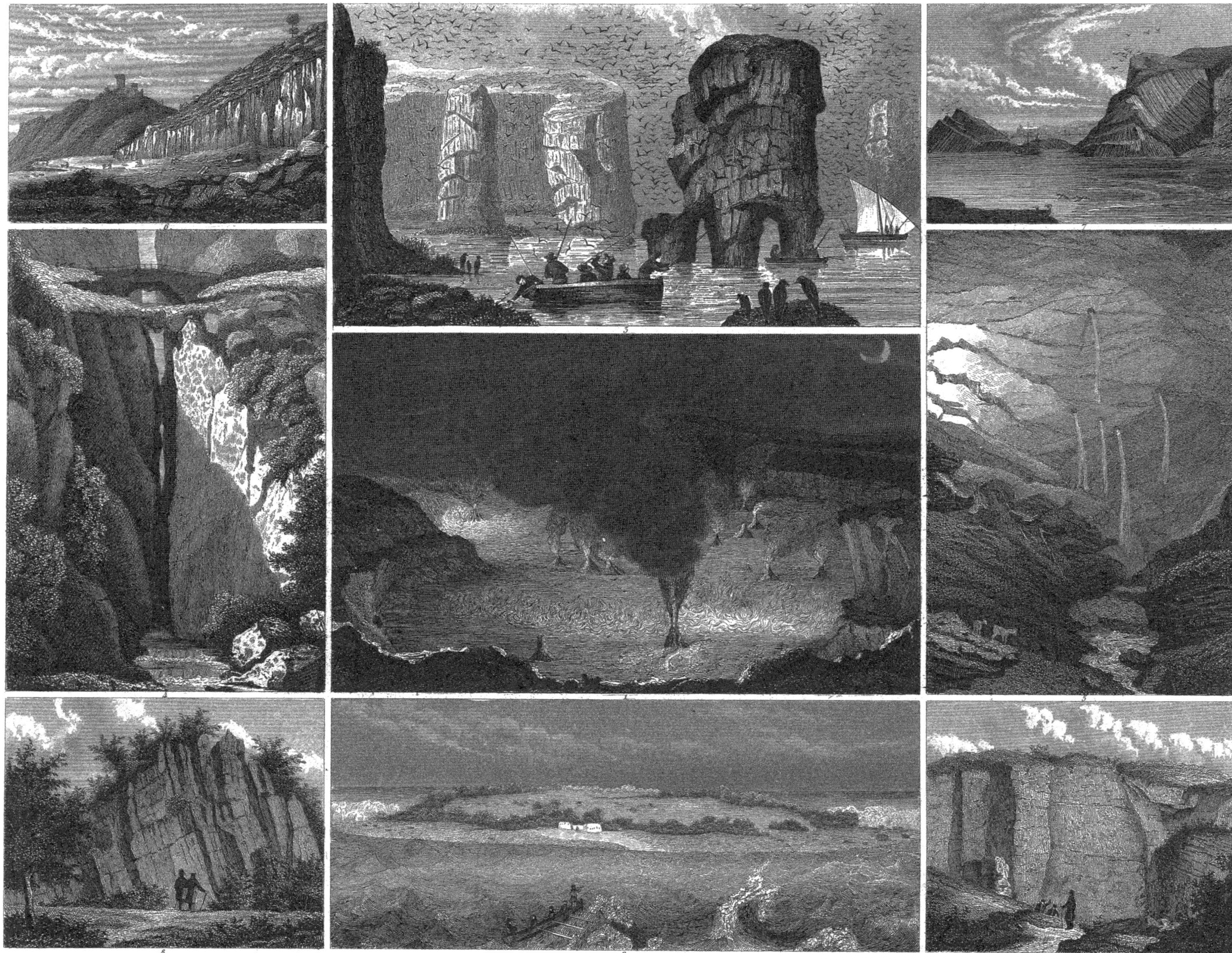

Plate 49 (Chemistry, Mineralogy, and Geology, Plate 20)

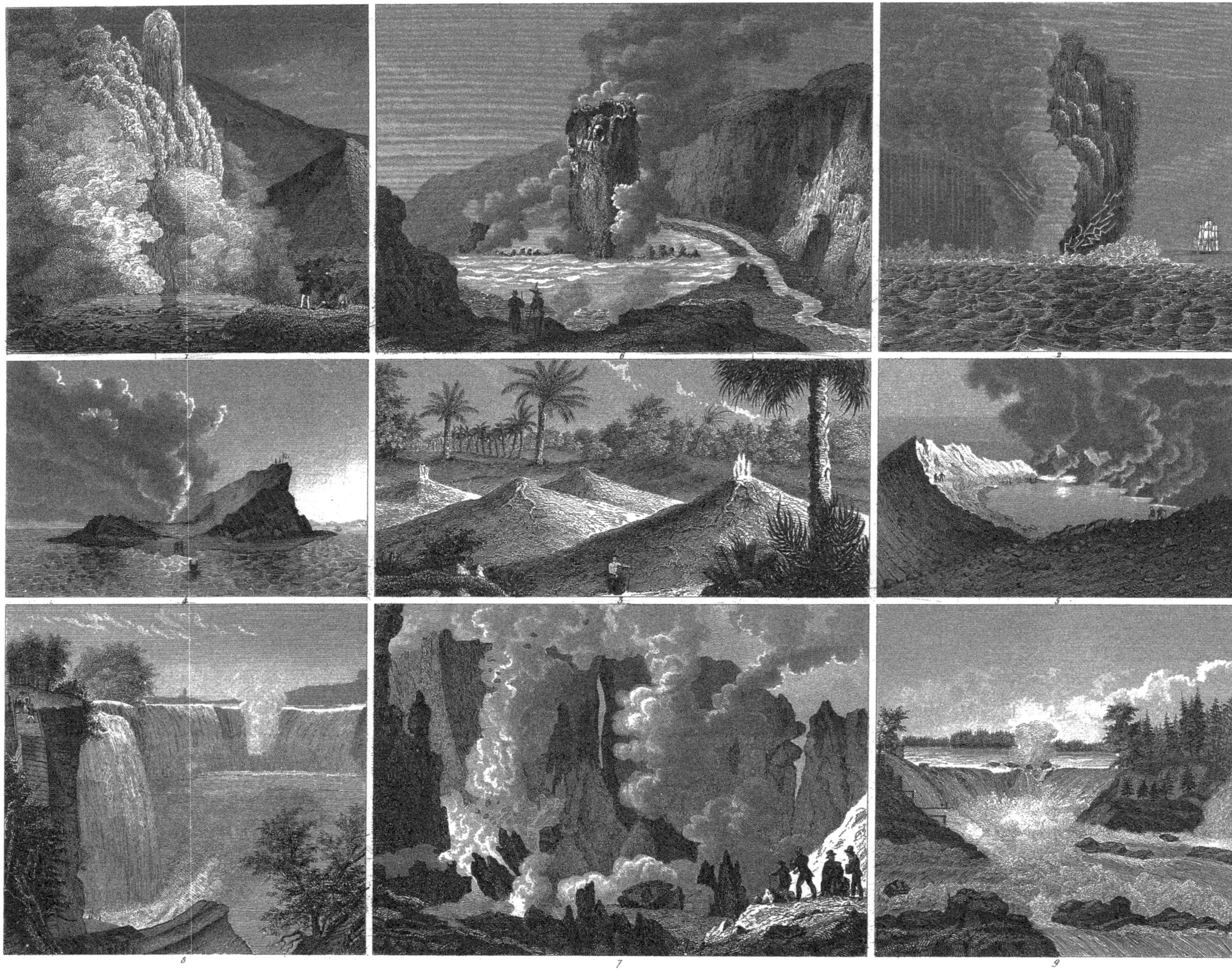

Plate 50 (Chemistry, Mineralogy and Geology, Plate 21)

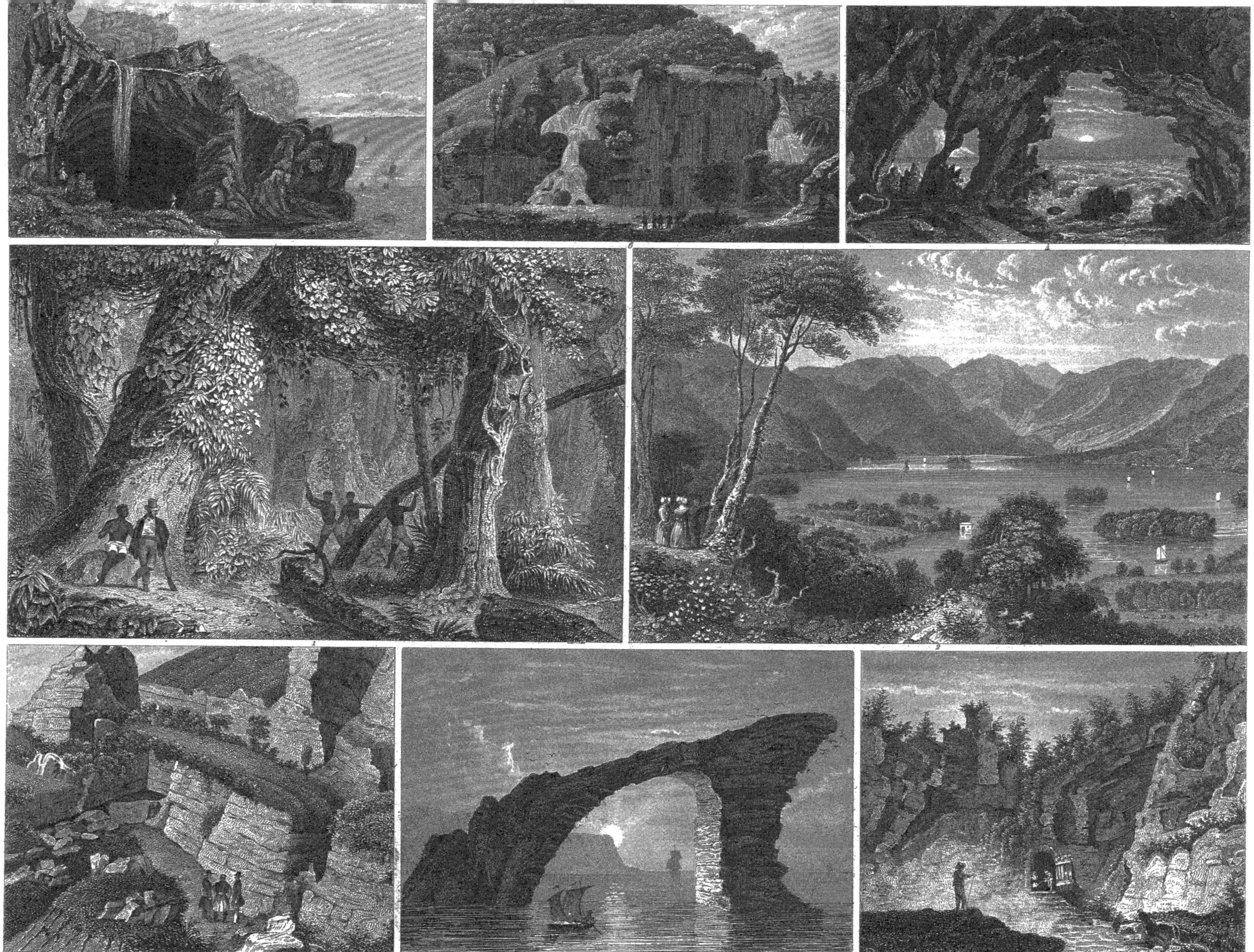

Plate 51 (Chemistry, Mineralogy, and Geology, Plate 22)

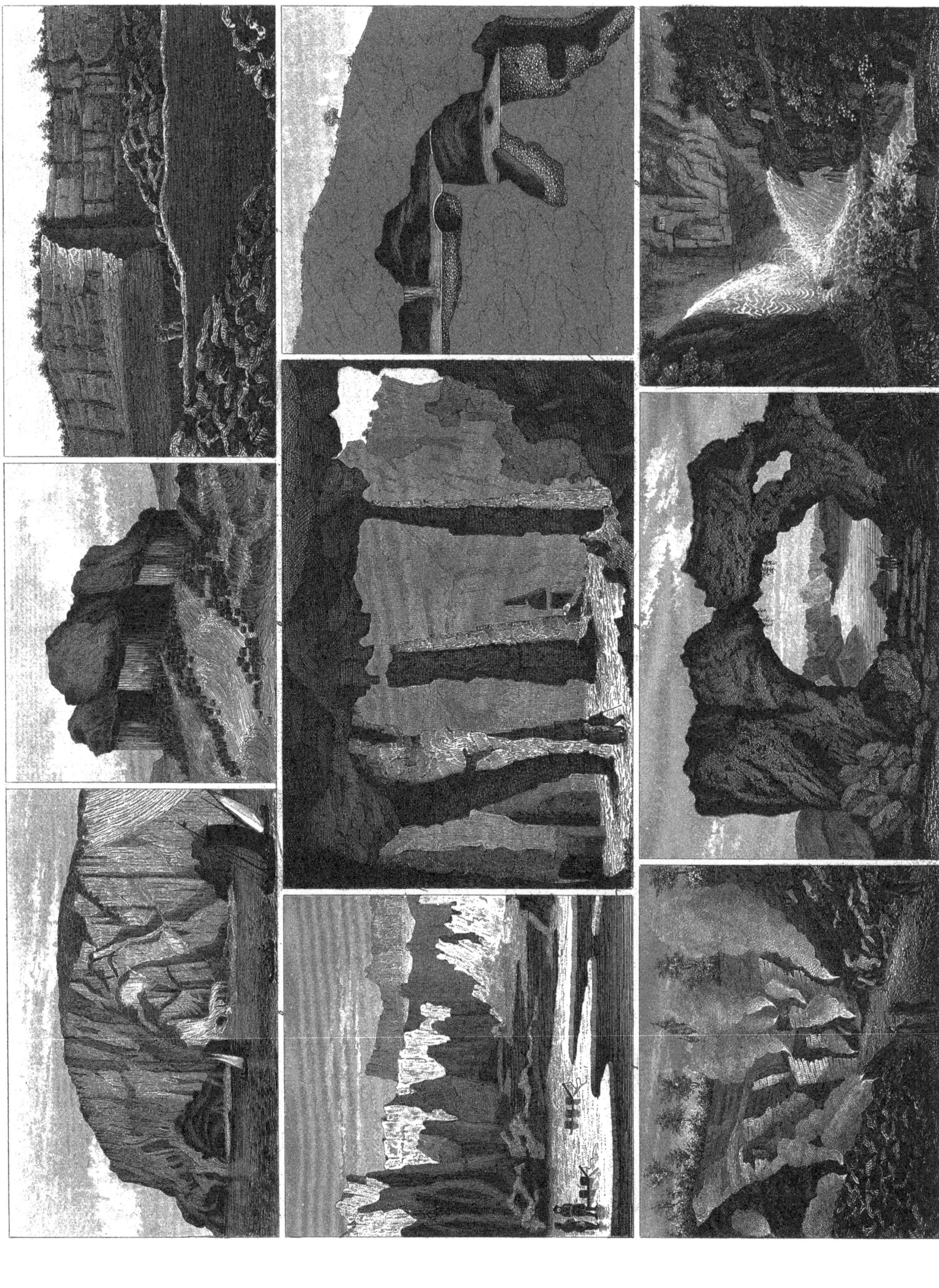

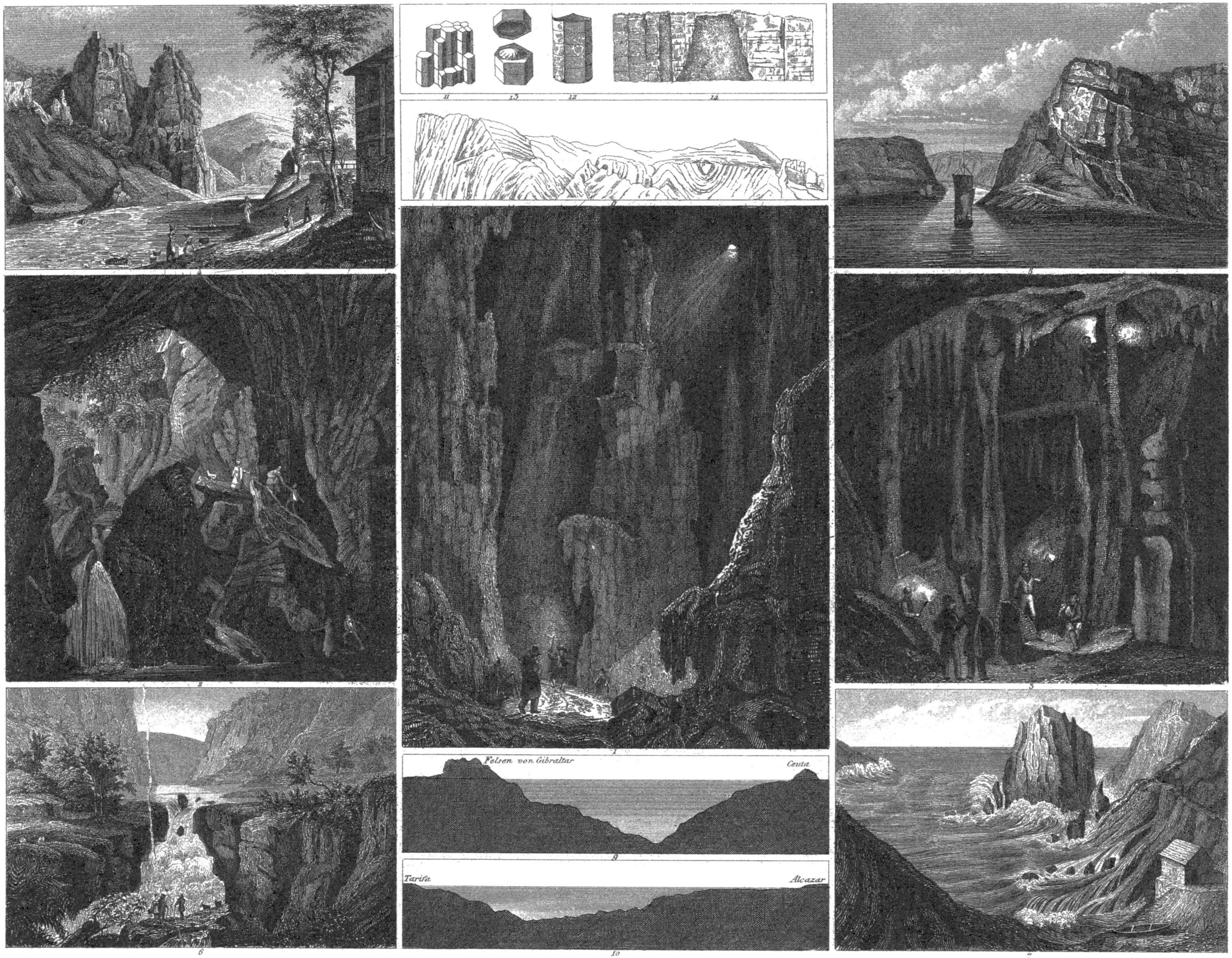

Plate 53 (Chemistry, Mineralogy, and Geology, Plate 24)

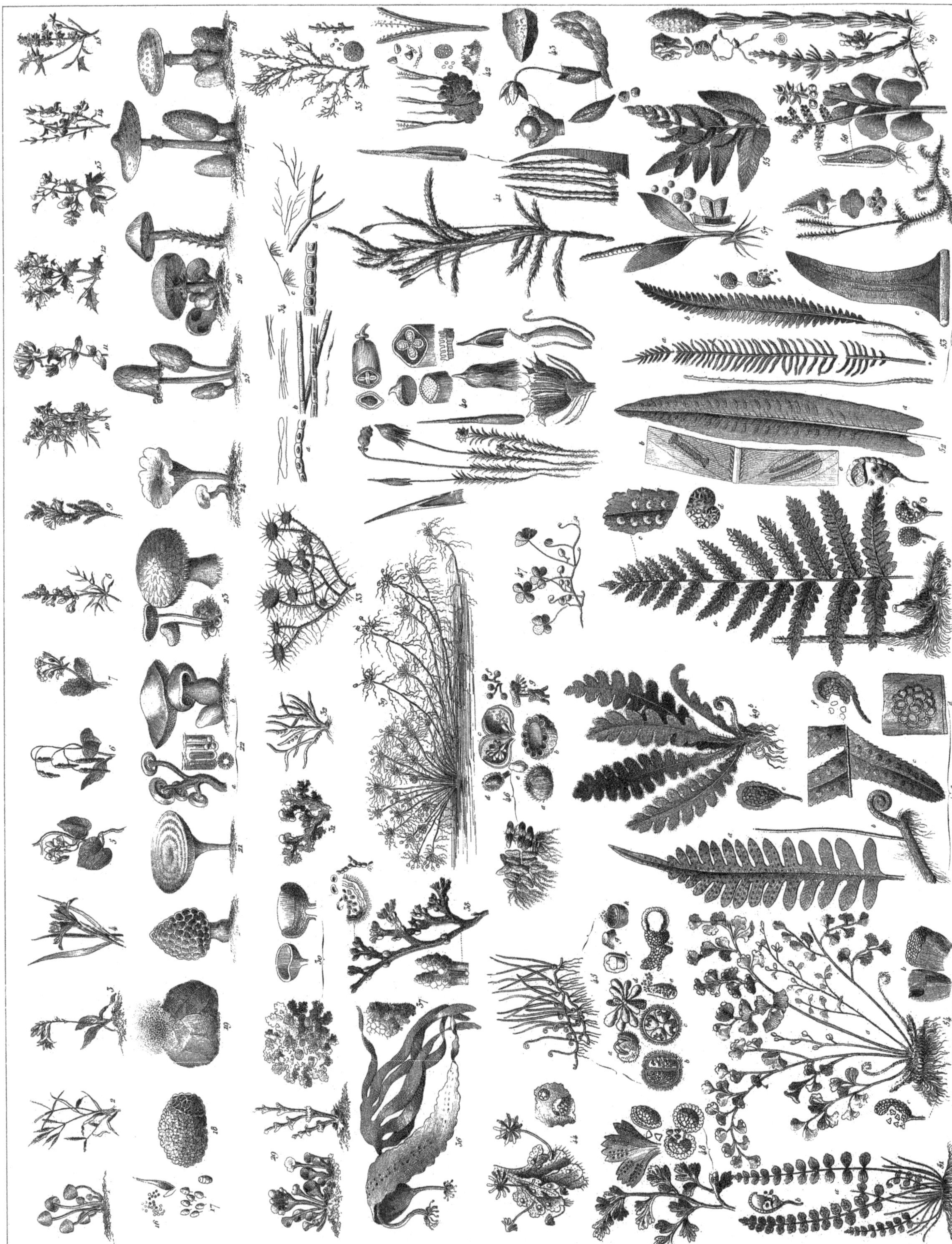

Plate 54 (Botany, Plate 1)

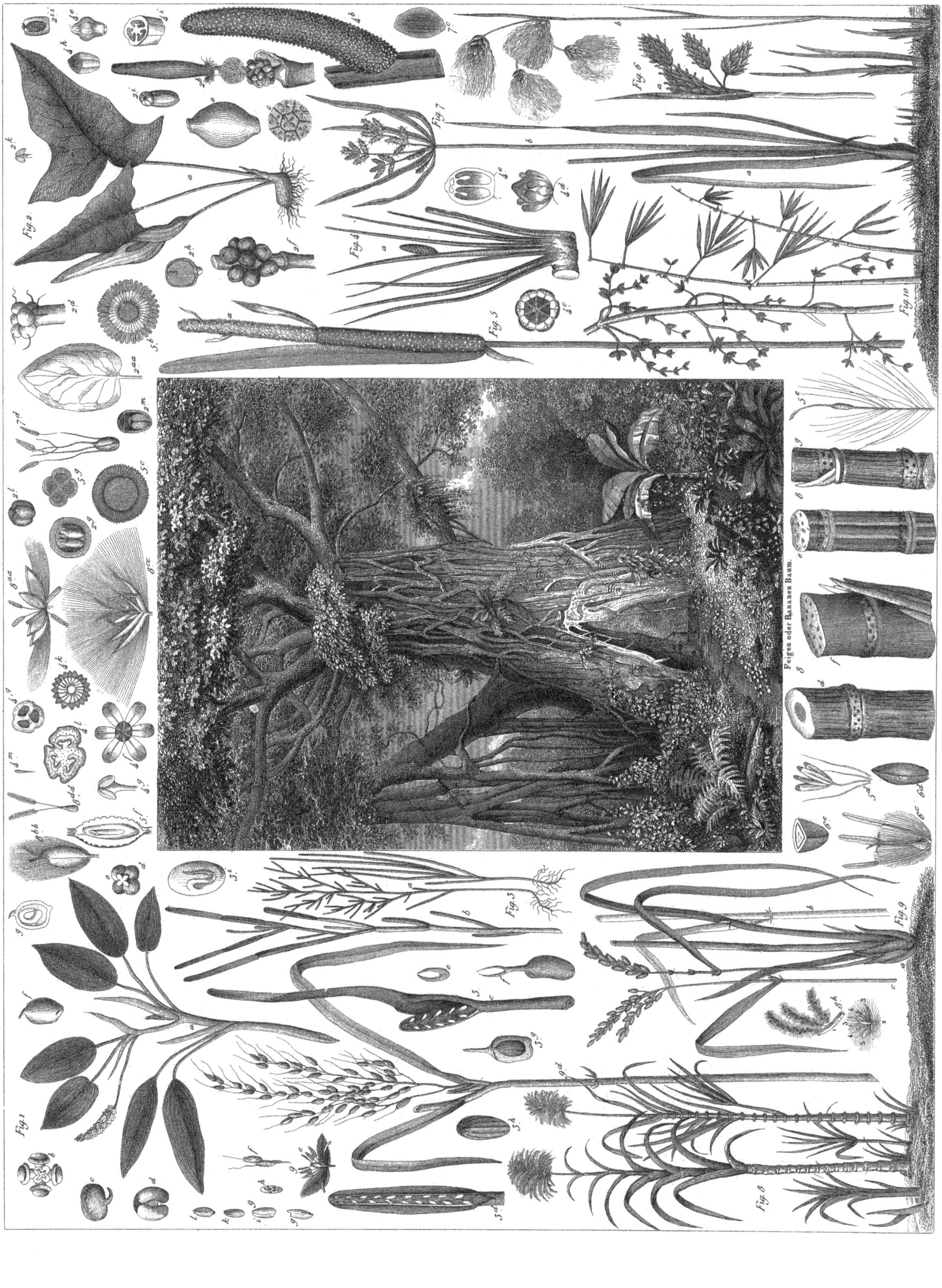

Plate 55 (Botany, Plate 2)

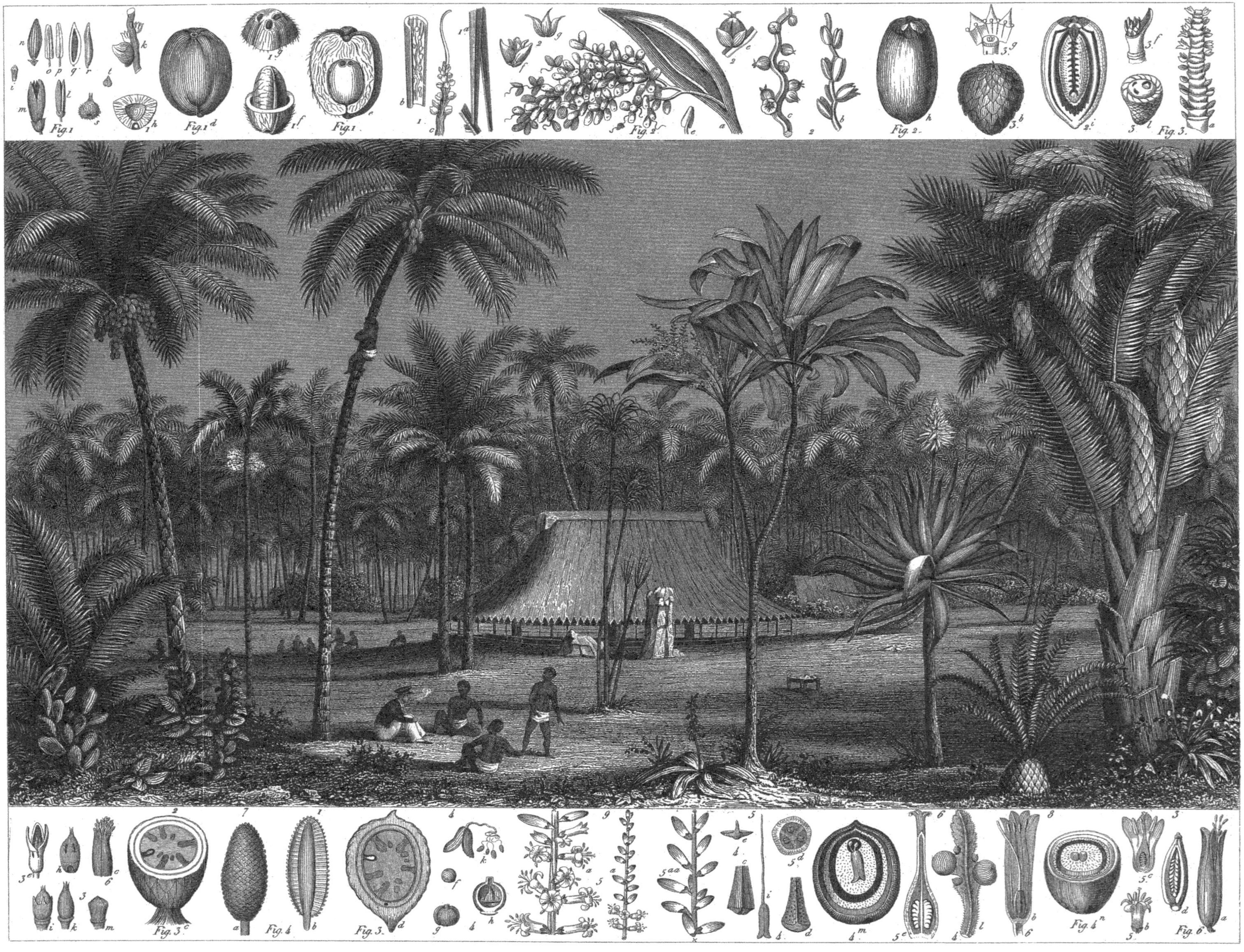

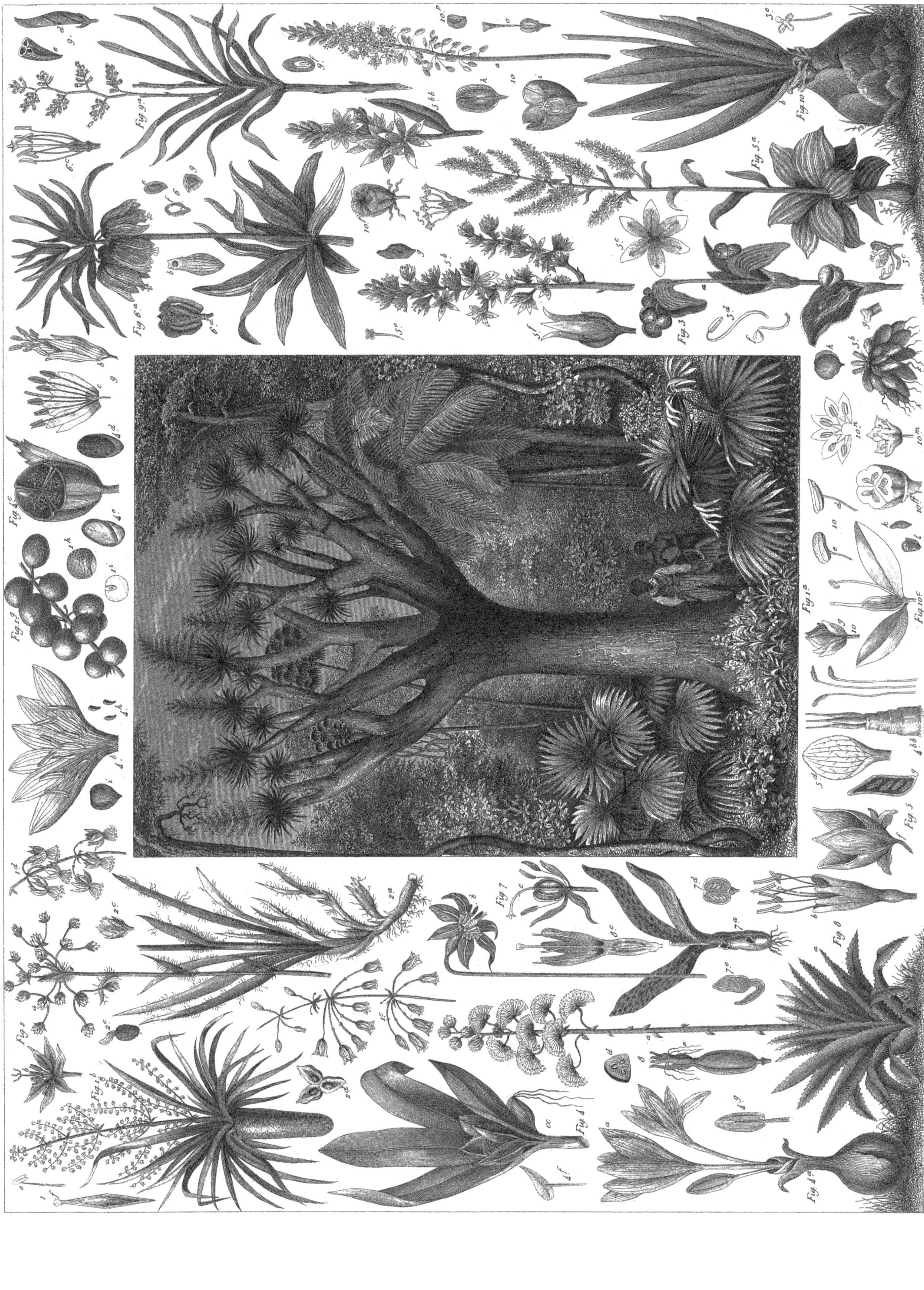

Plate 57 (Botany, Plate 4)

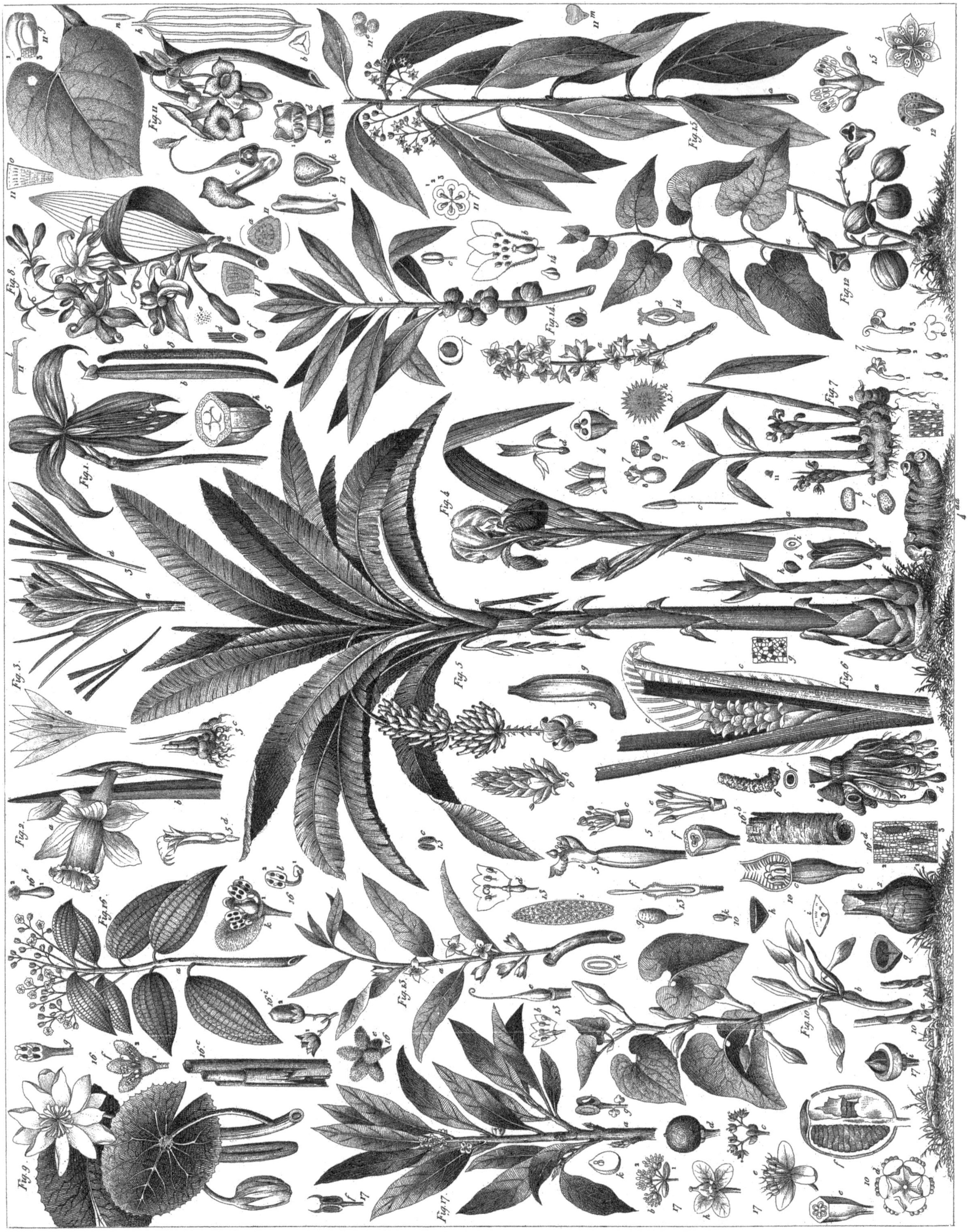

Plates 58 and 59 (Botany, Plates 5 and 6)

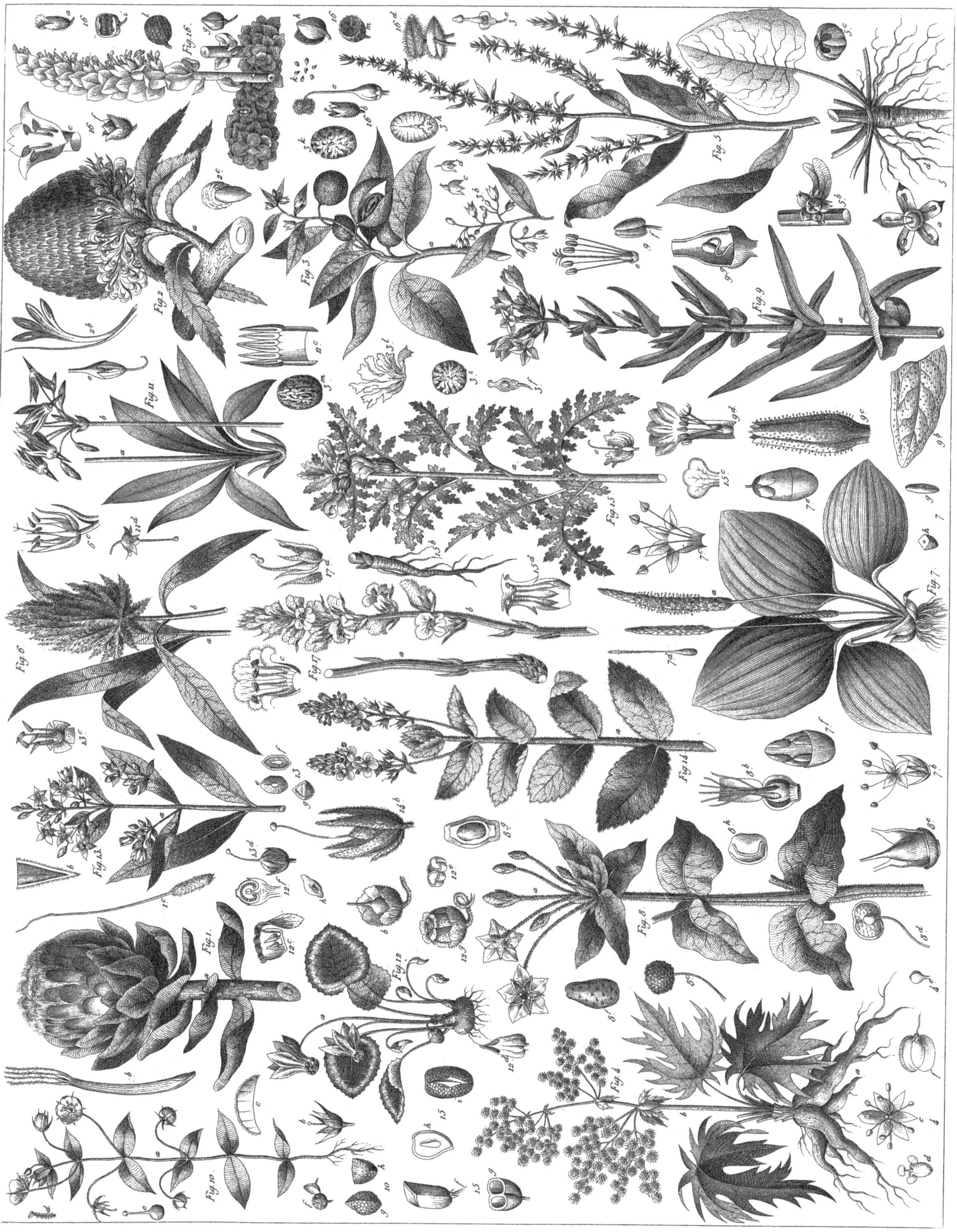

Plates 60 and 61 (Botany, Plates 7 and 8)

Plate 62 (Botany, Plate 9)

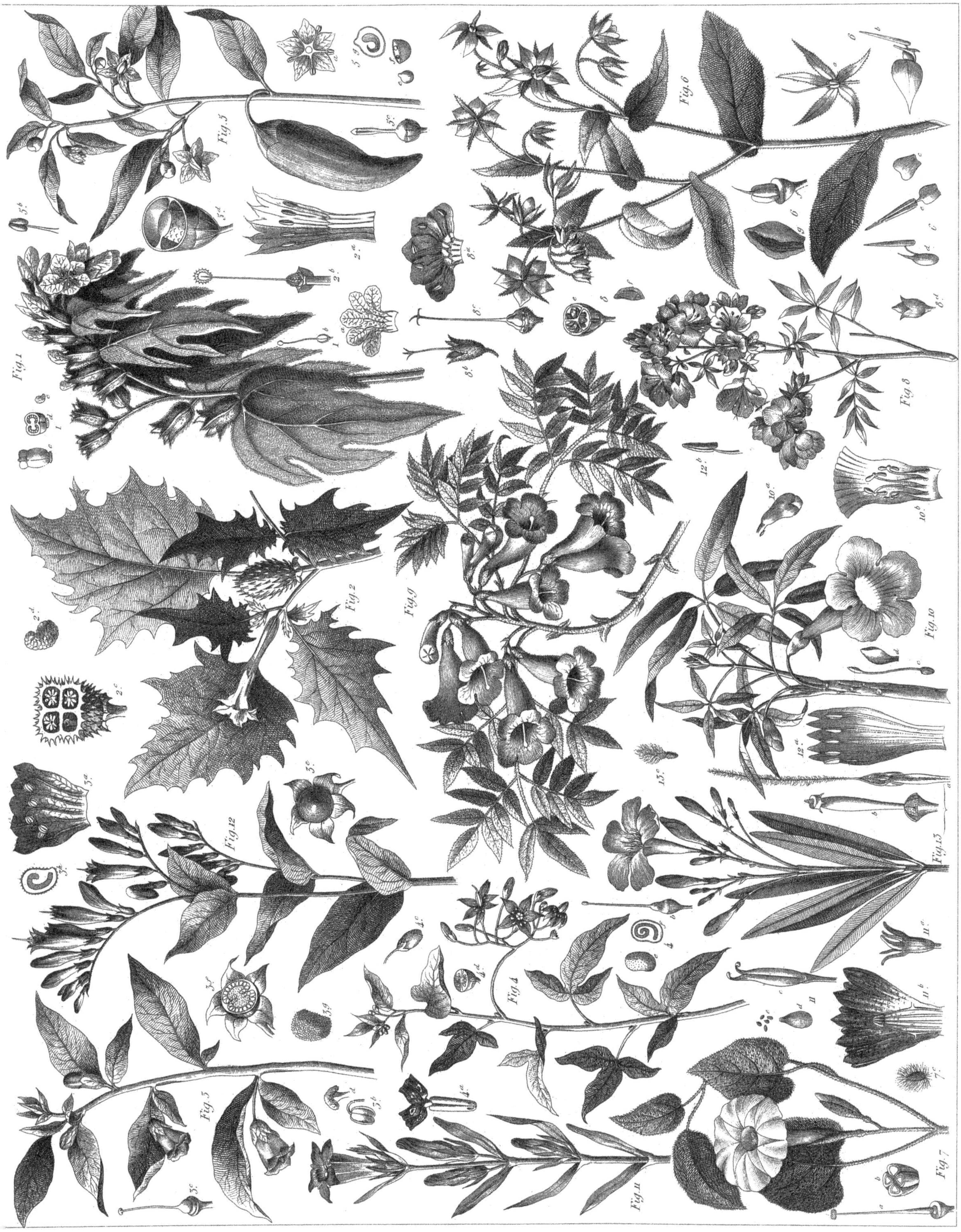

Plate 63 (Botany, Plate 10)

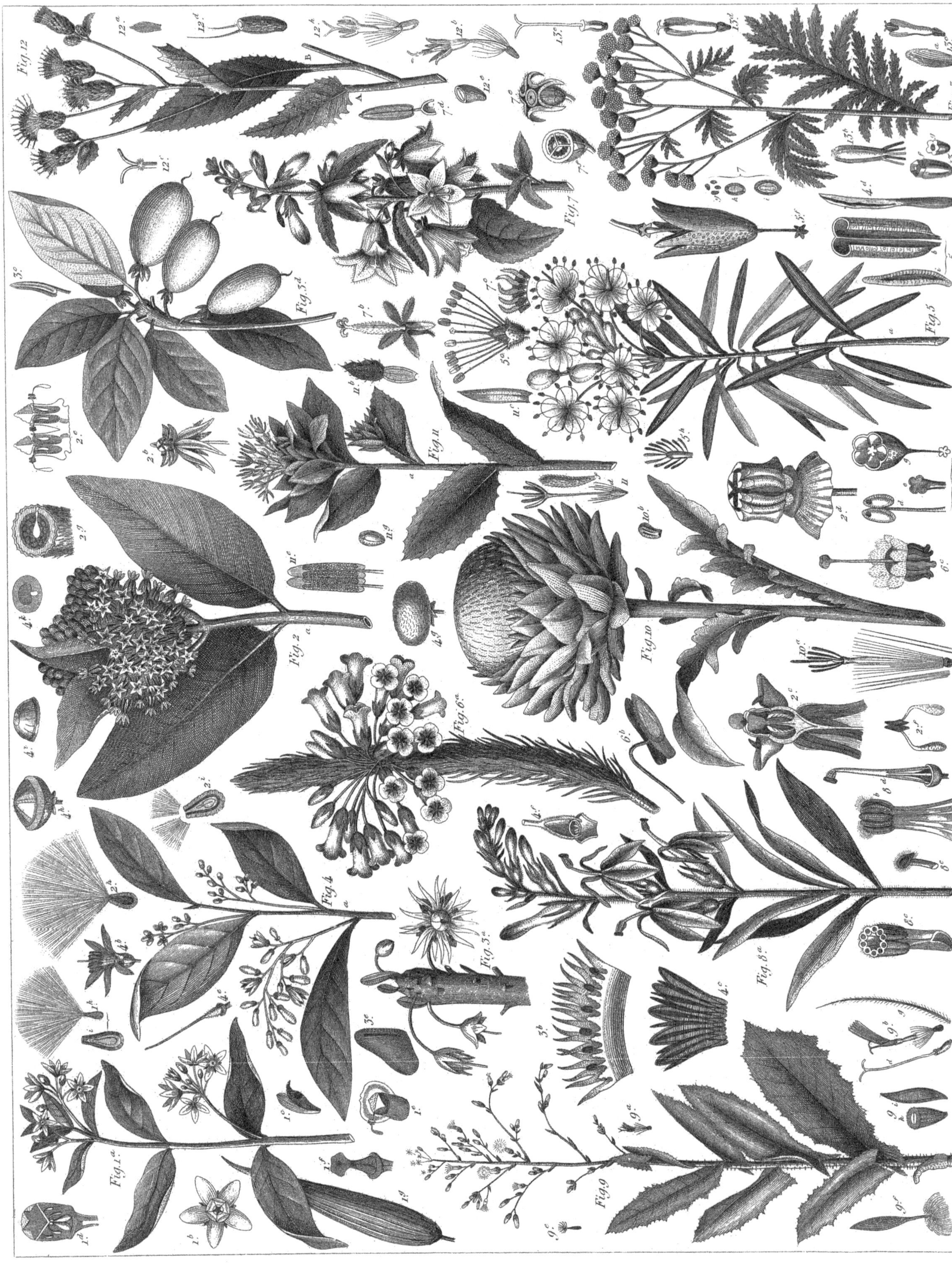

Plate 64 (Botany, Plate 11)

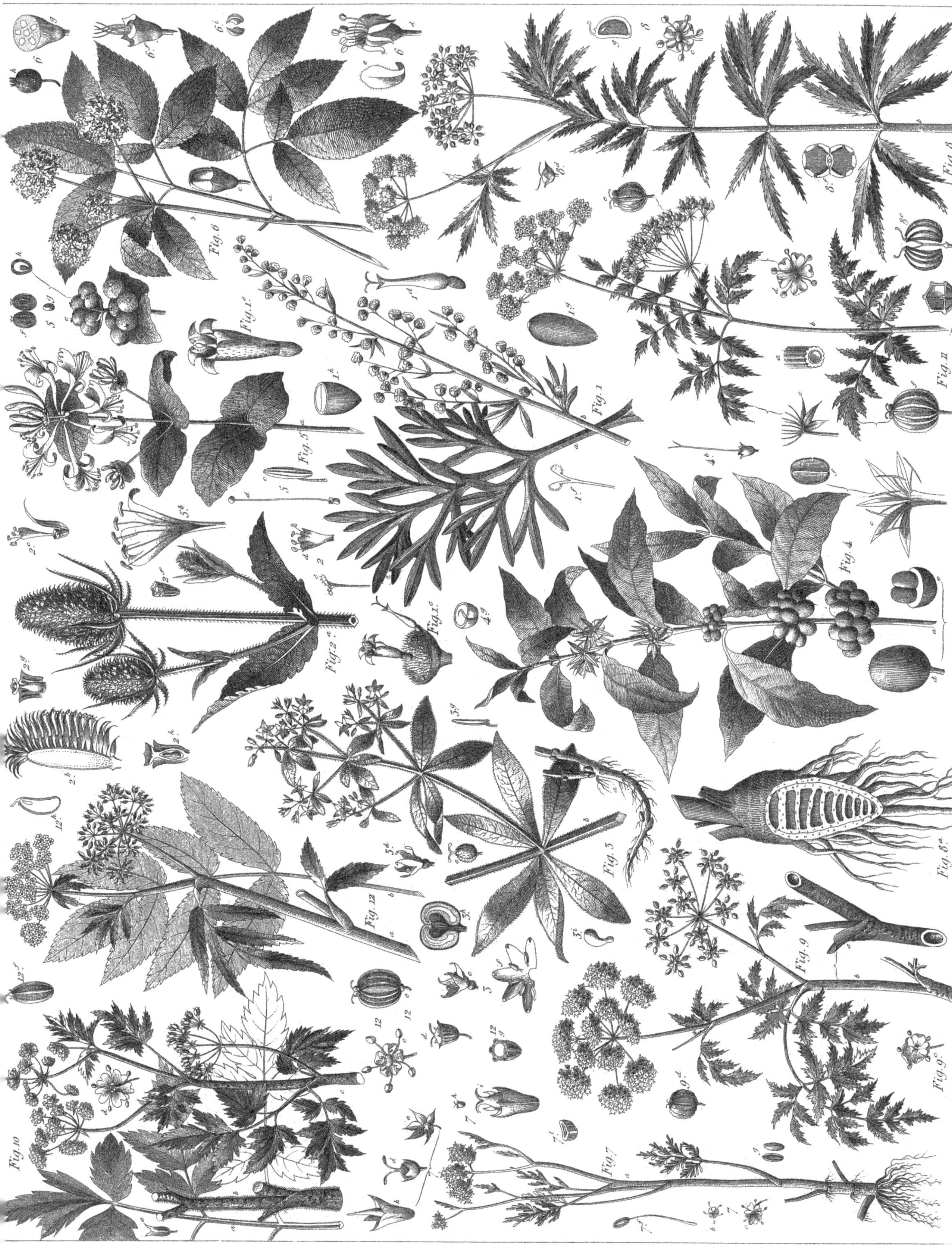

Plate 65 (Botany, Plate 12)

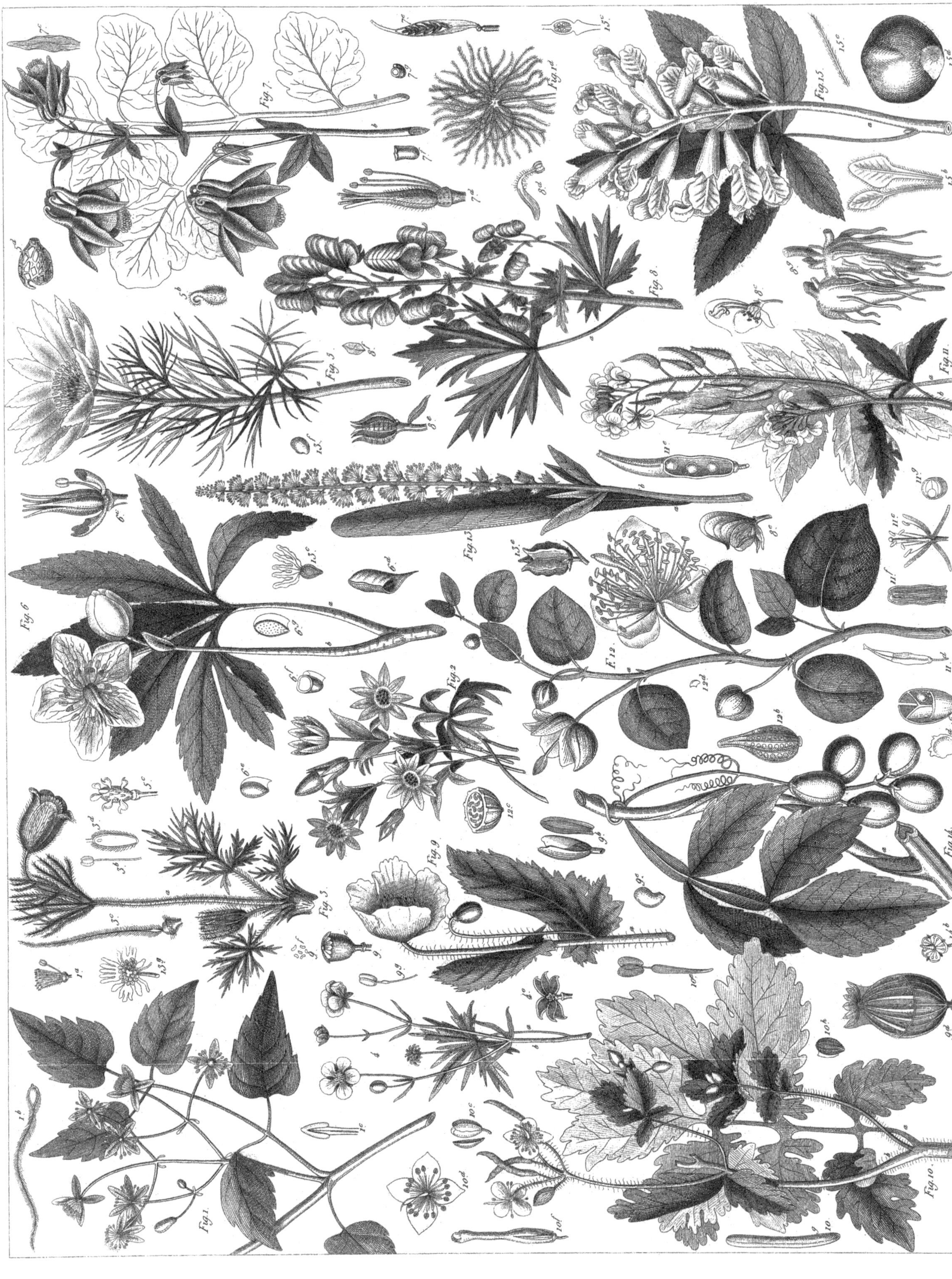

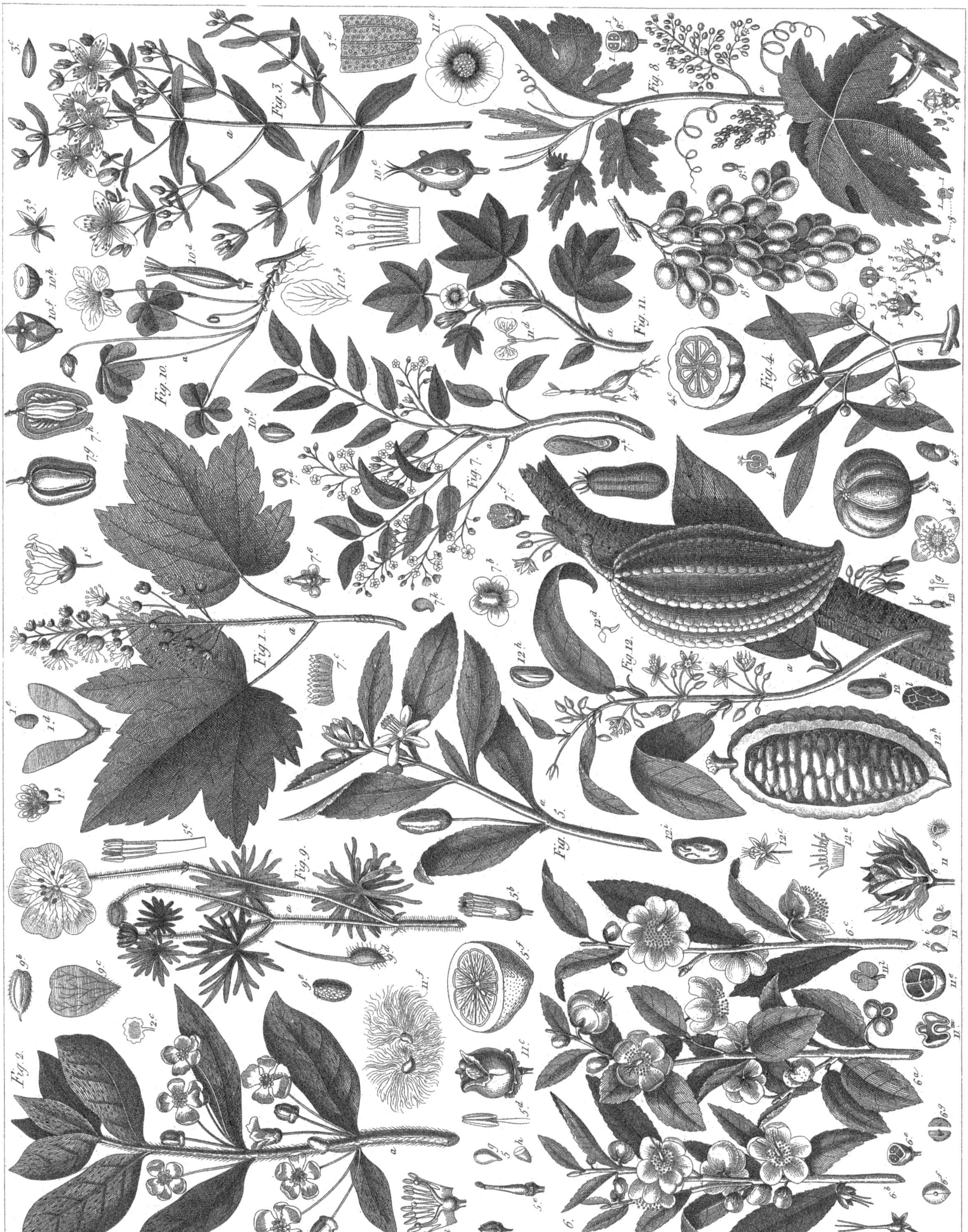

Plate 67 (Botany, Plate 14)

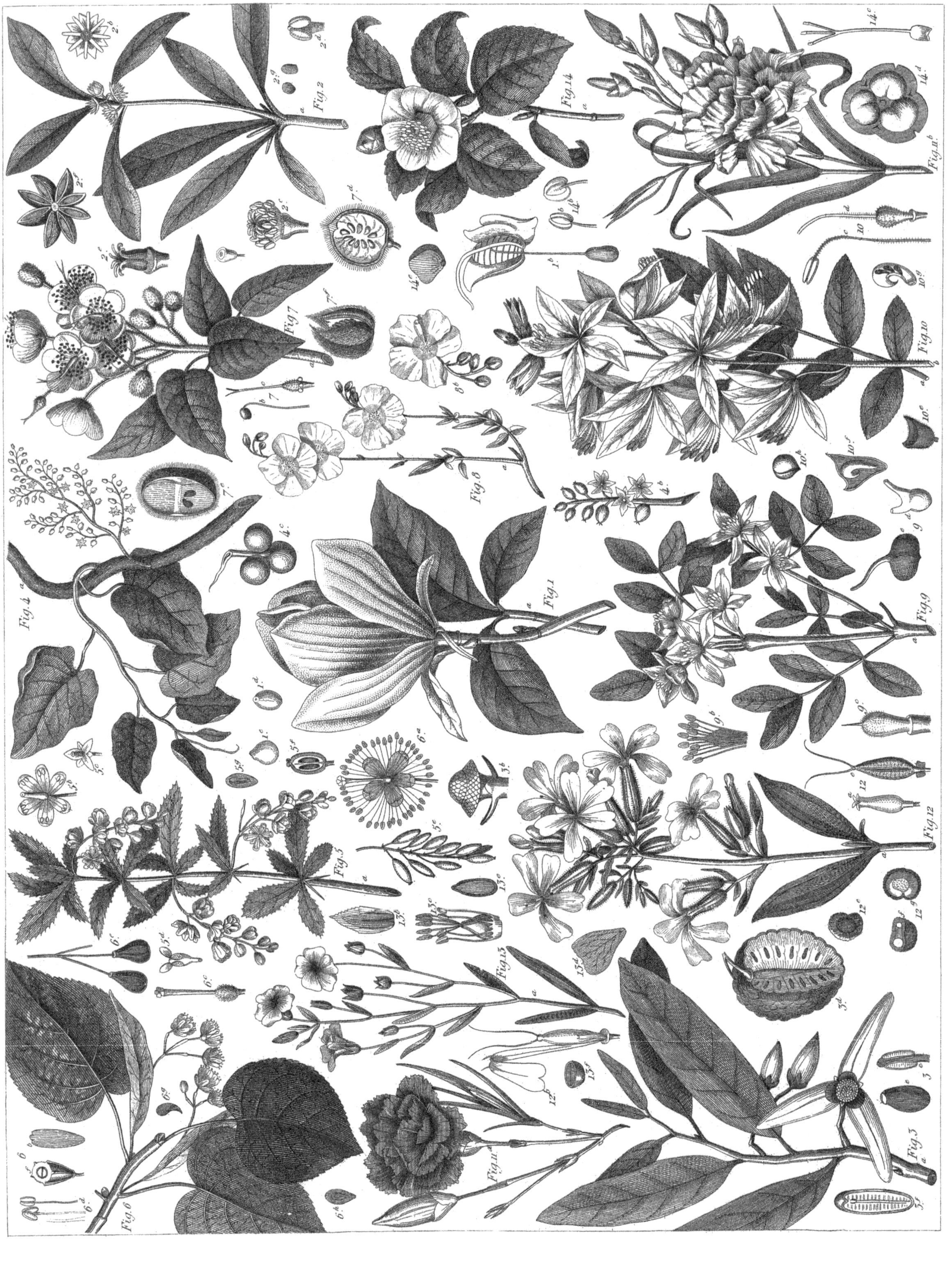

Plate 68 (Botany, Plate 15)

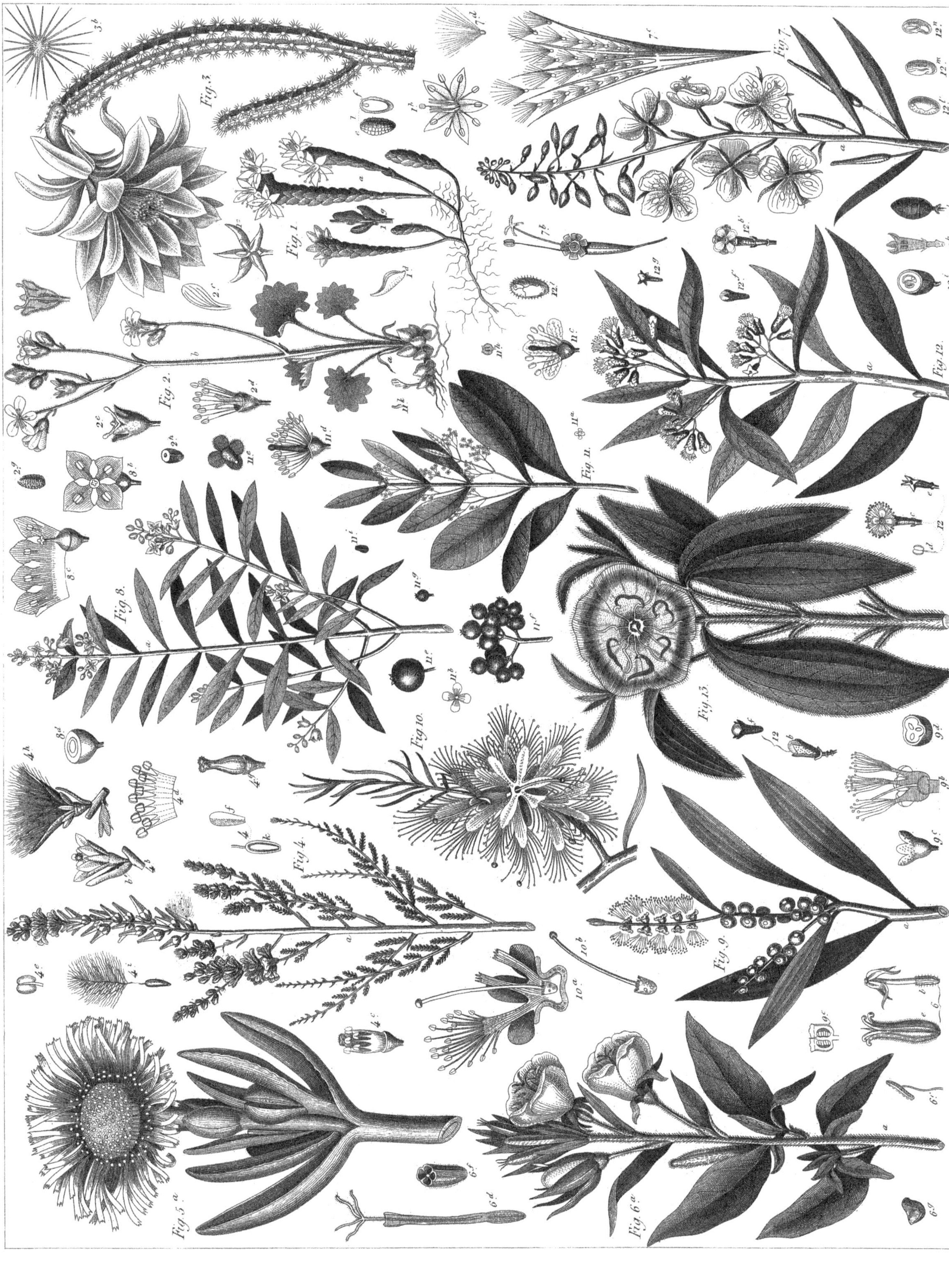

Plate 69 (Botany, Plate 16)

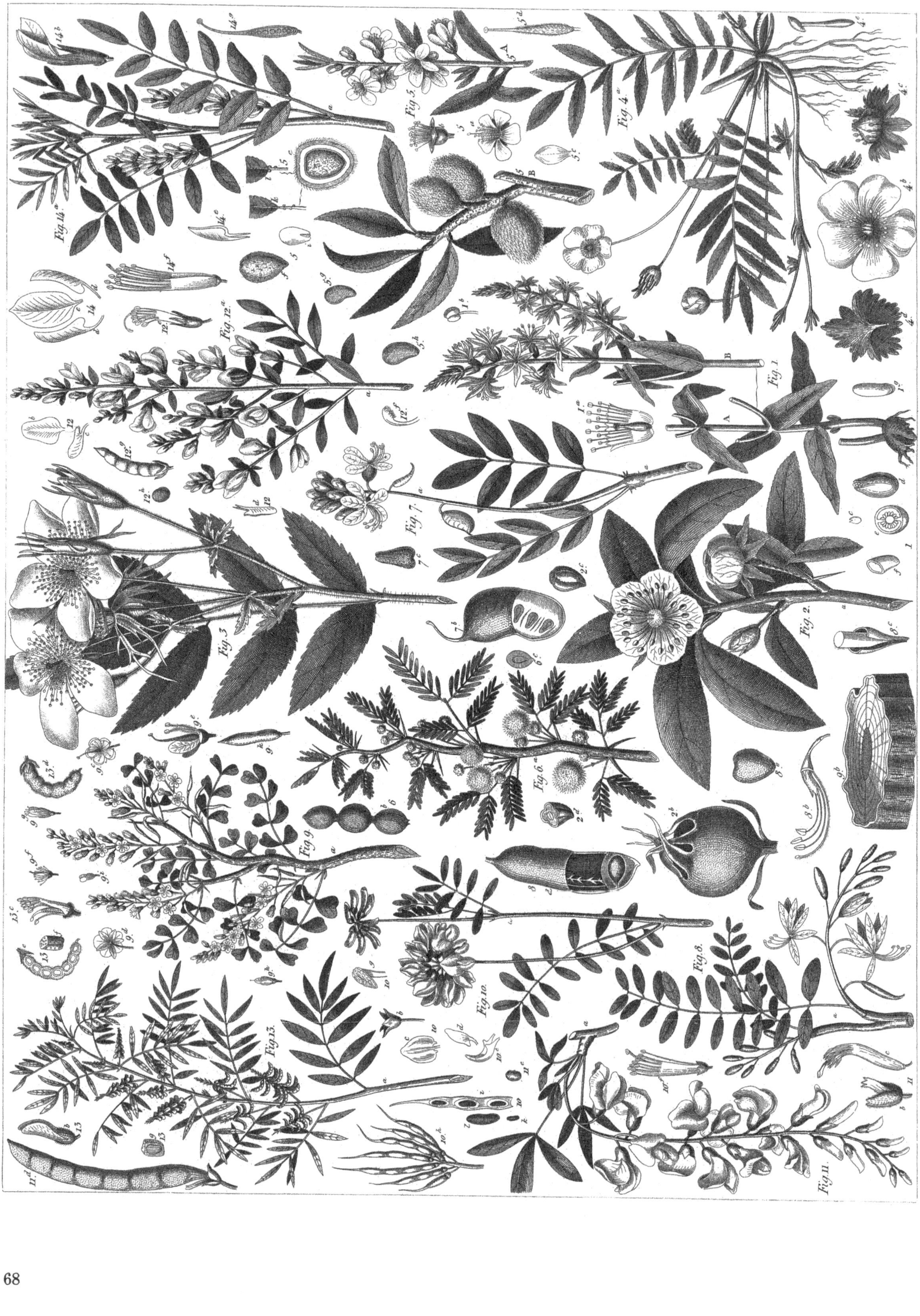

Plate 70 (Botany, Plate 17)

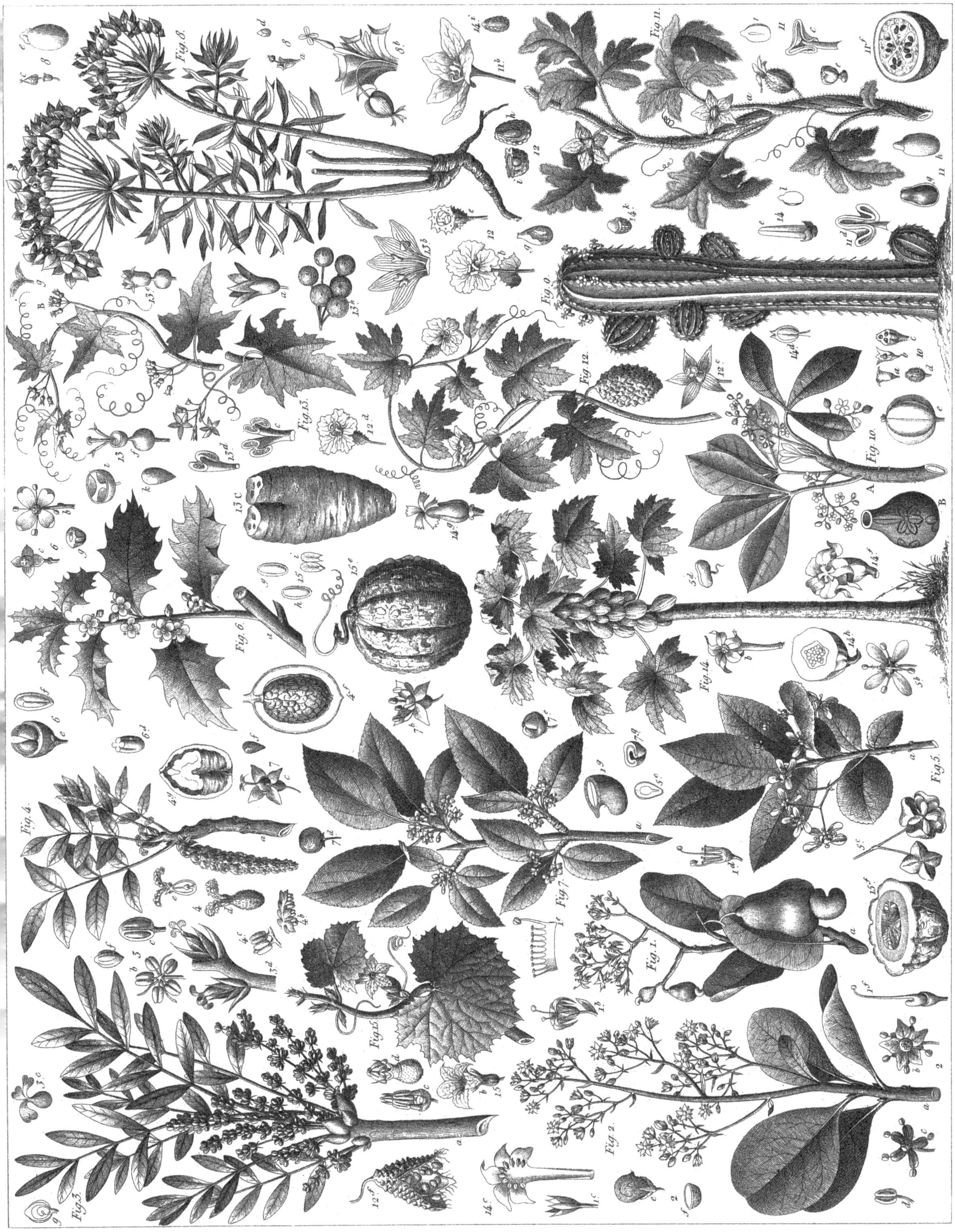

Plate 71 (Botany, Plate 18)

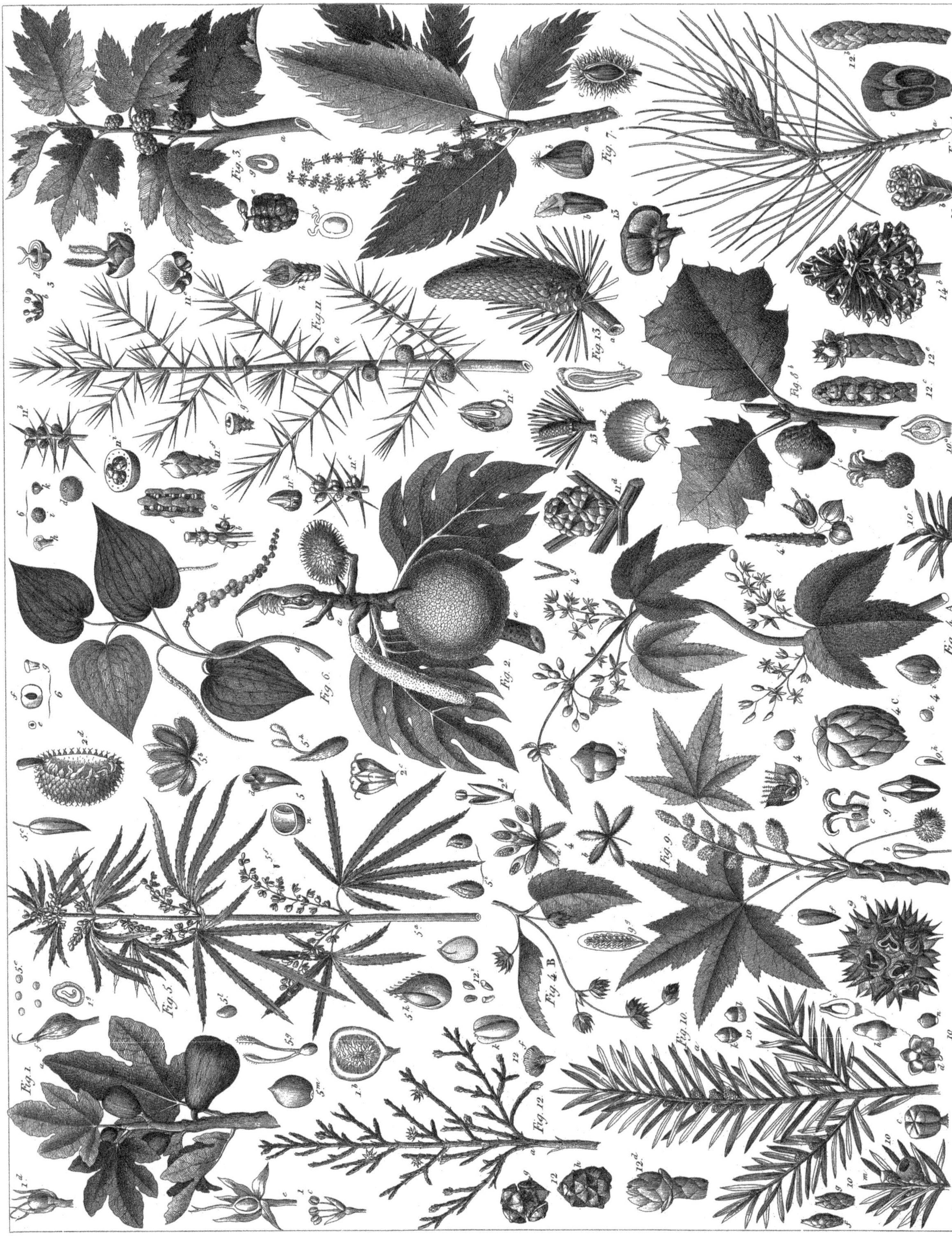

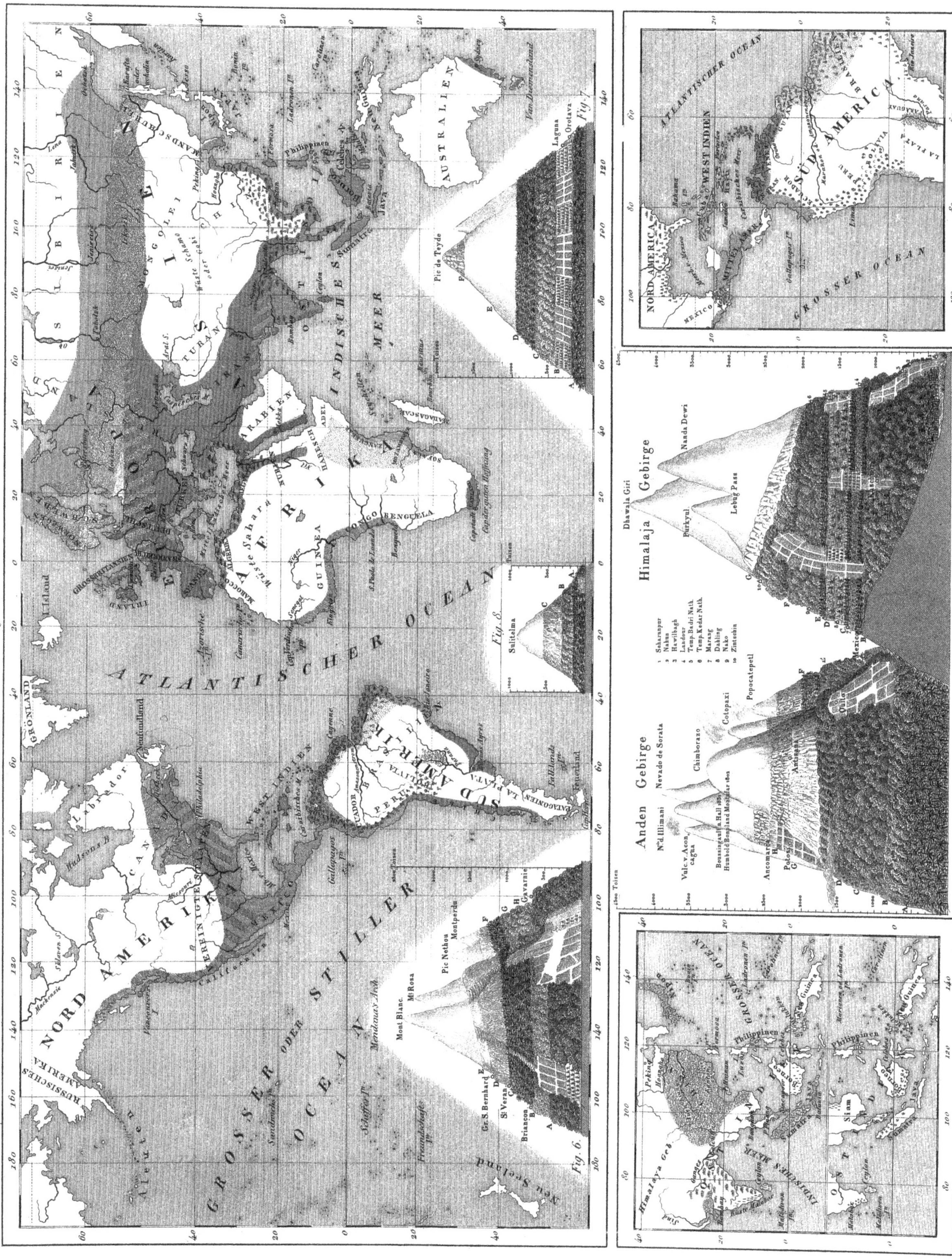

Plate 73 (Botany, Plate 20)

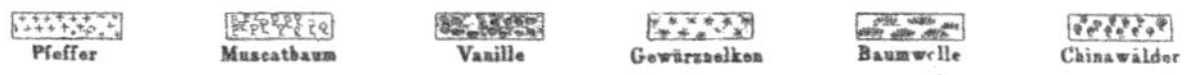

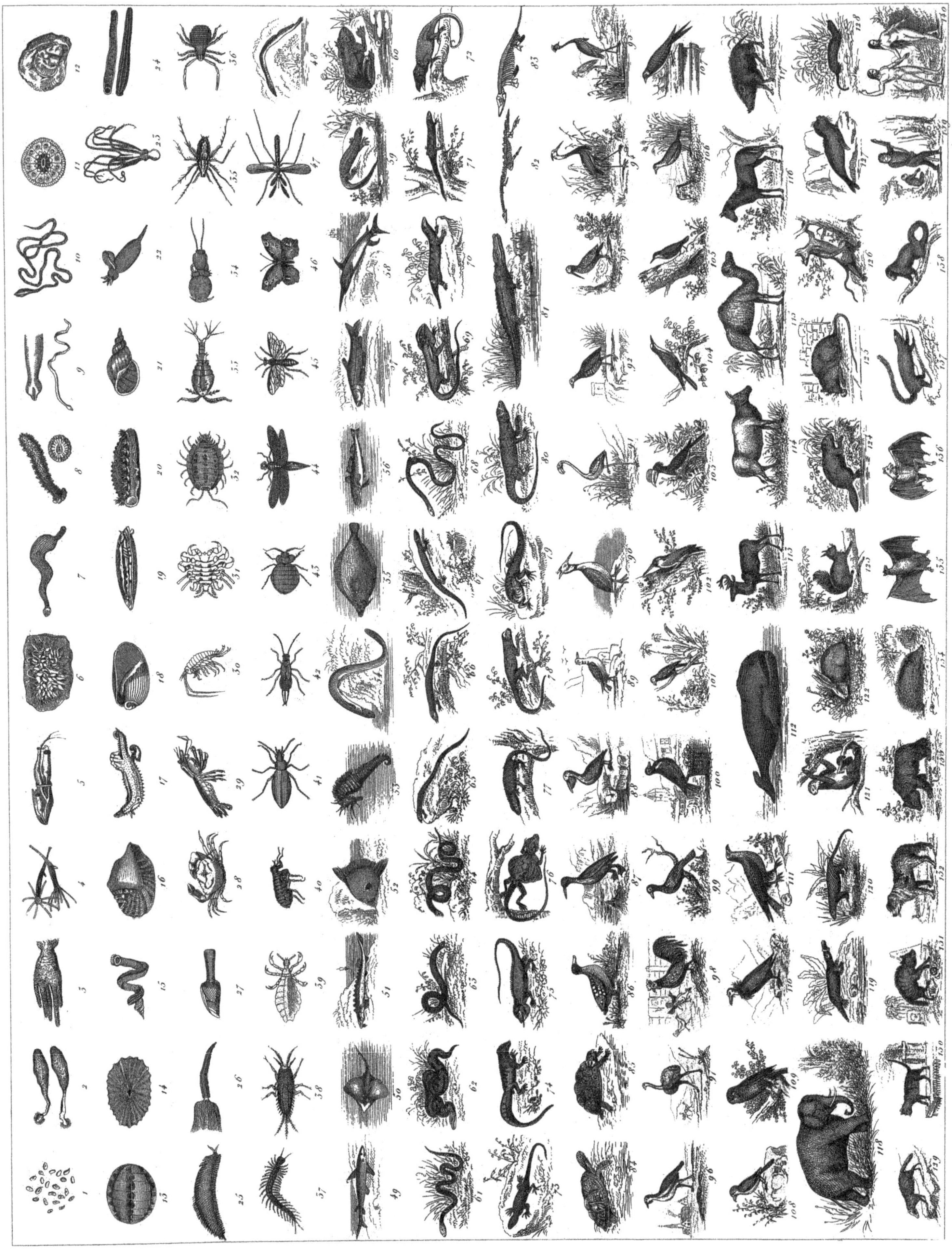

Plate 74 (Zoology, Plate 1)

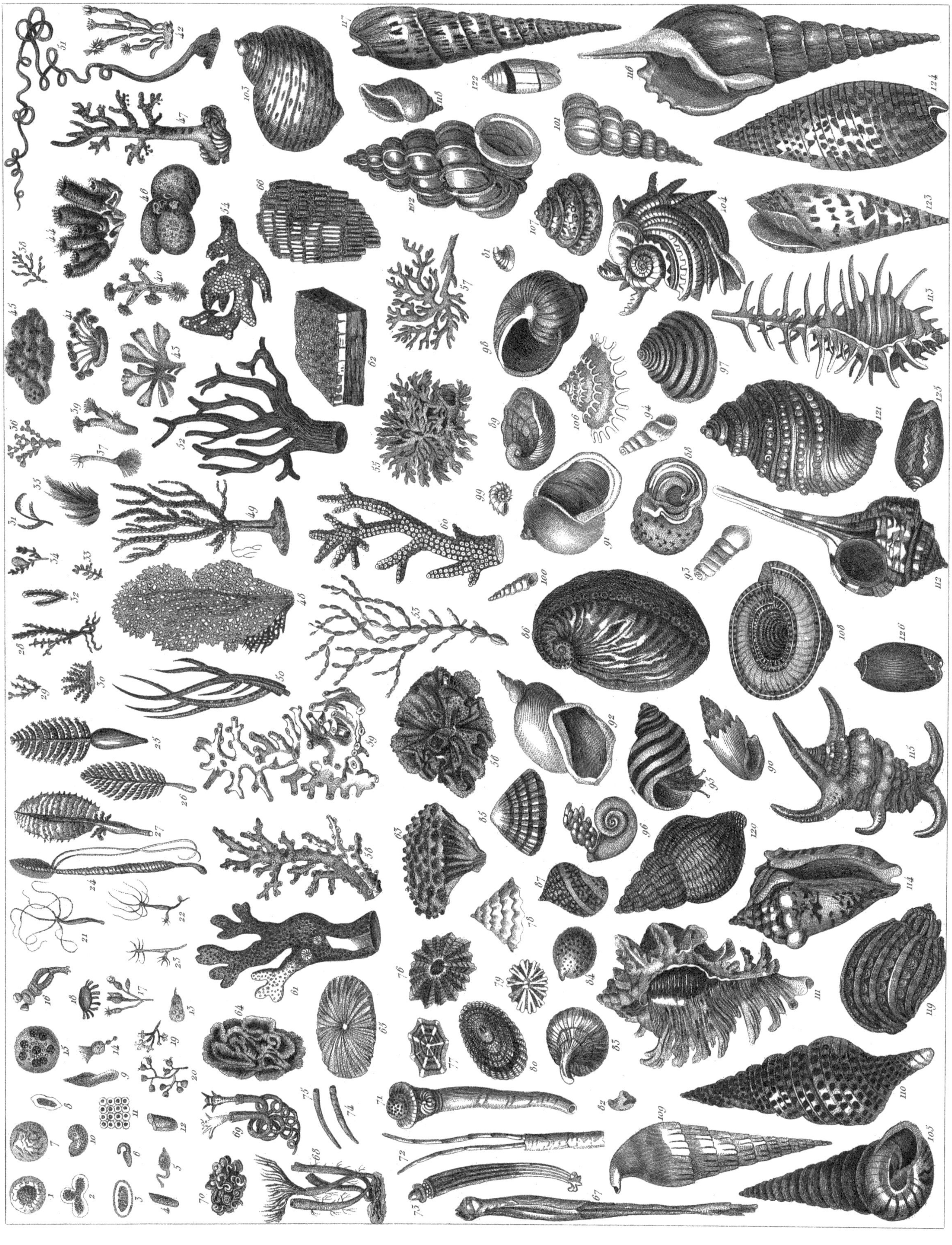

Plate 75 (Zoology, Plate 2)

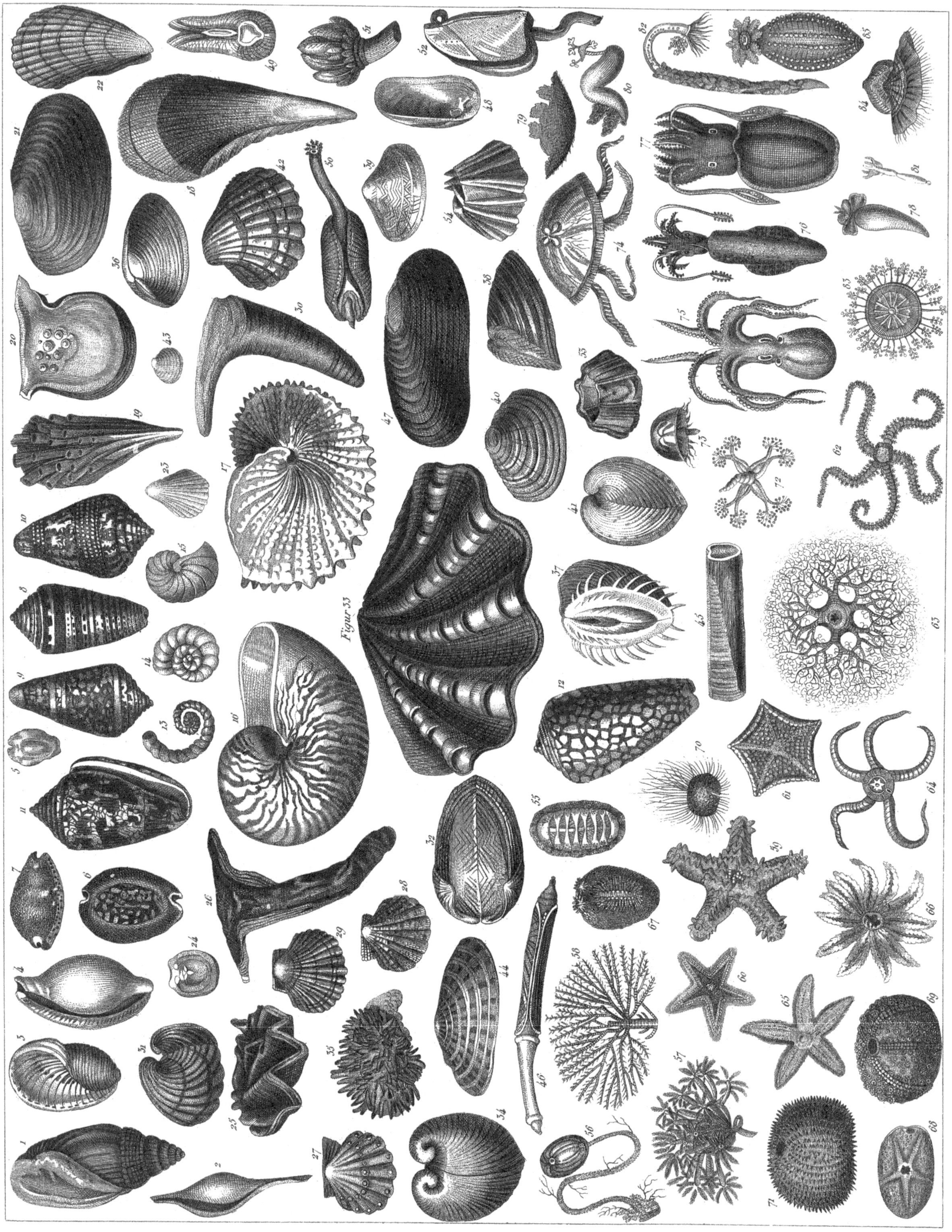

Plate 76 (Zoology, Plate 2)

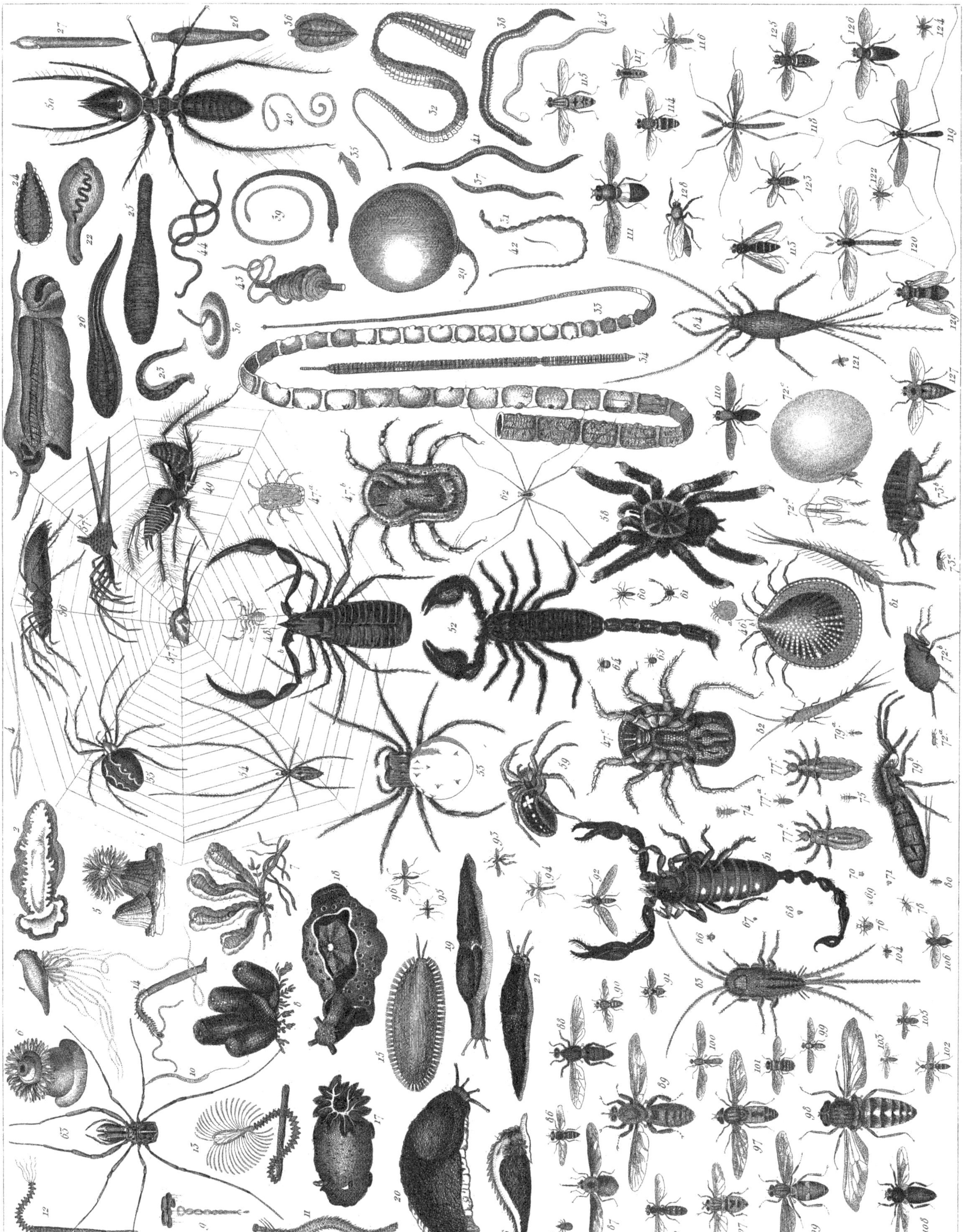

Plate 77 (Zoology, Plate 4)

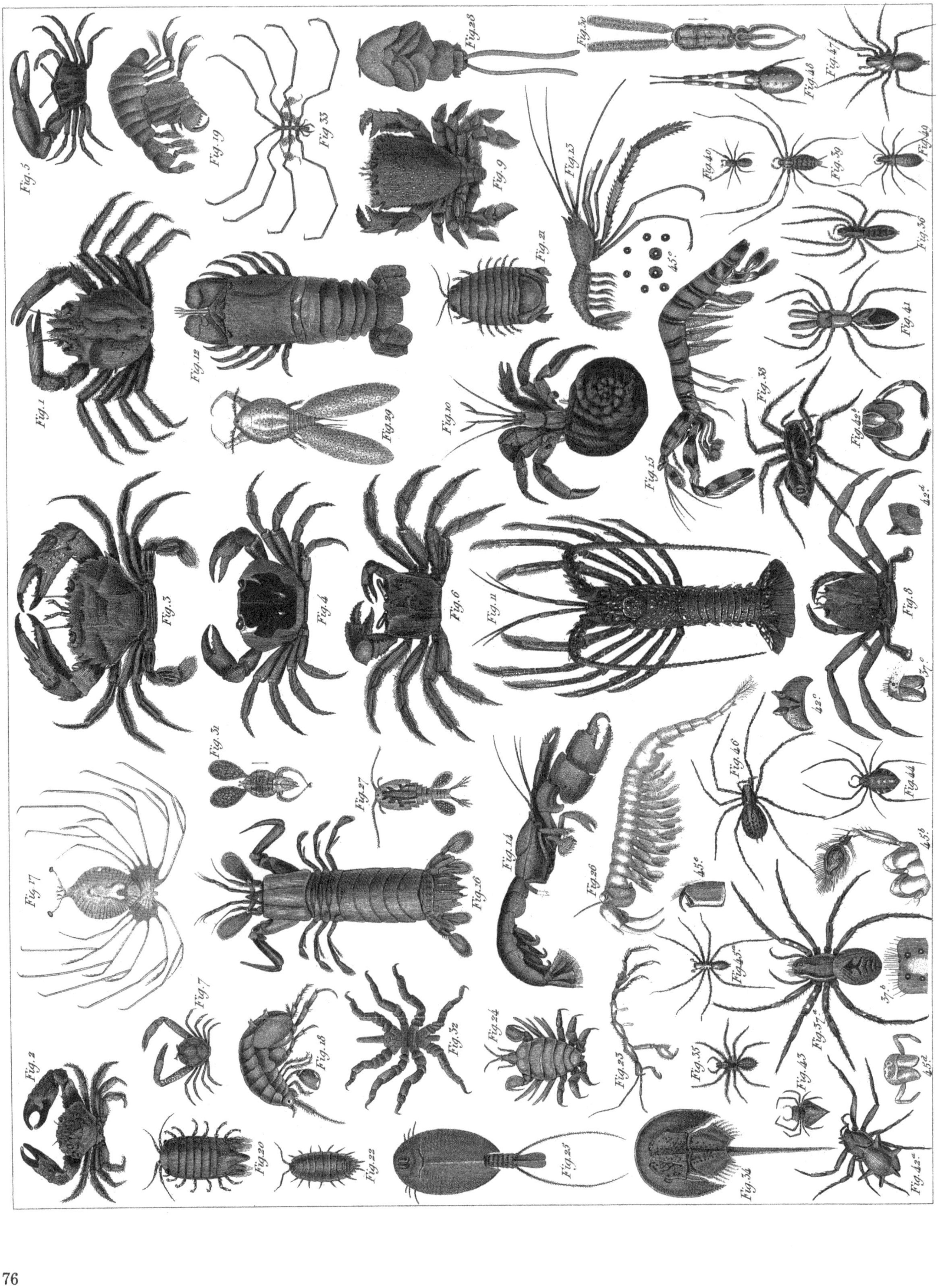

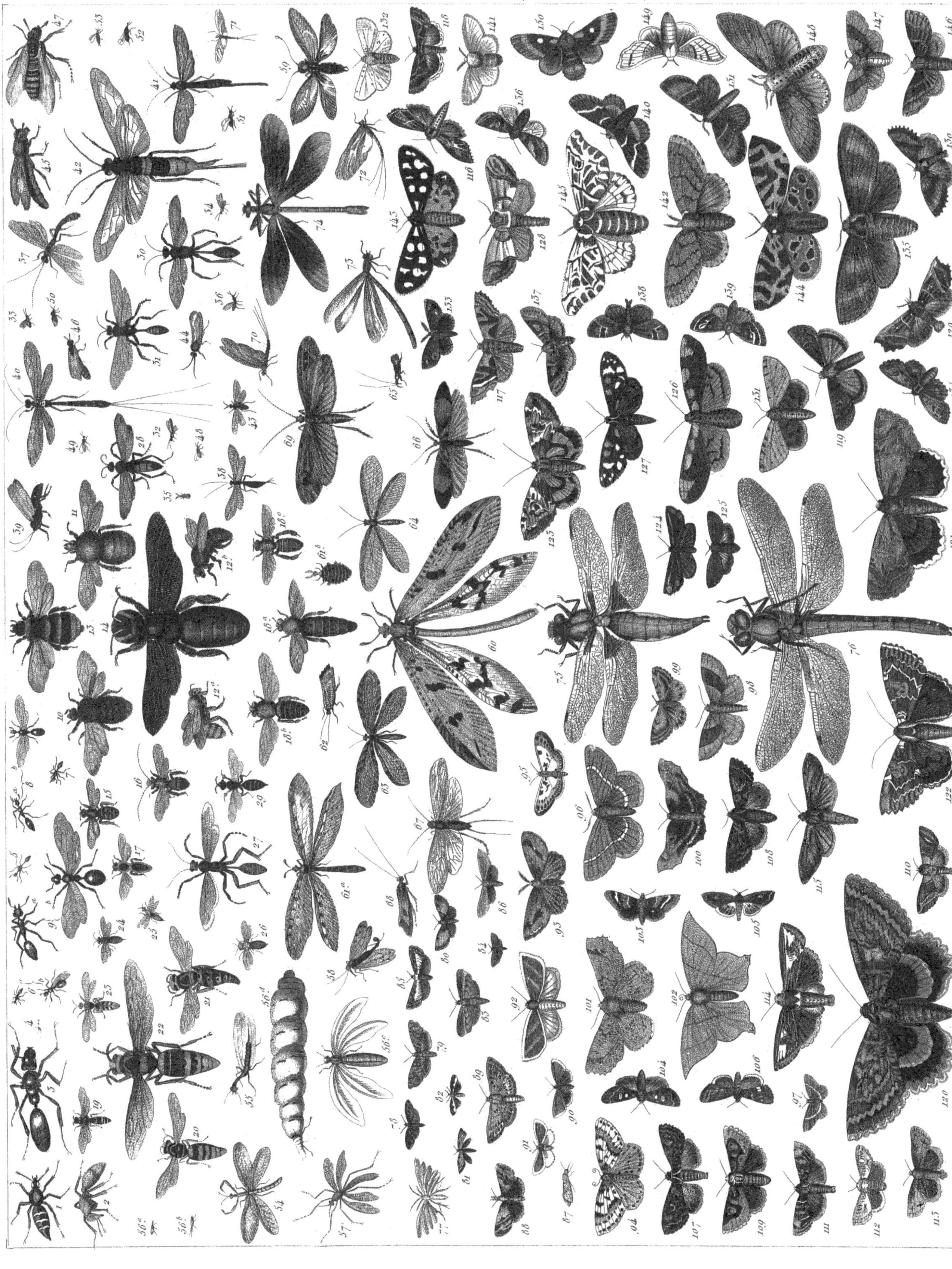

Plate 79 (Zoology, Plate 6)

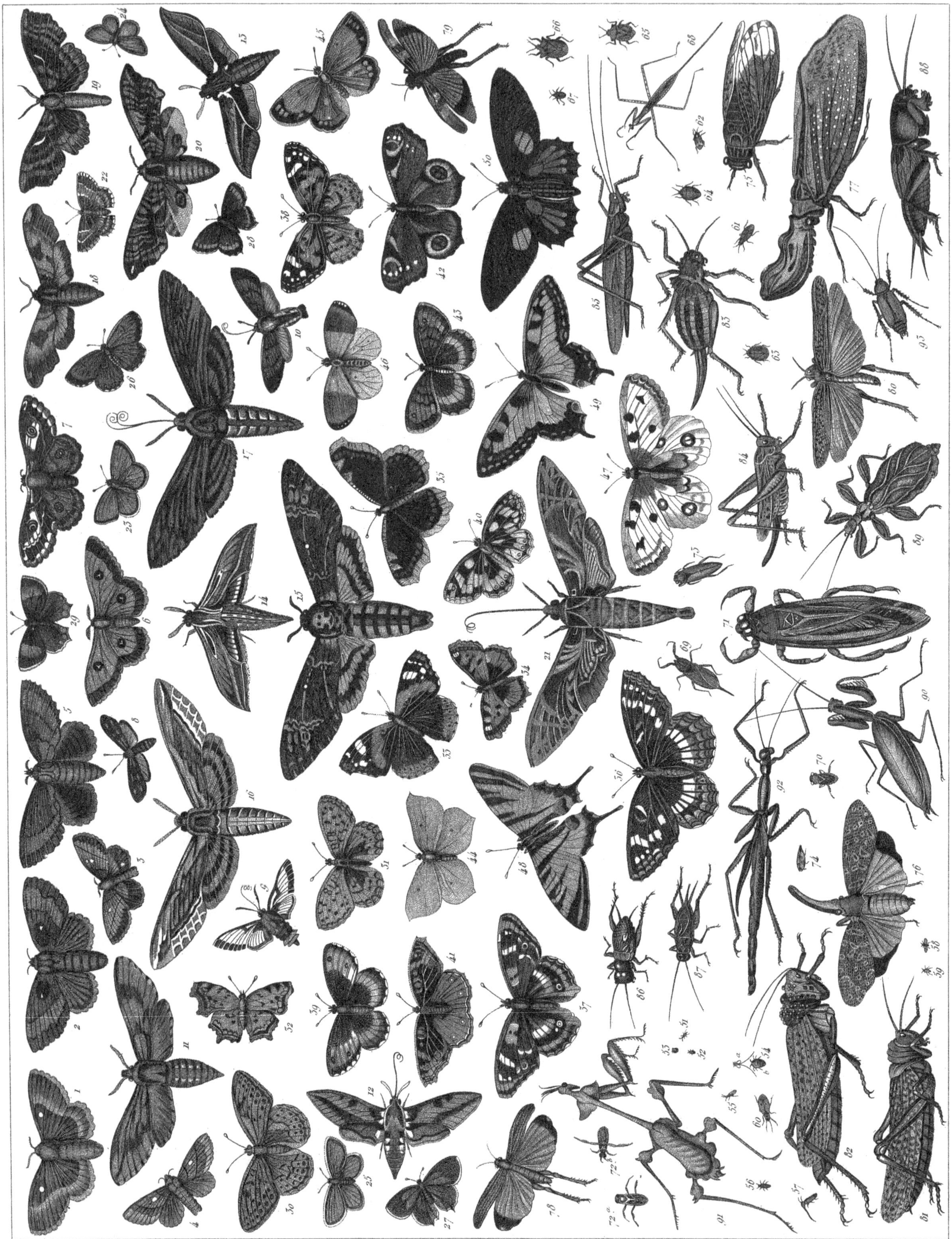

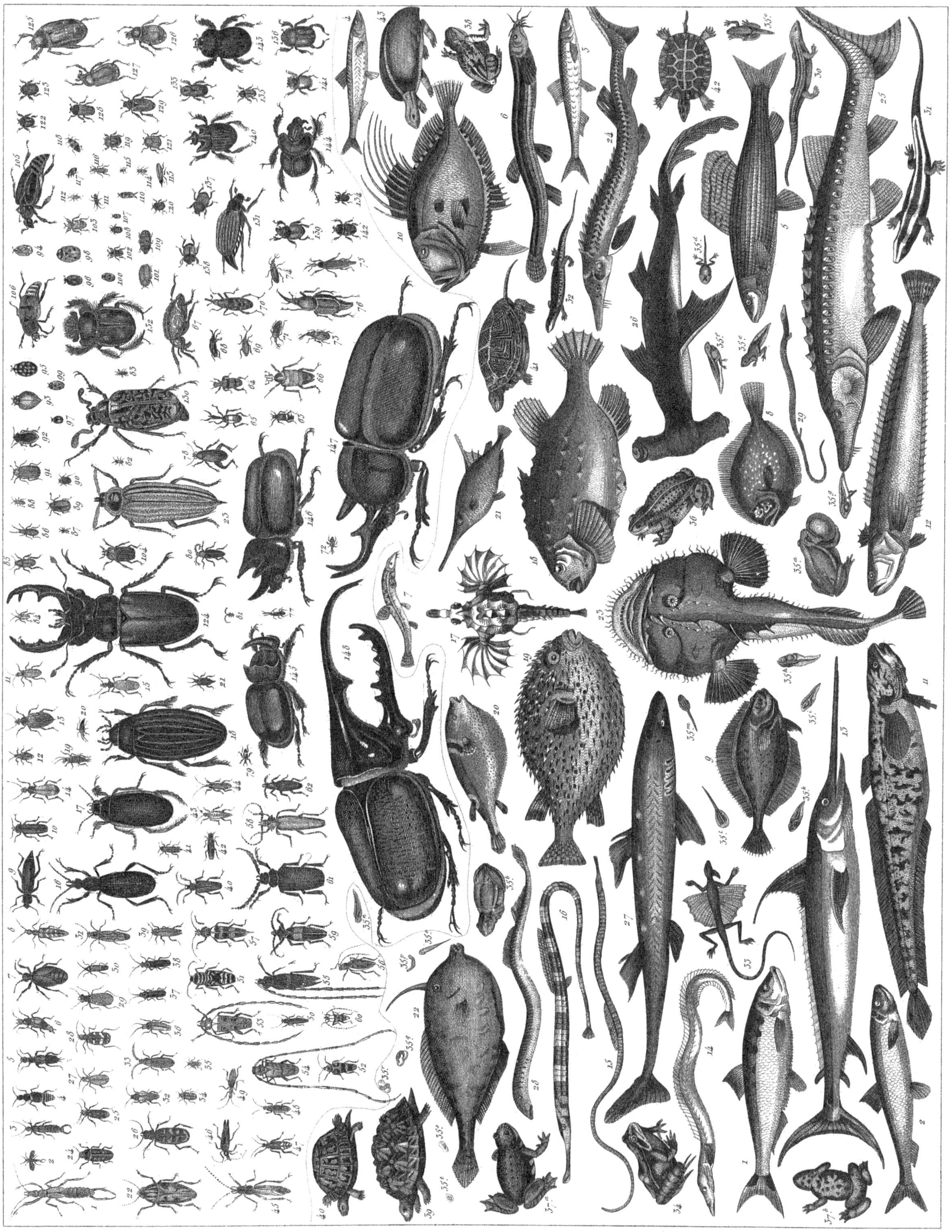

Plate 81 (Zoology, Plate 8)

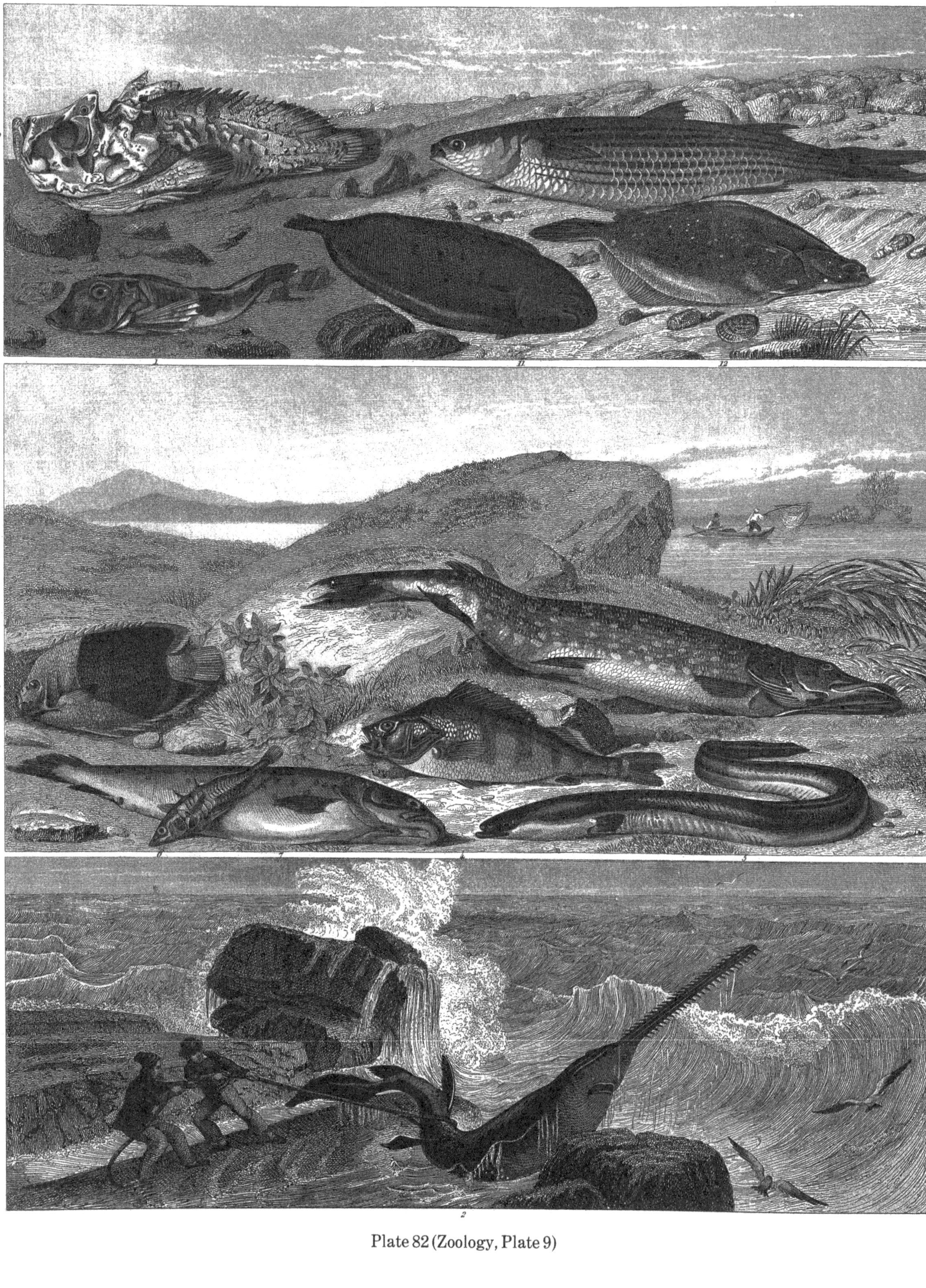

Plate 82 (Zoology, Plate 9)

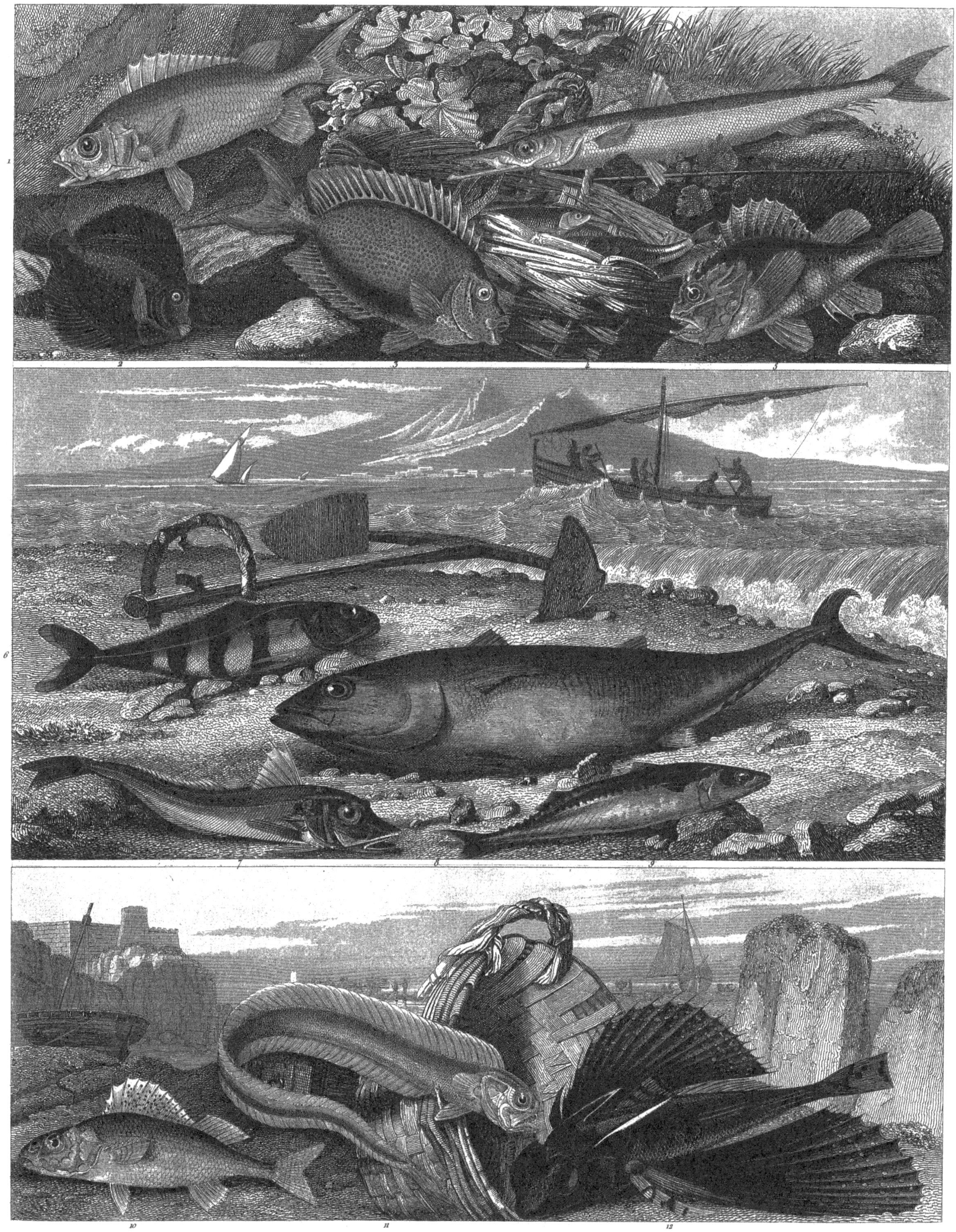

Plate 83 (Zoology, Plate 10)

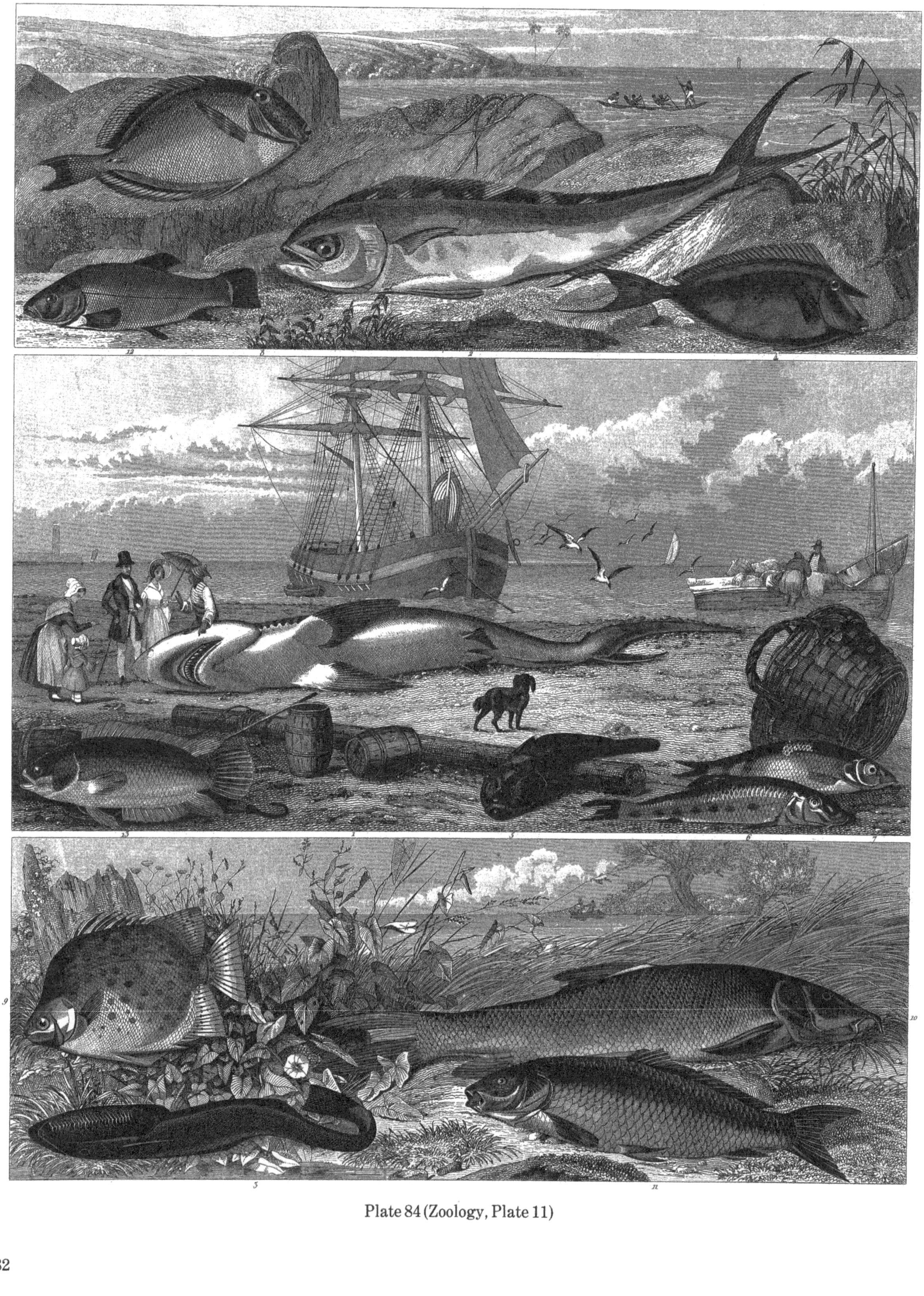

Plate 84 (Zoology, Plate 11)

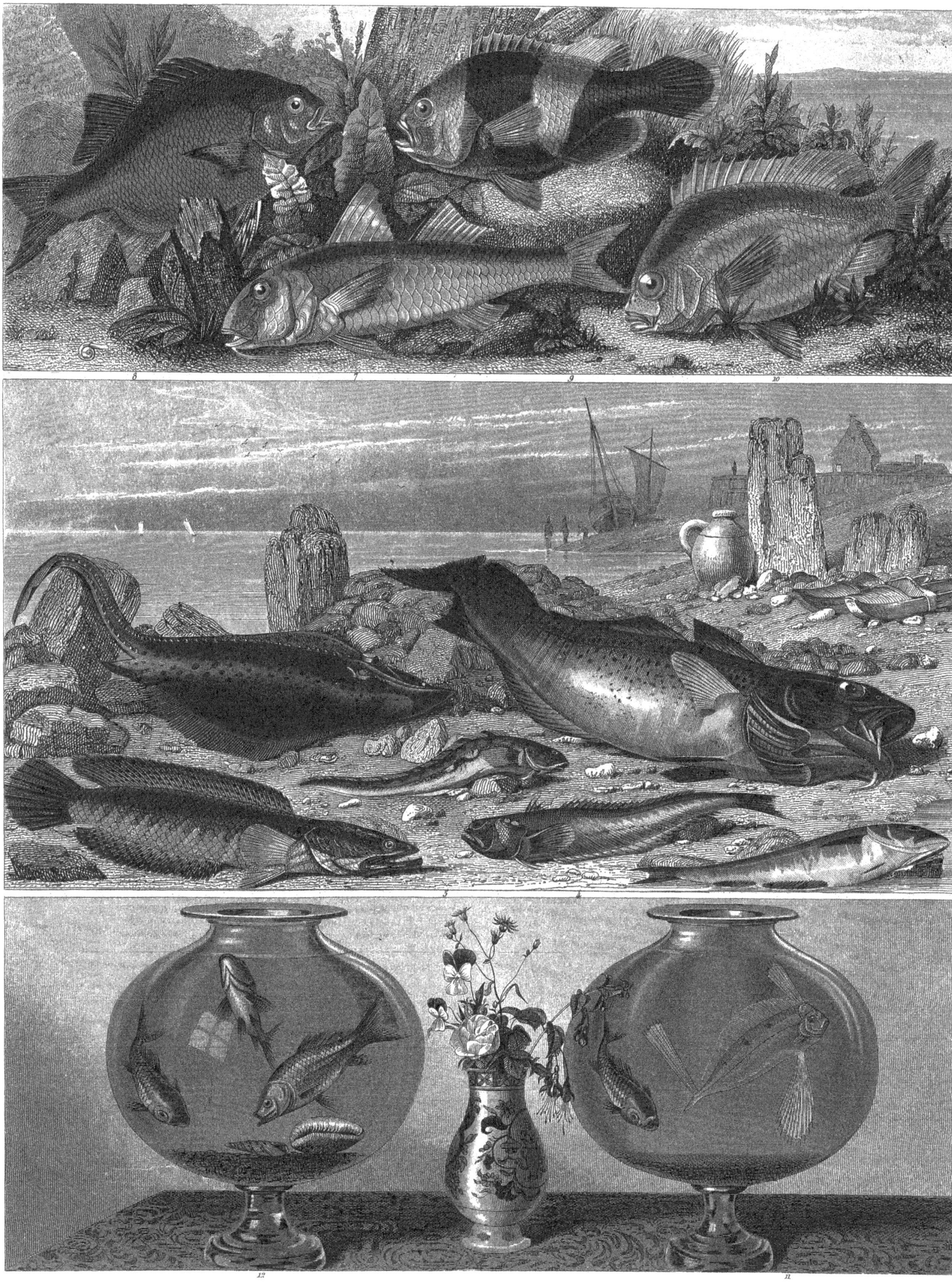

Plate 85 (Zoology, Plate 12)

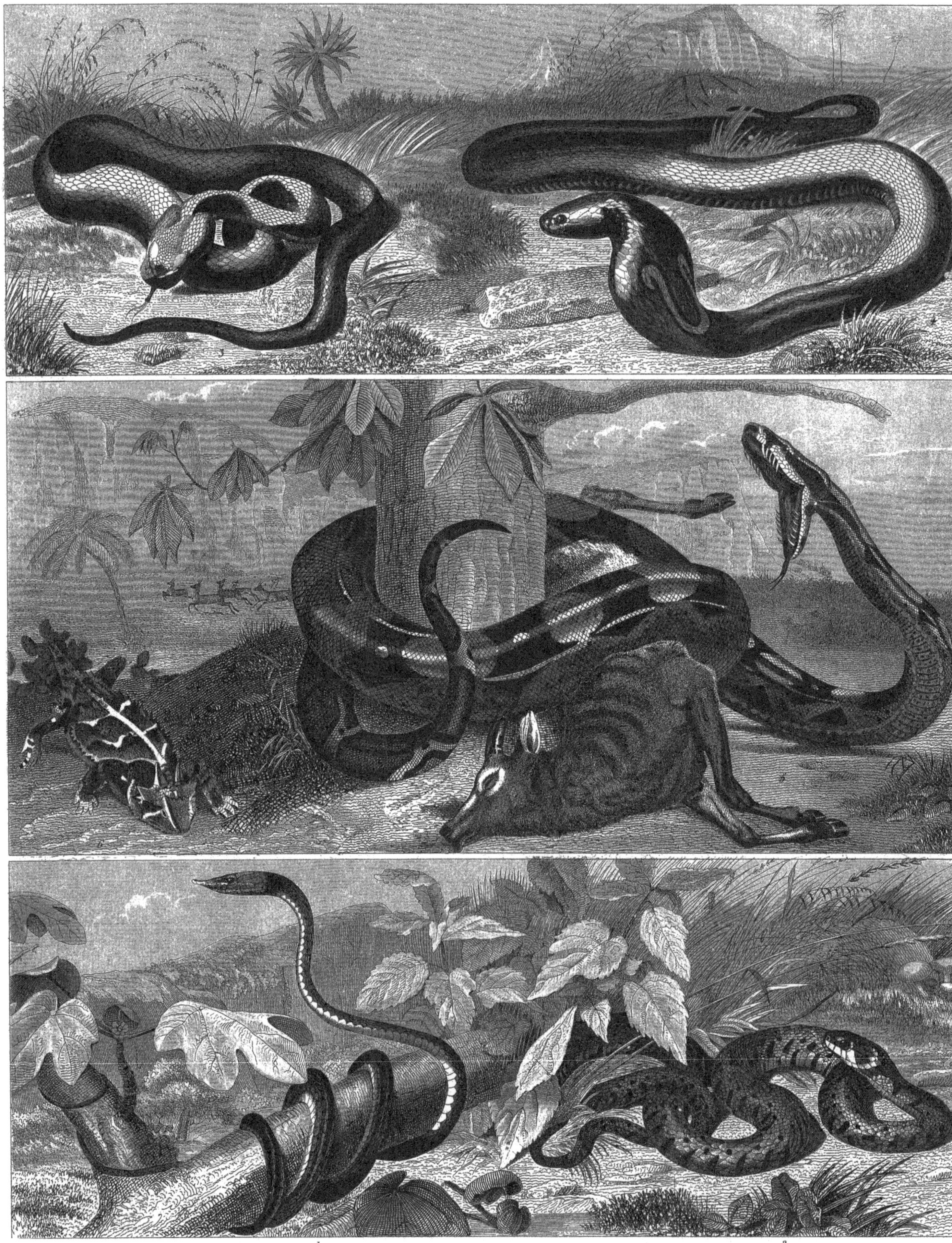

1 *2*

Plate 86 (Zoology, Plate 13)

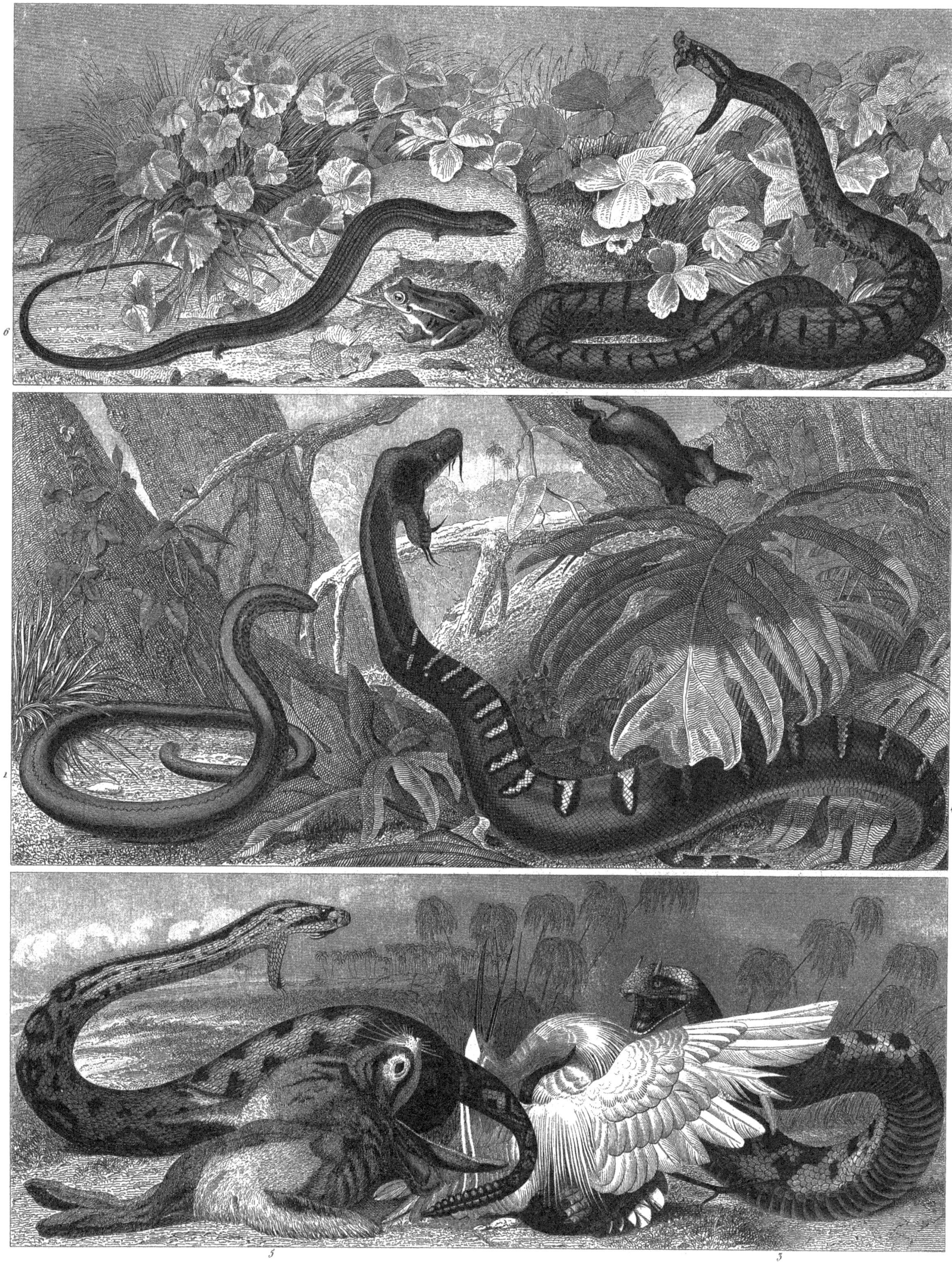

Plate 87 (Zoology, Plate 14)

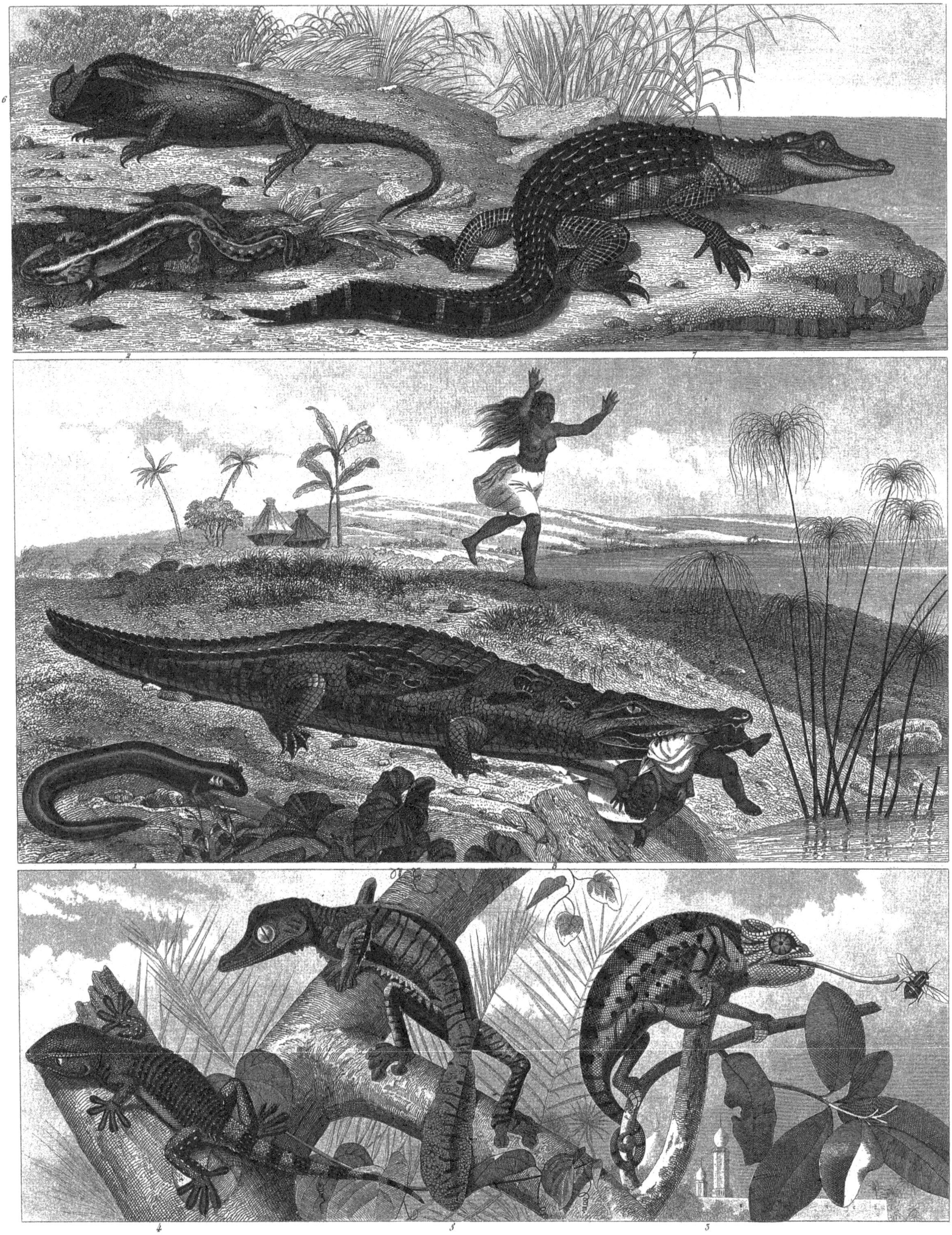

Plate 88 (Zoology, Plate 15)

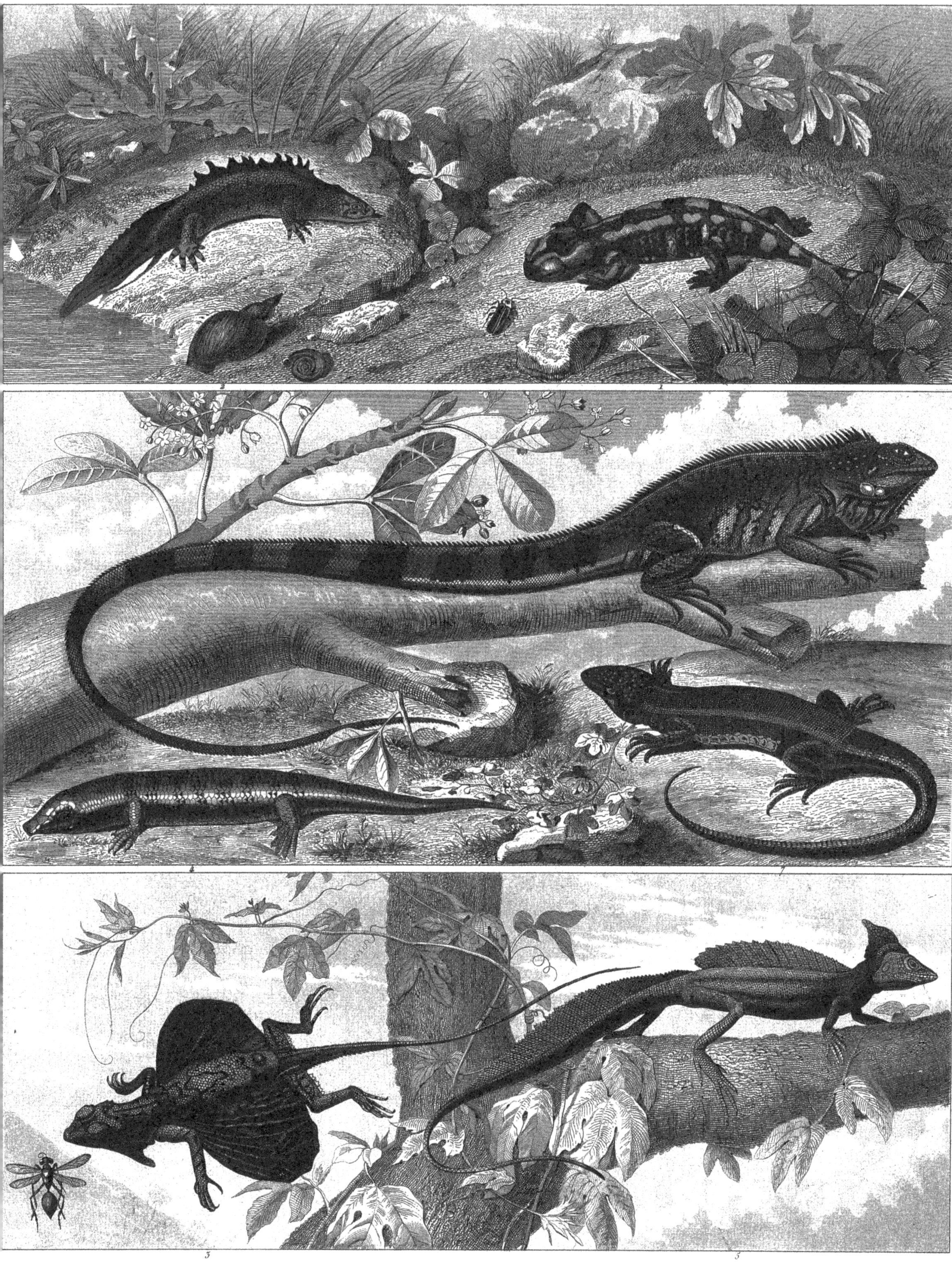

Plate 89 (Zoology, Plate 16)

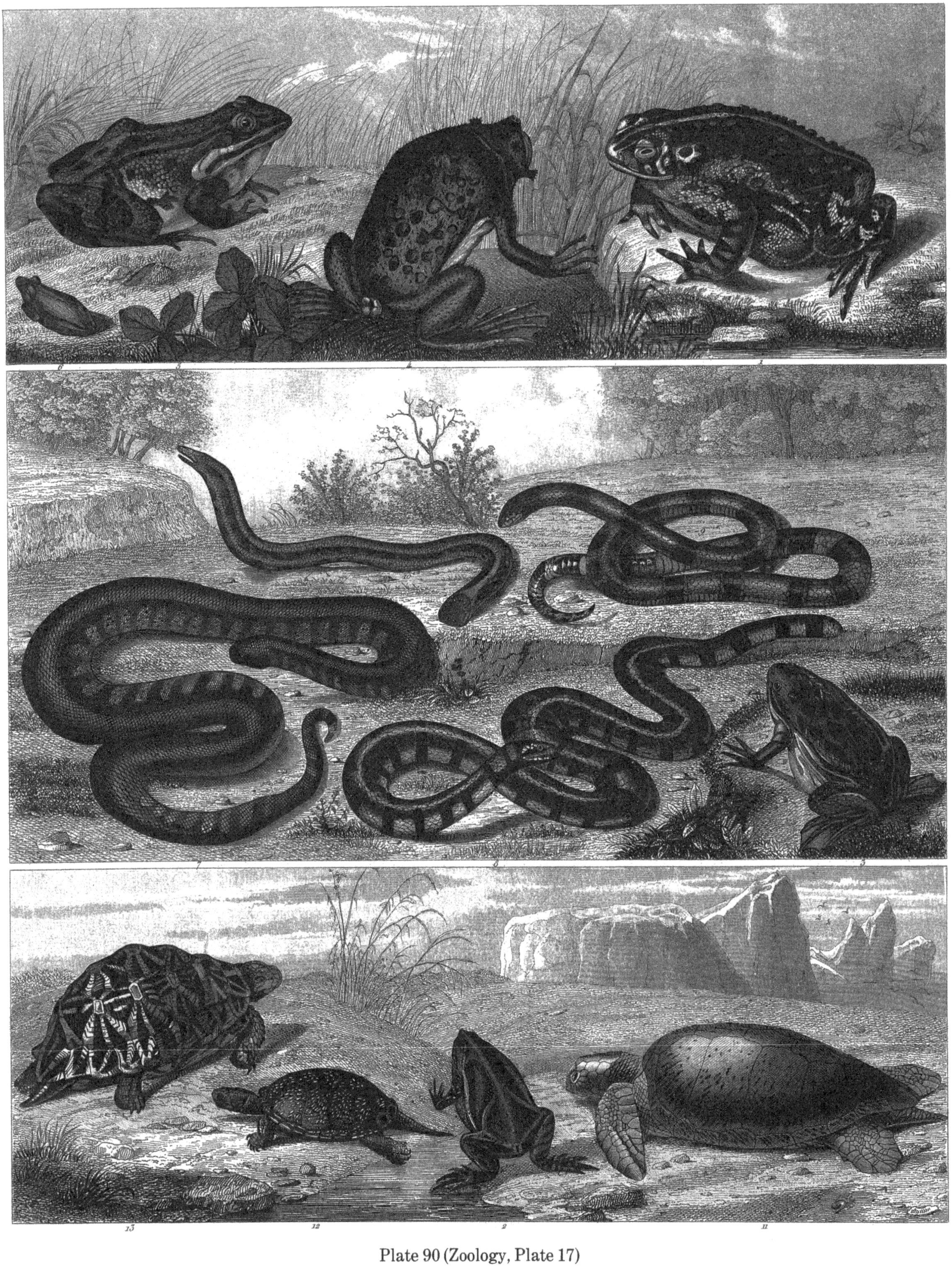

Plate 90 (Zoology, Plate 17)

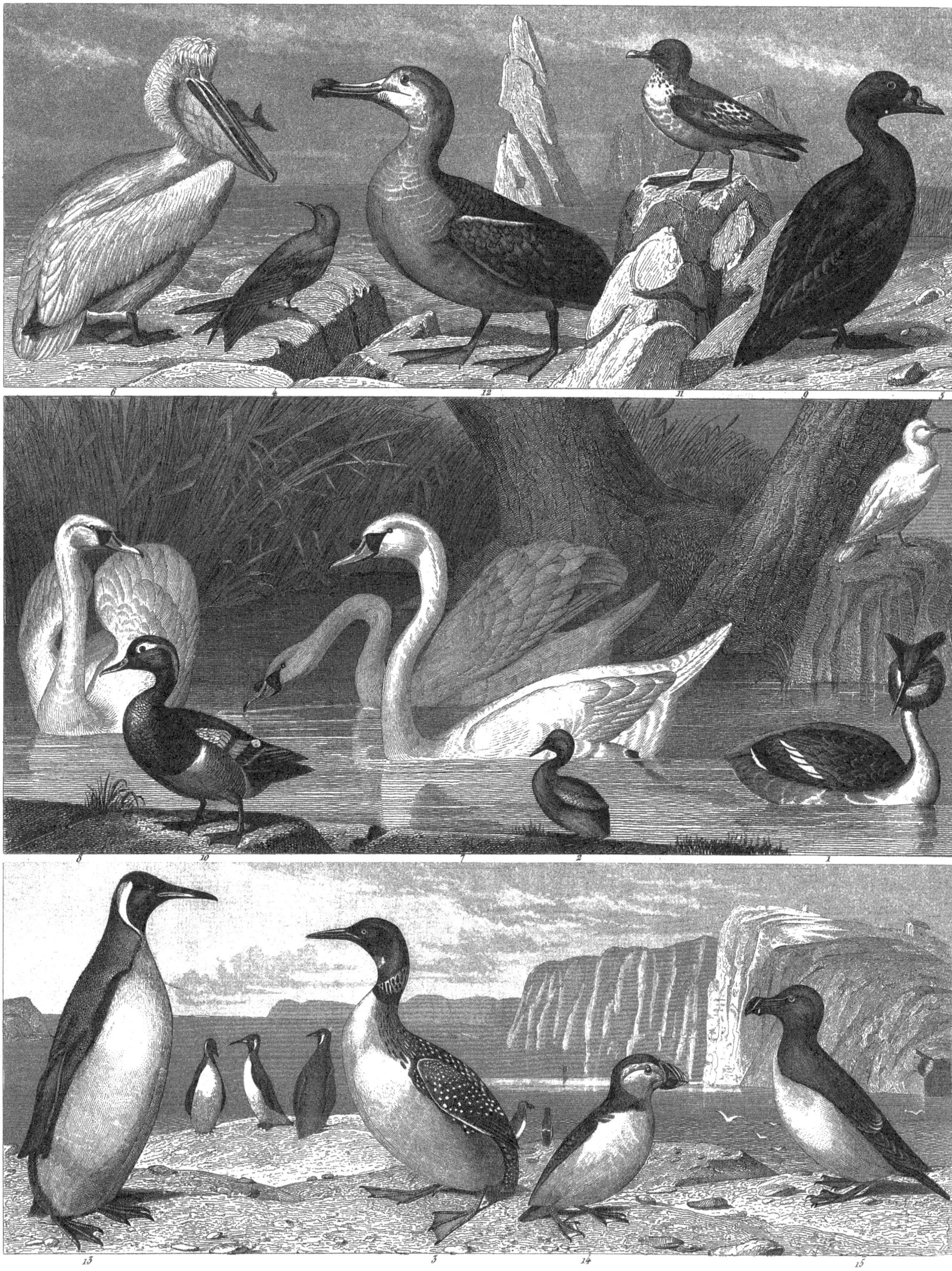

Plate 91 (Zoology, Plate 18)

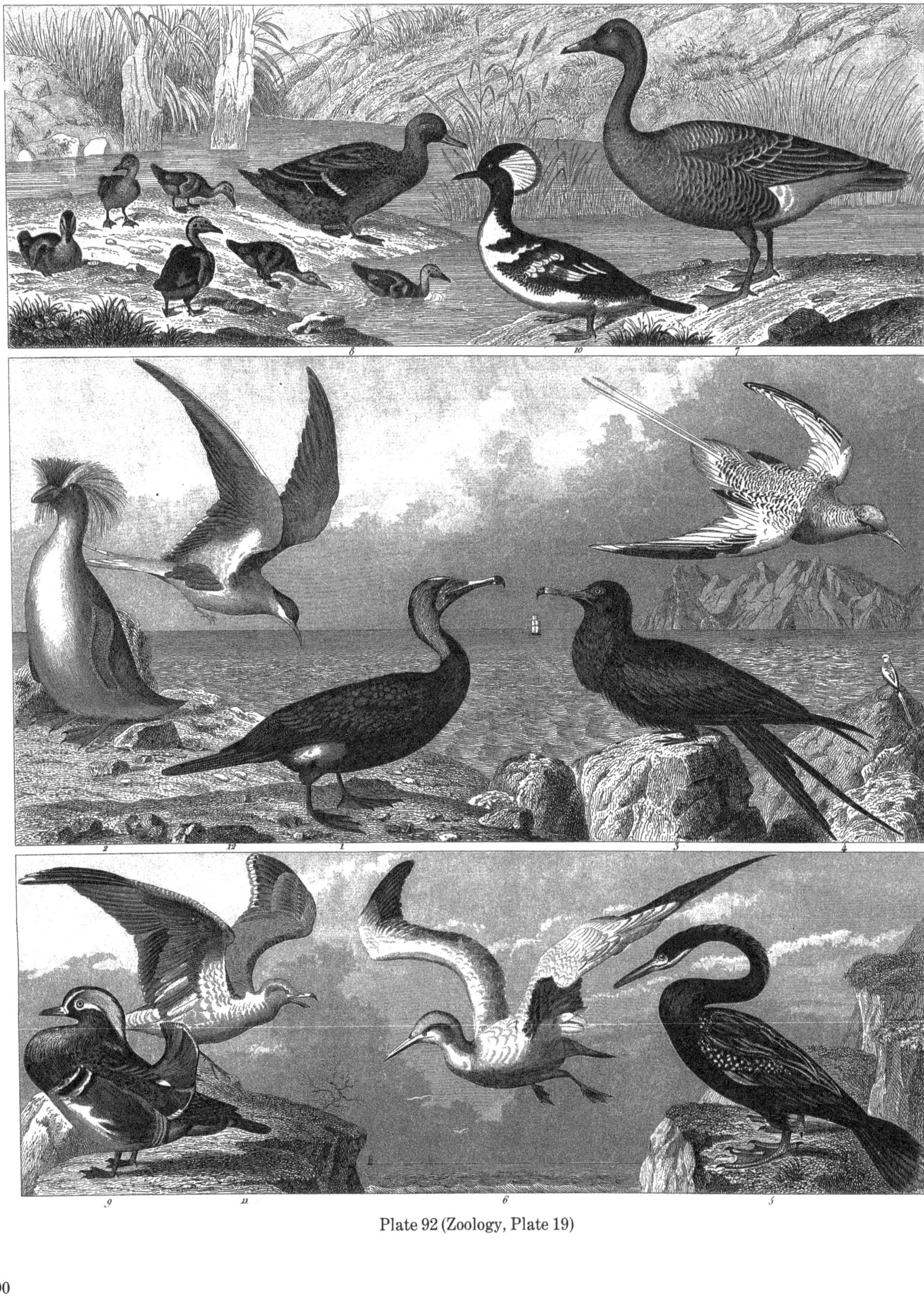

Plate 92 (Zoology, Plate 19)

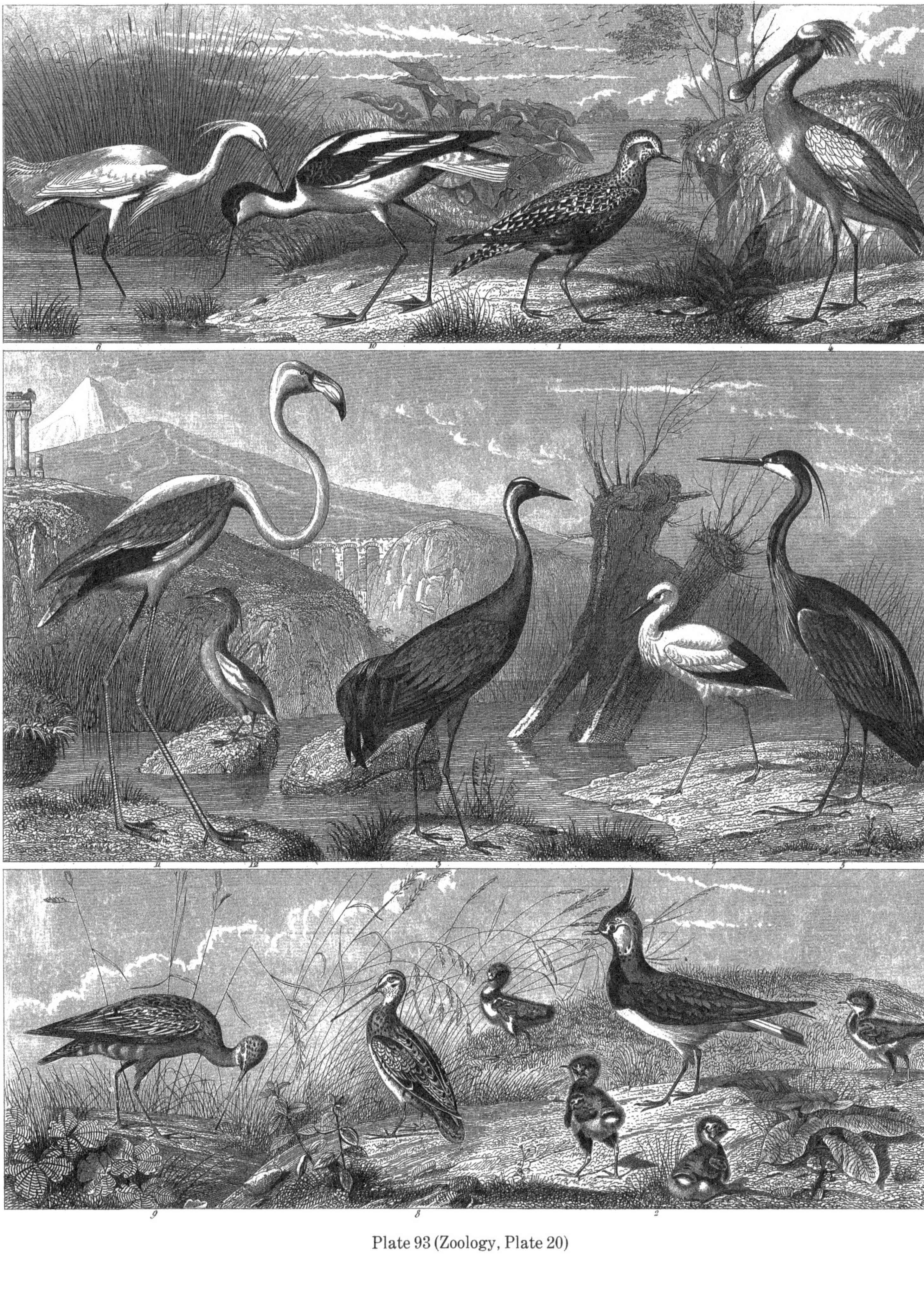

Plate 93 (Zoology, Plate 20)

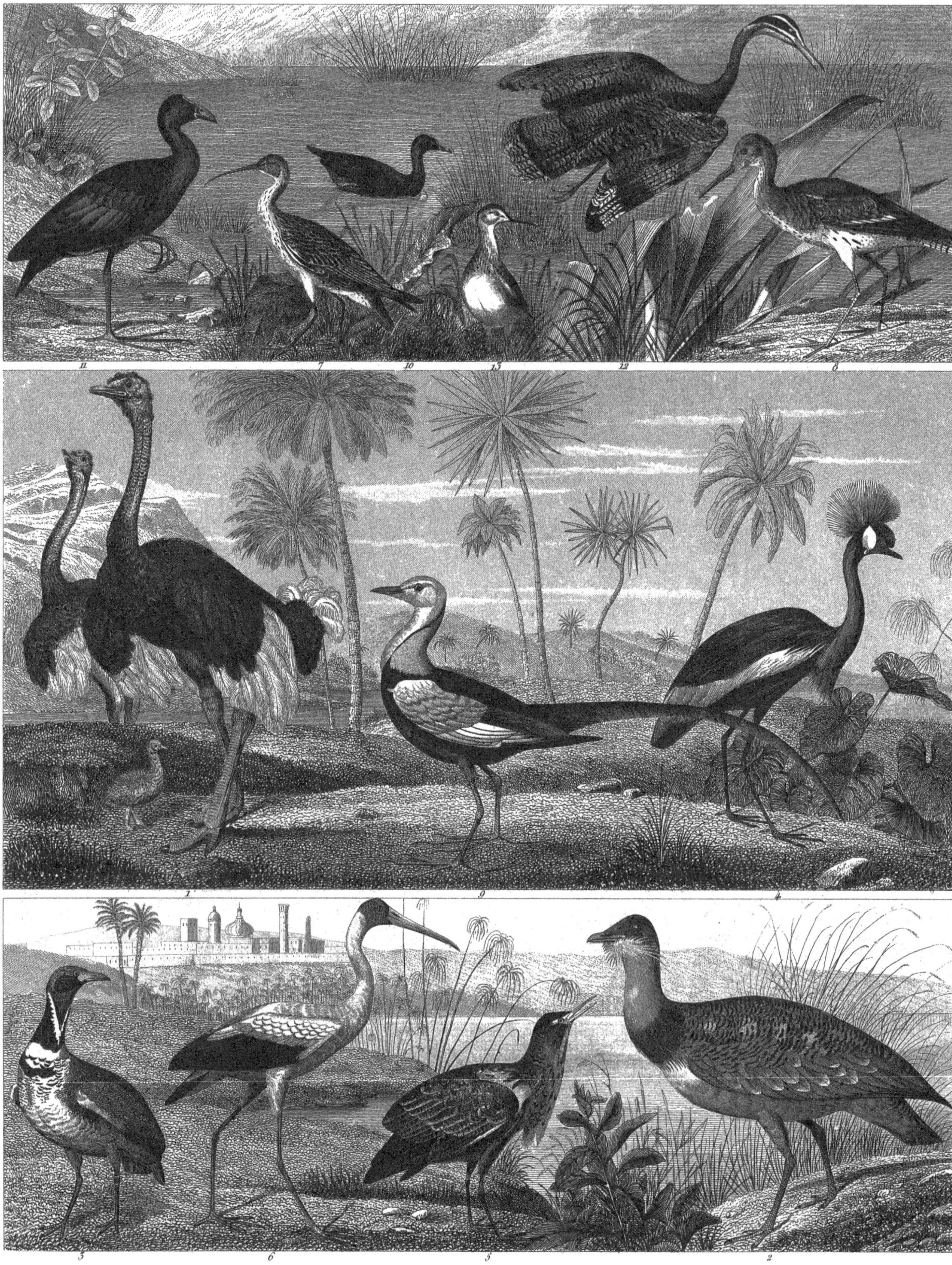

Plate 94 (Zoology, Plate 21)

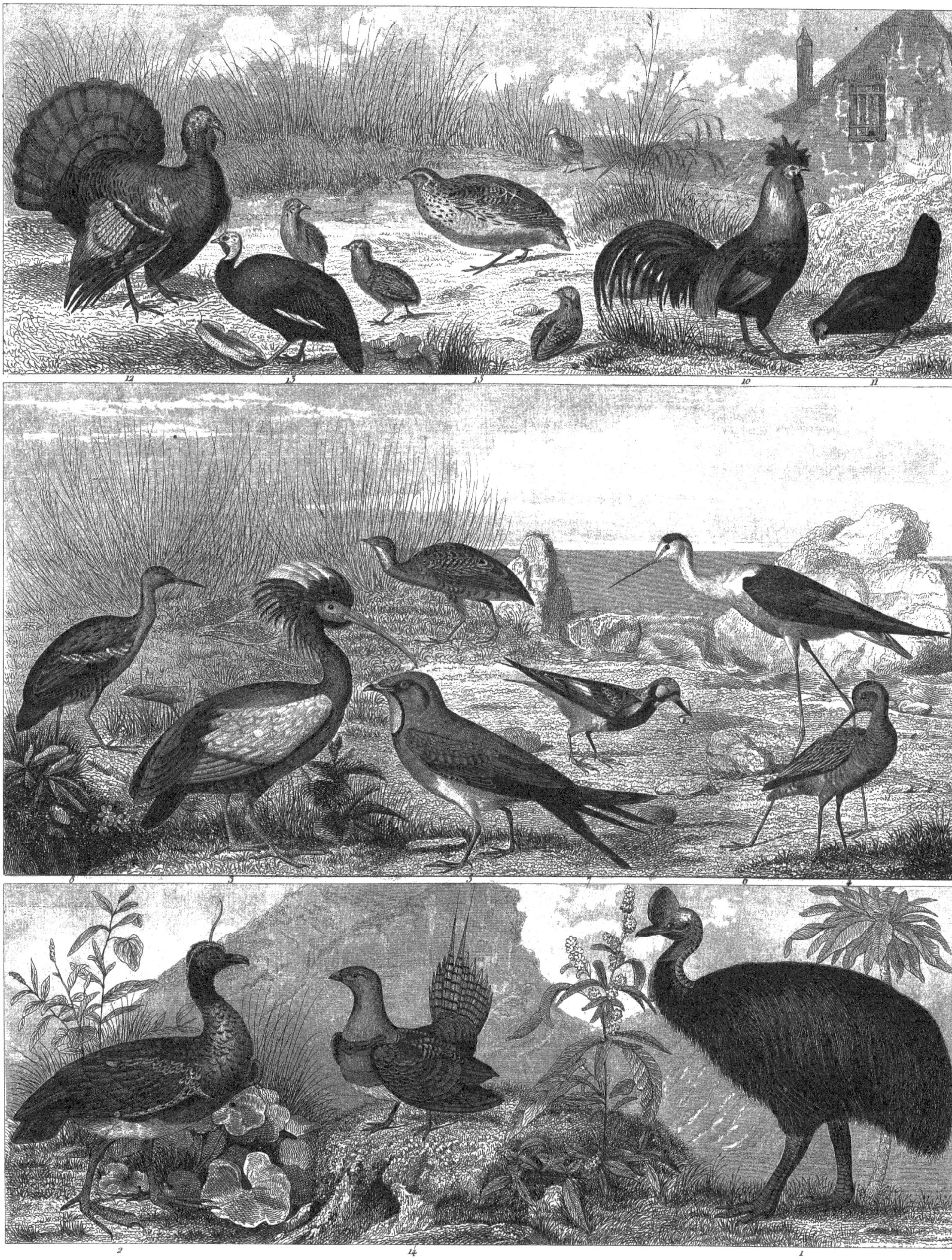

Plate 95 (Zoology, Plate 22)

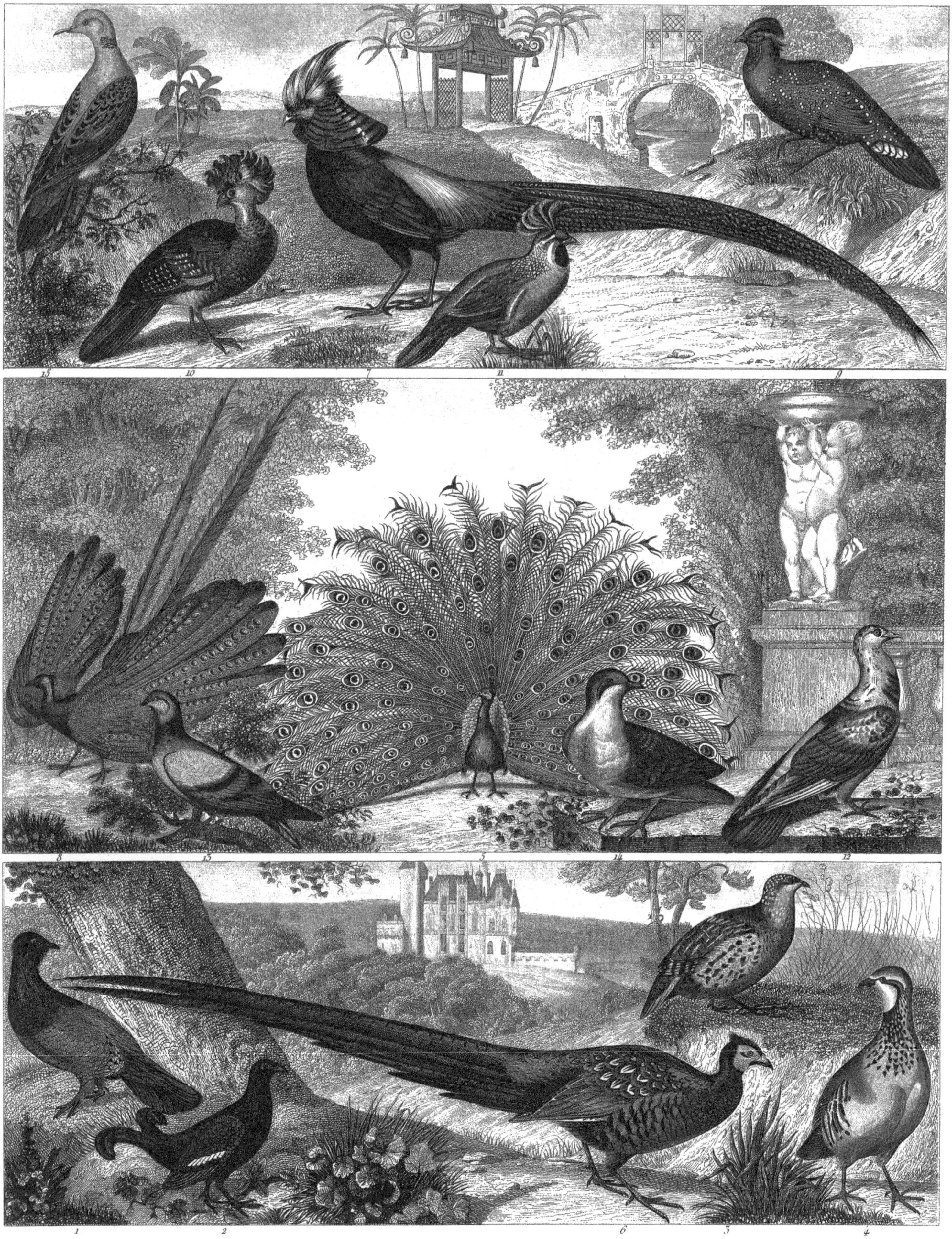

Plate 96 (Zoology, Plate 23)

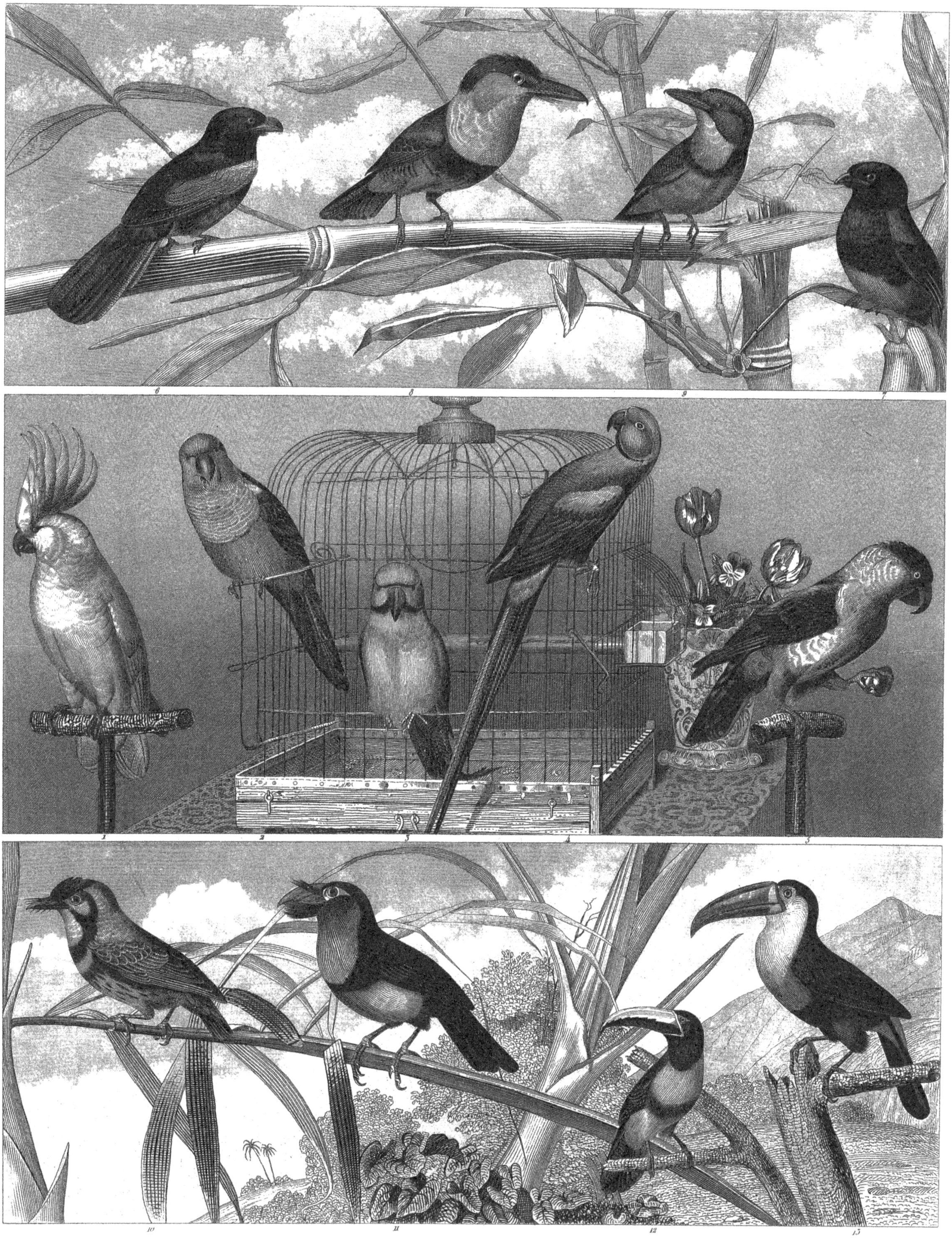

Plate 97 (Zoology, Plate 24)

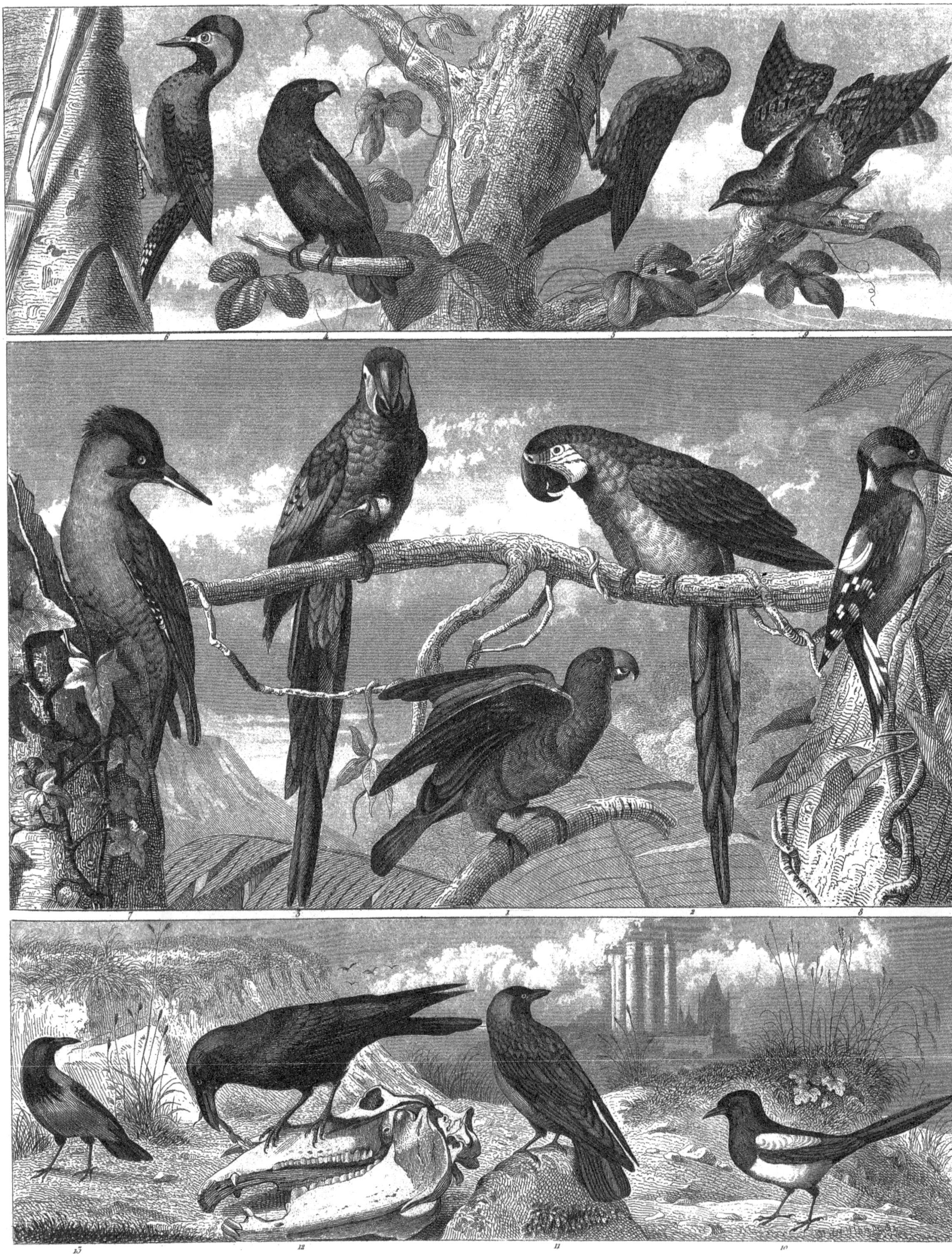

Plate 98 (Zoology, Plate 25)

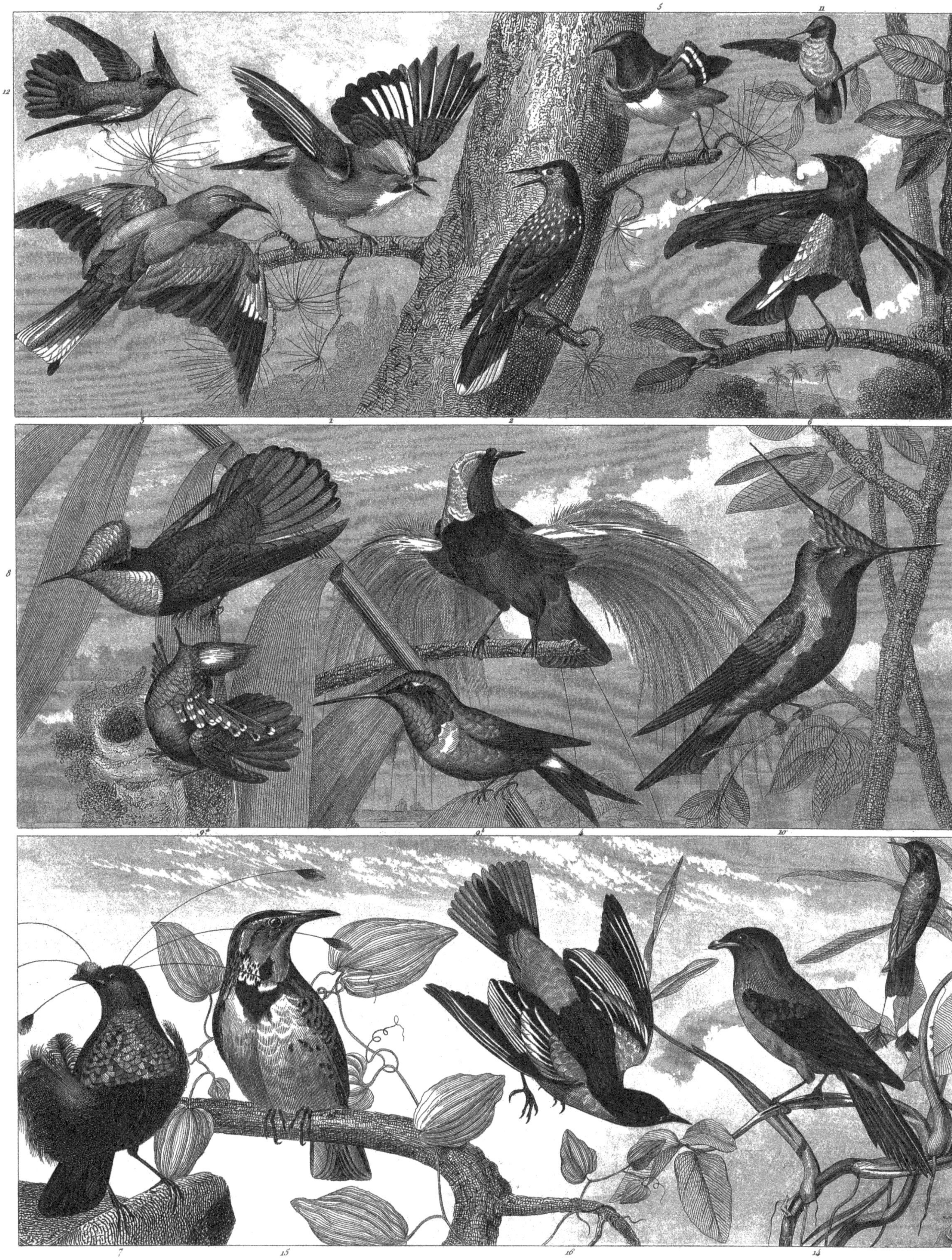

Plate 99 (Zoology, Plate 26)

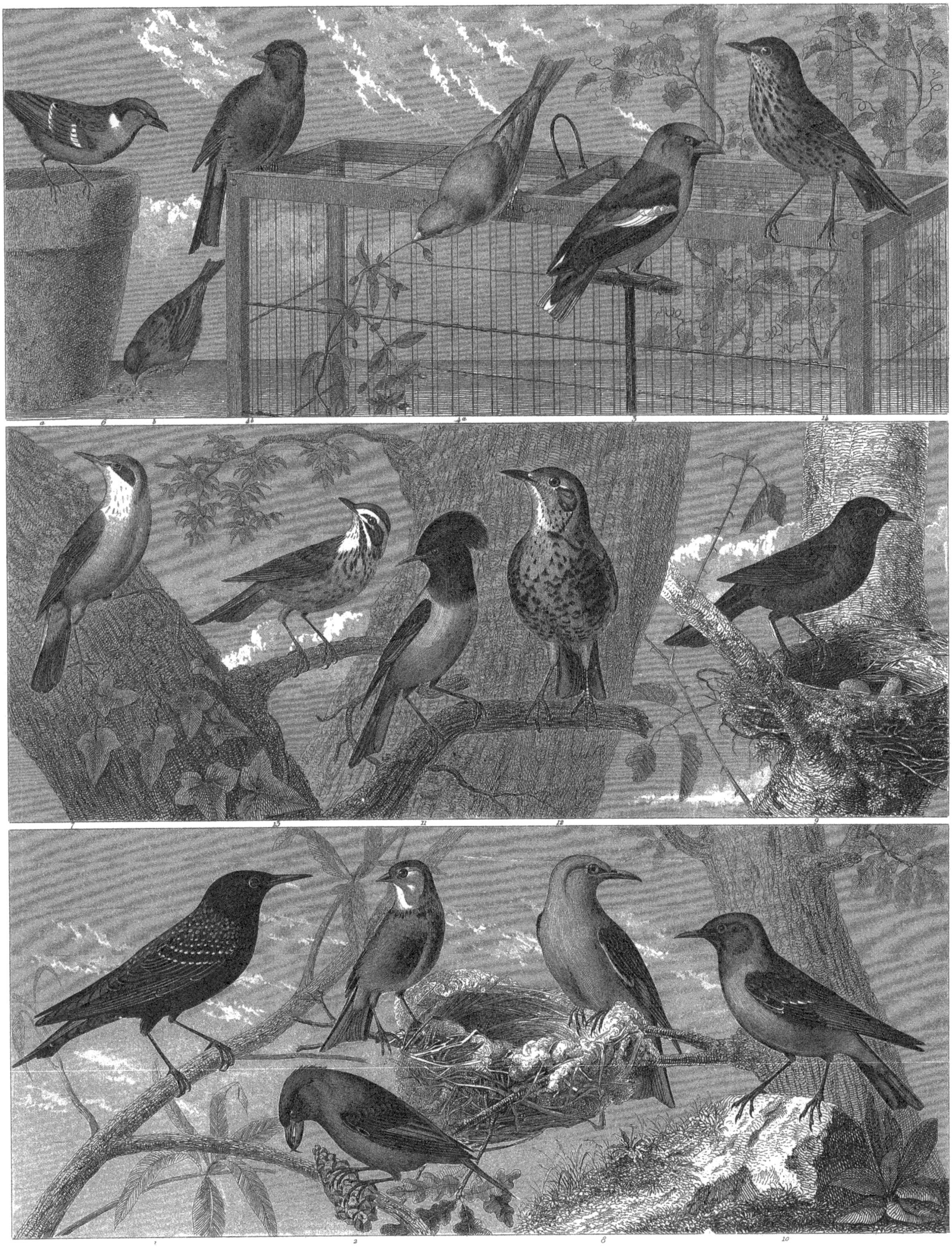

Plate 100 (Zoology, Plate 27)

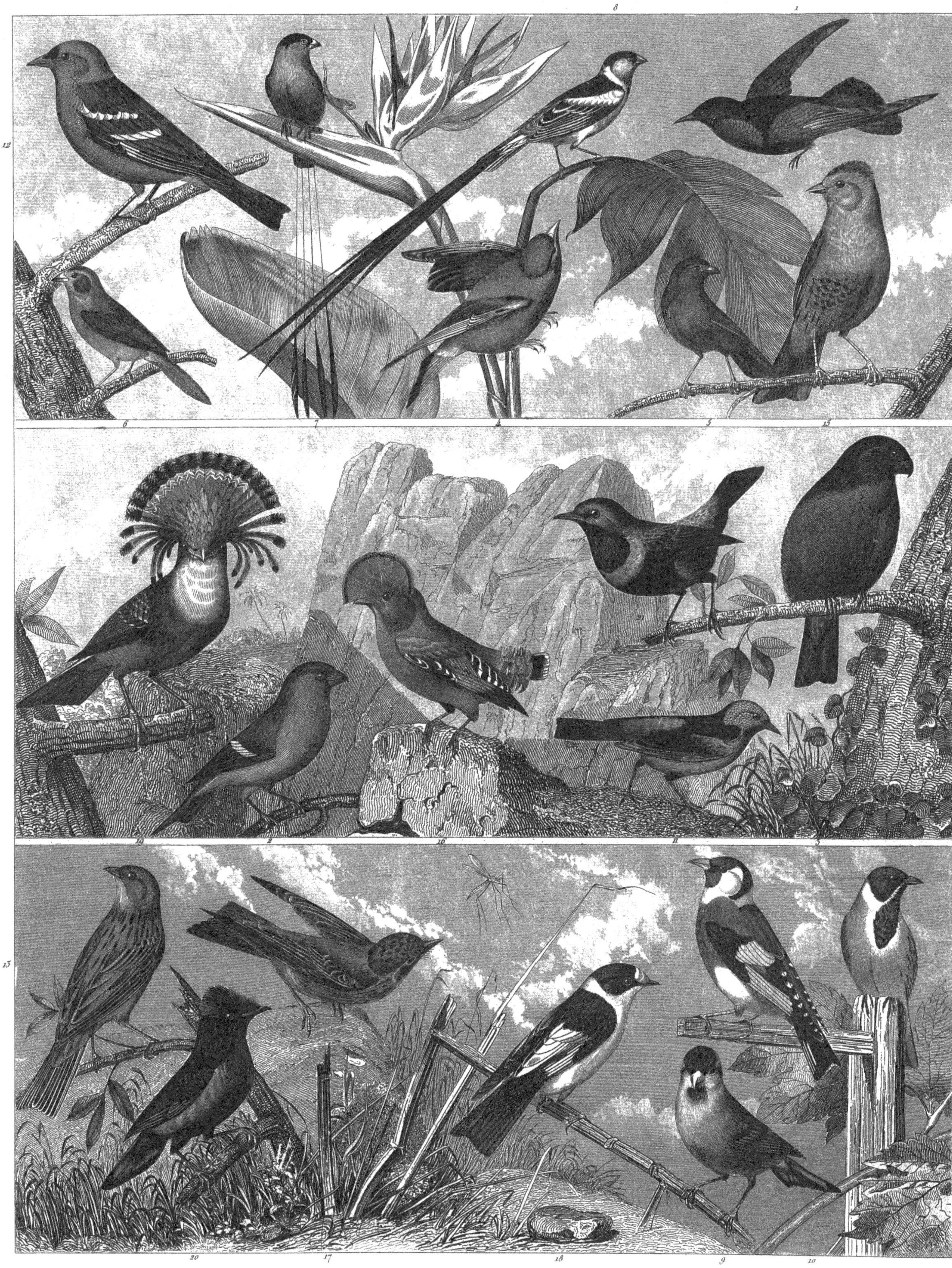

Plate 101 (Zoology, Plate 28)

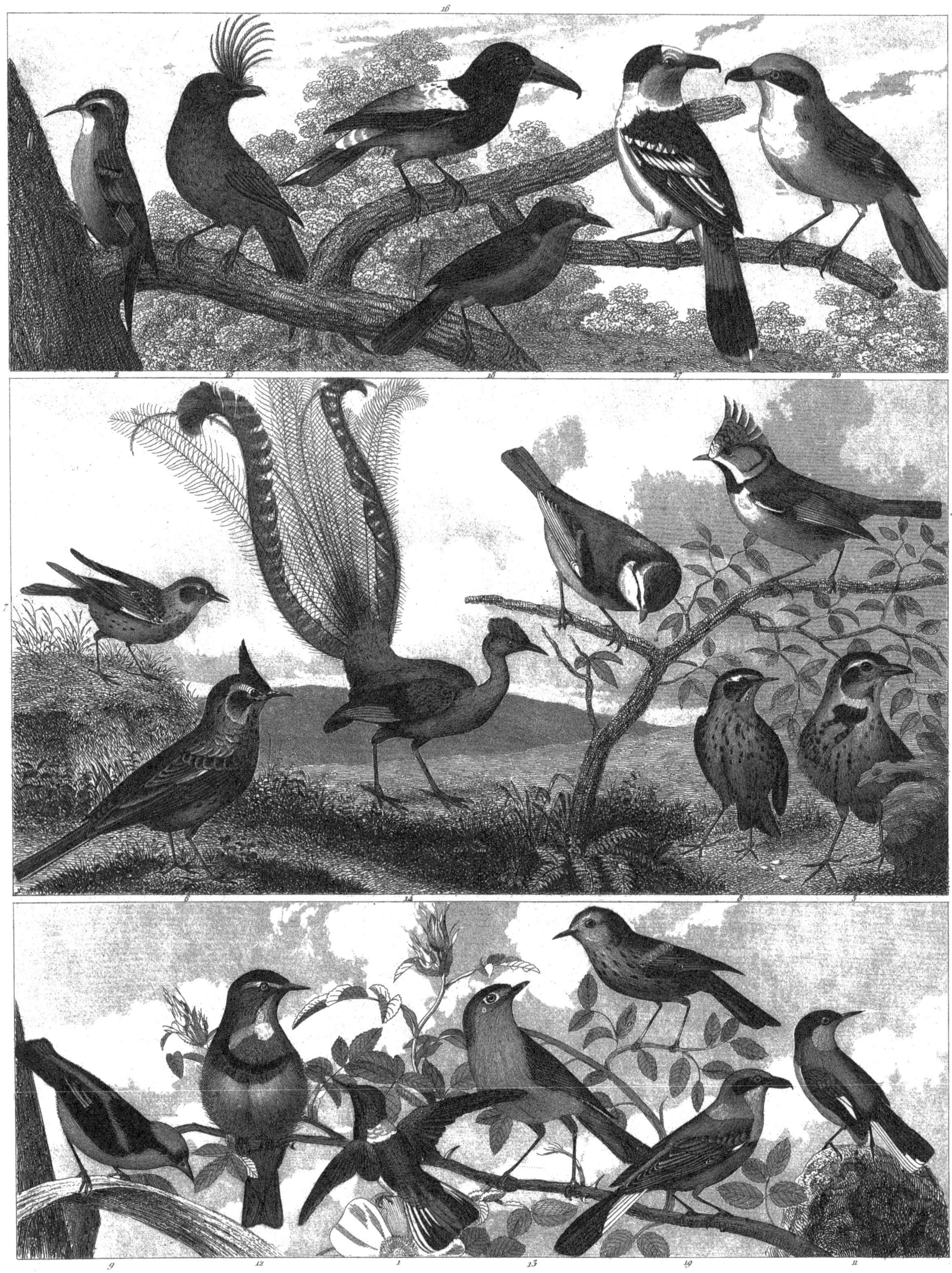

Plate 102 (Zoology, Plate 29)

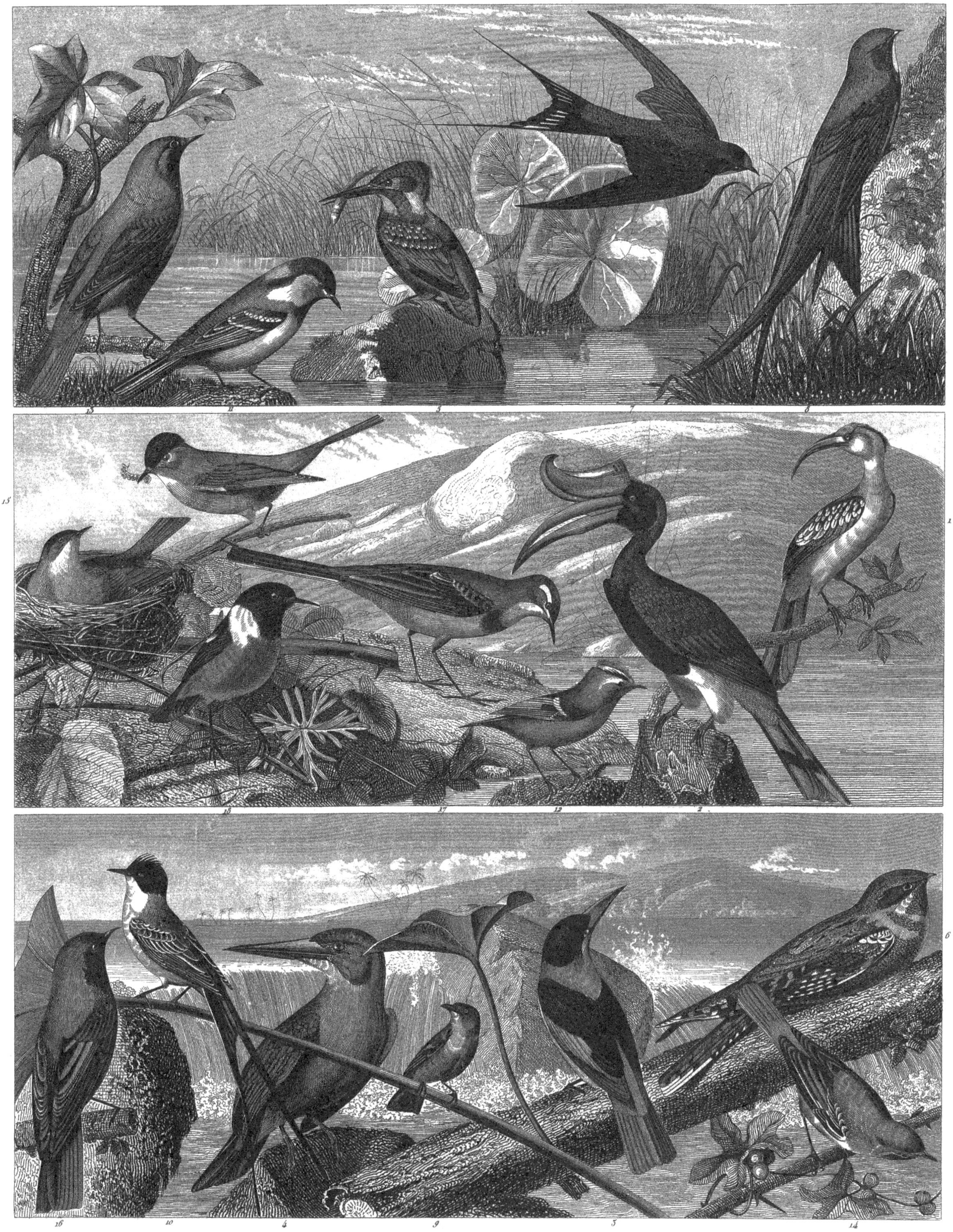

Plate 103 (Zoology, Plate 30)

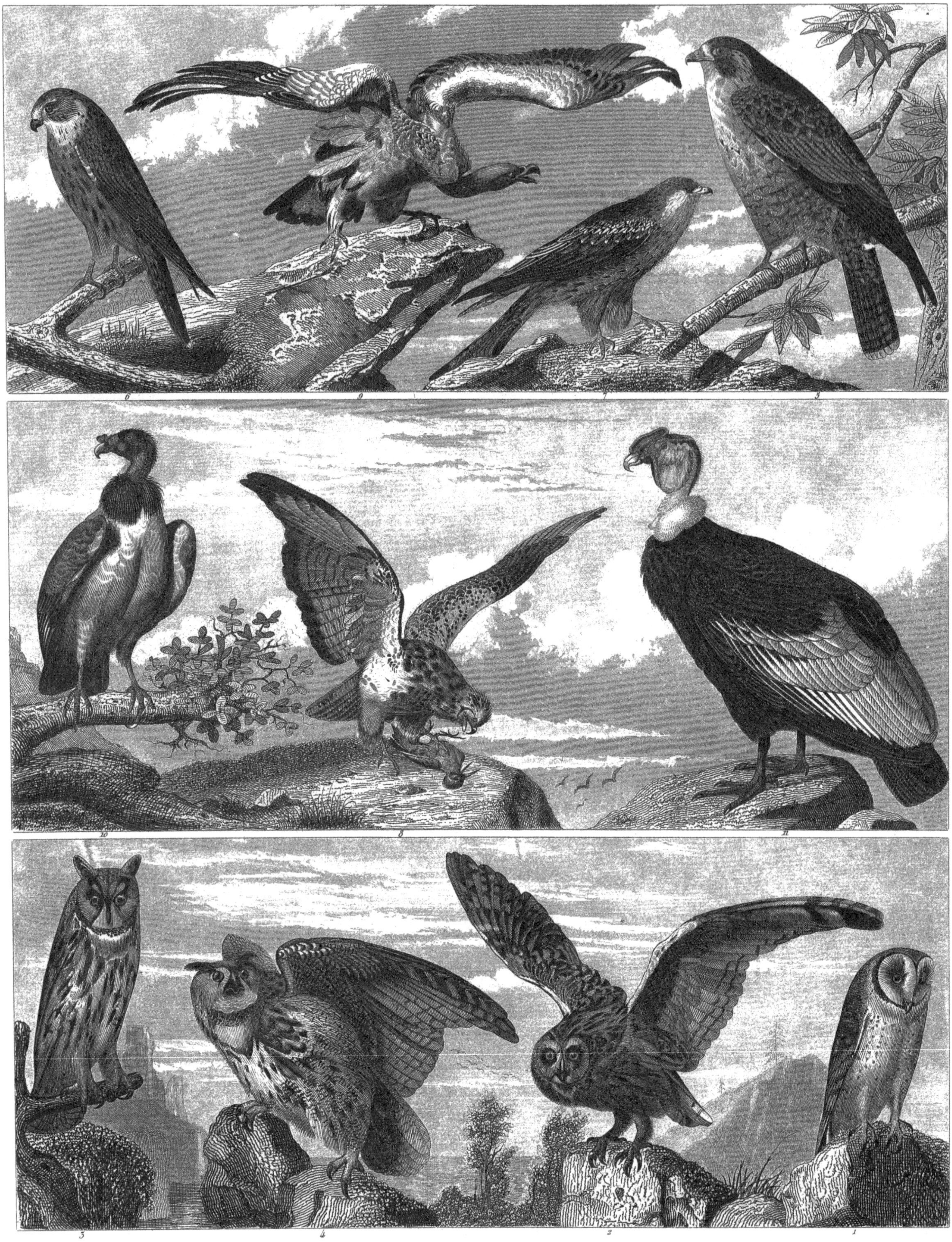

Plate 104 (Zoology, Plate 31)

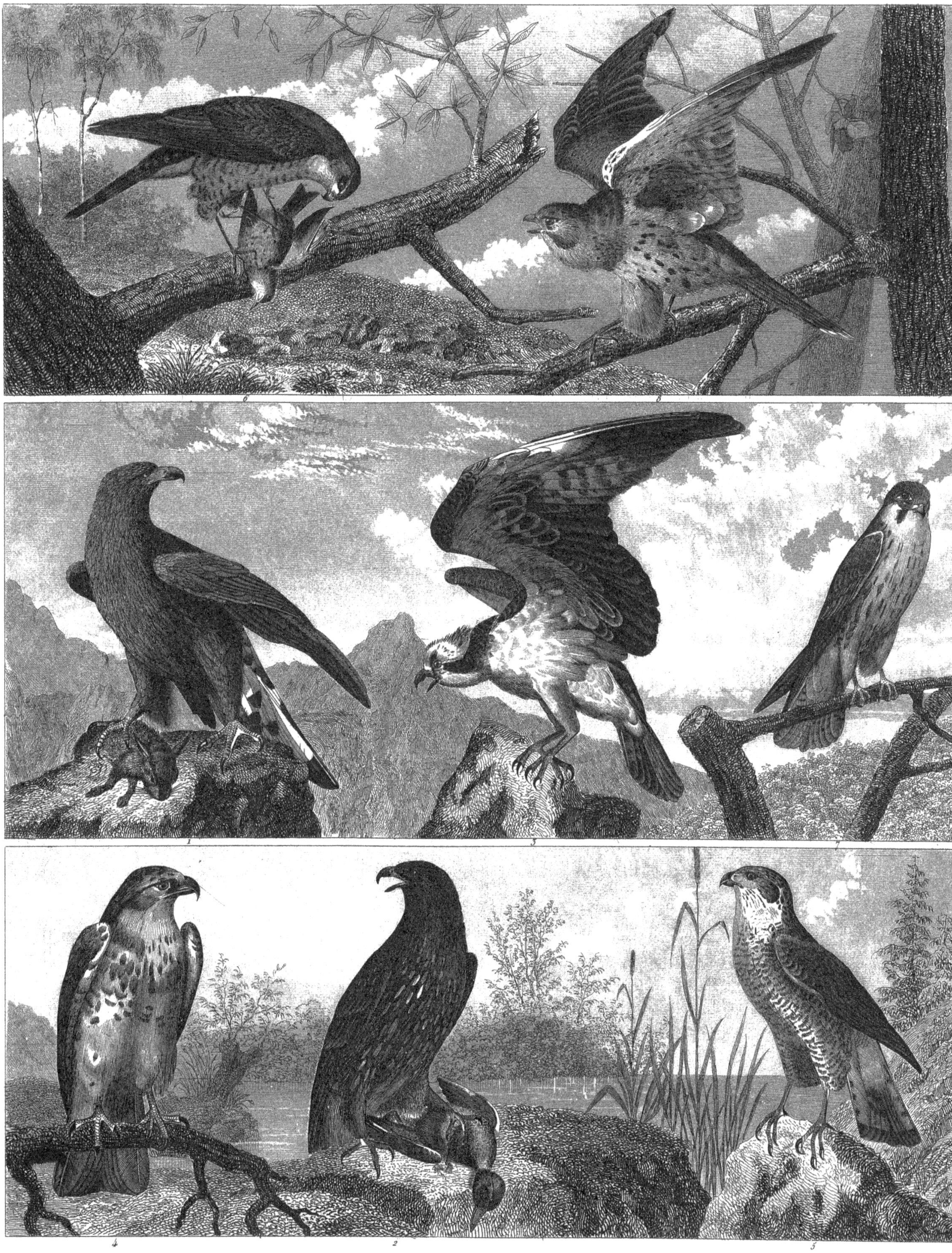

Plate 105 (Zoology, Plate 32)

Plate 105^{b} occurs between Plate 105 and Plate 106.

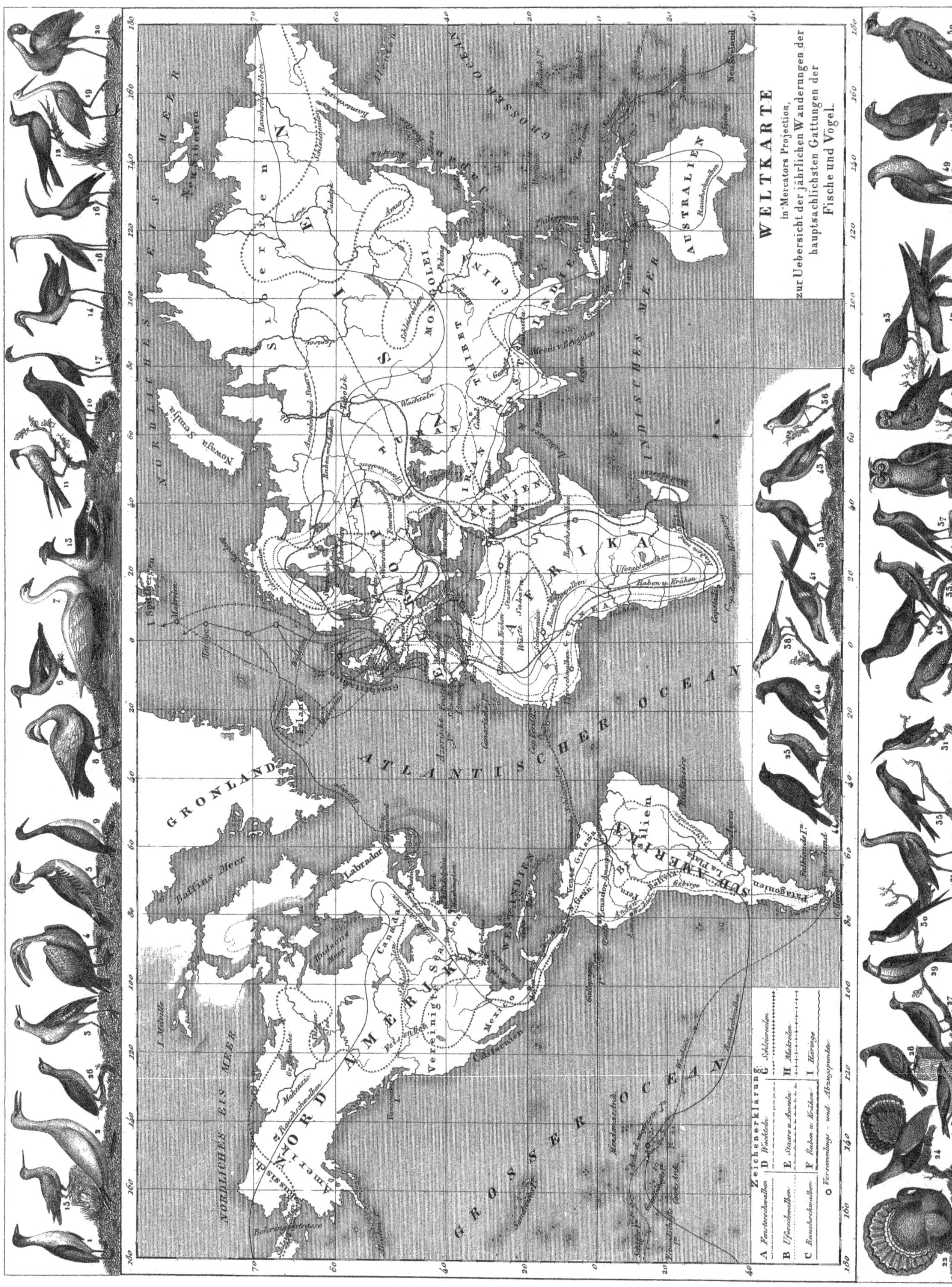
WELTKARTE
in Mercators Projection,
zur Uebersicht der jährlichen Wanderungen der
hauptsächlichsten Gattungen der
Fische und Vögel.
Zeichenerklärung
A Fensterschwalben
B Uferschwalben
C Rauchschwalben
D Wachteln
E Staare u. Amseln
F Raben u. Krähen
G Schleiereulen
H Makrelen
I Heringe
o Versammlung- und Abzugspunkte
NORDLICHES EIS MEER
NÖRDLICHES EIS MEER
ATLANTISCHER OCEAN
GROSSER OCEAN
INDISCHES MEER
NORD AMERIKA
SÜD AMERIKA
AFRIKA
AUSTRALIEN
GRONLAND
Nowaja Semlja
Labrador
Baffins Meer
Hudsons Meer
WESTINDIEN
Brasilien
Vereinigte Staaten
MONGOLEI
ARABIEN
TIBET
IRAN
CHINA
INDIEN
Neu Seeland

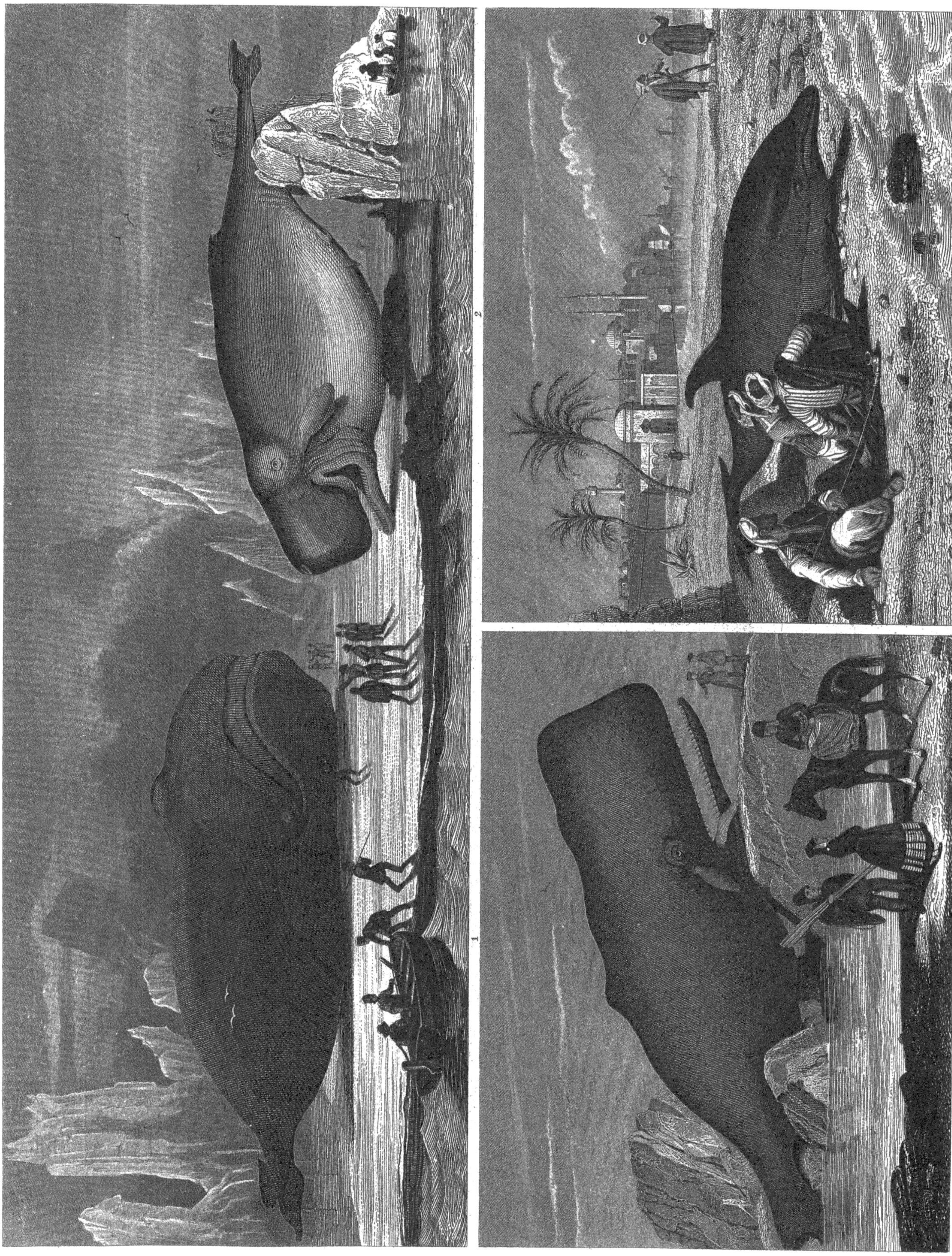

Plate 106 (Zoology, Plate 34)

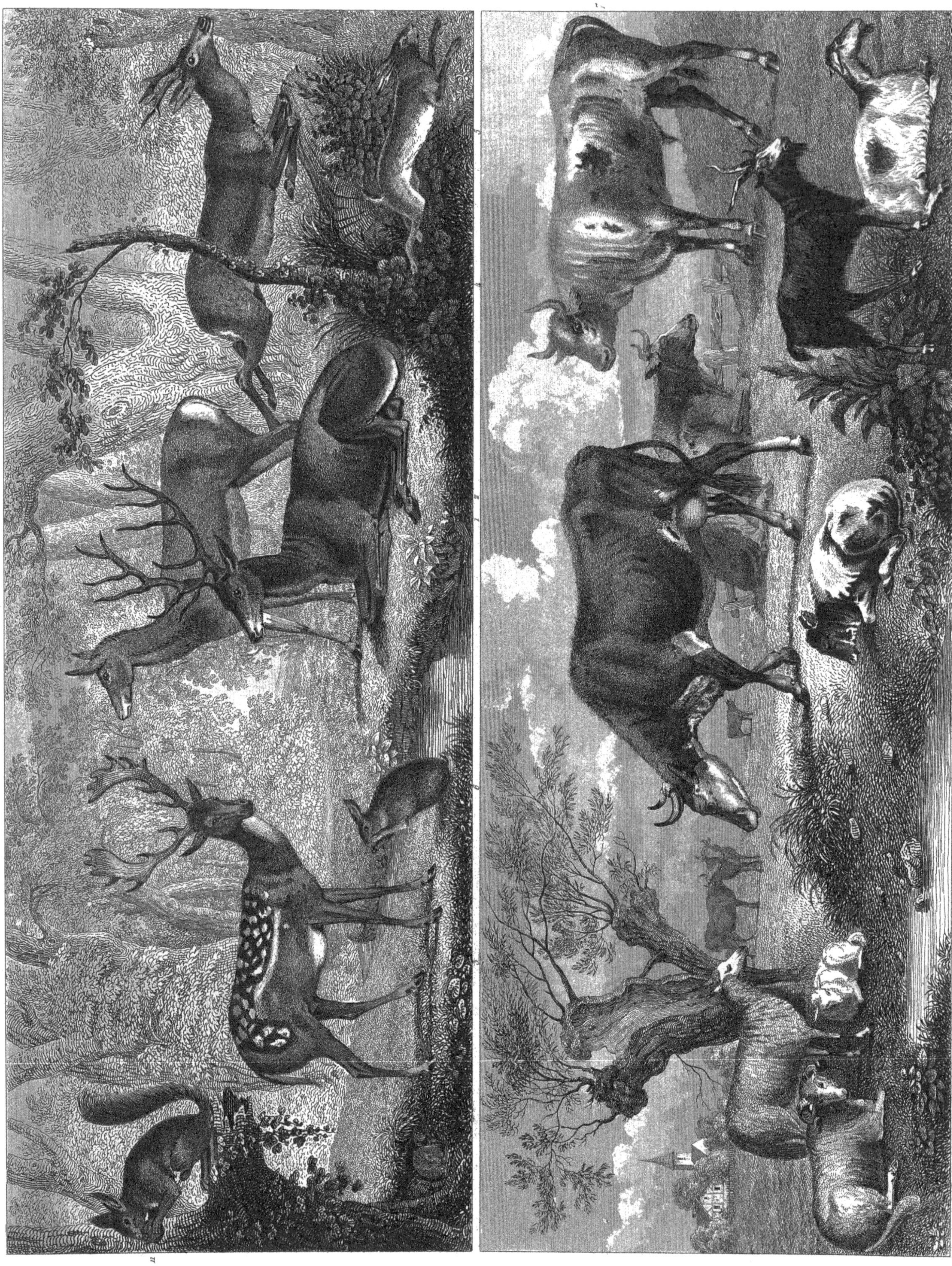

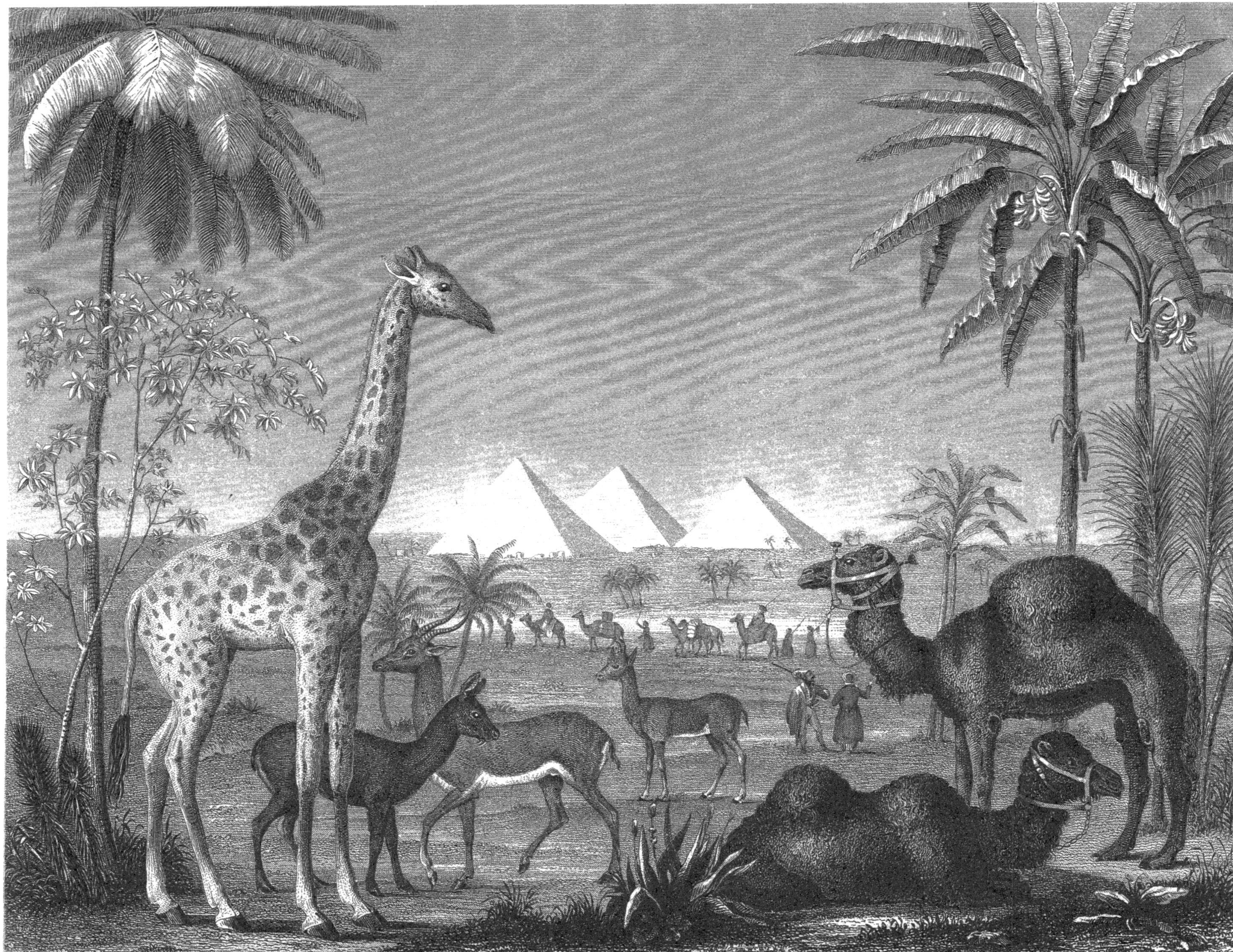

Plate 108 (Zoology, Plate 36)

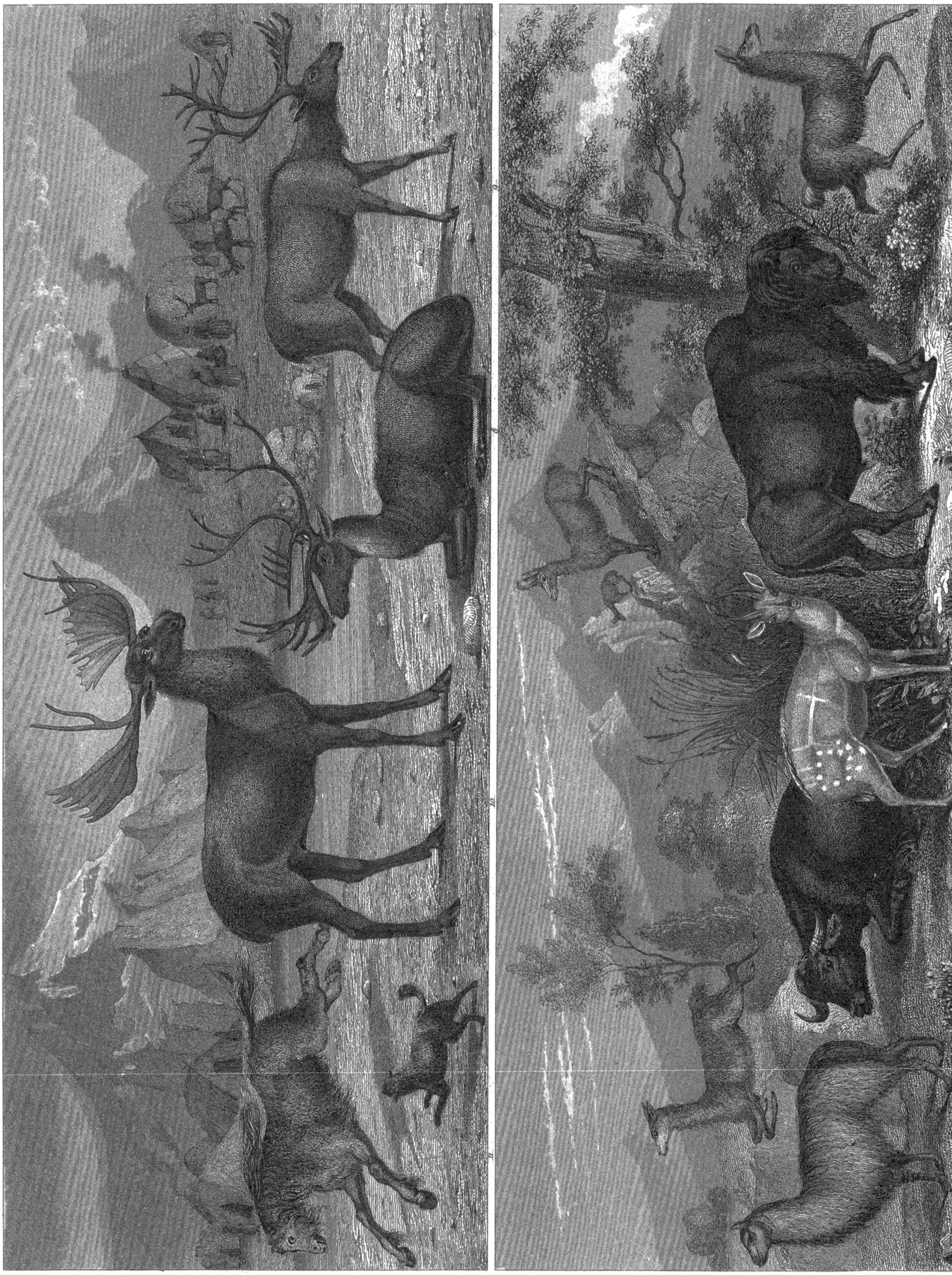

Plate 100 (Zoology, Plate 27)

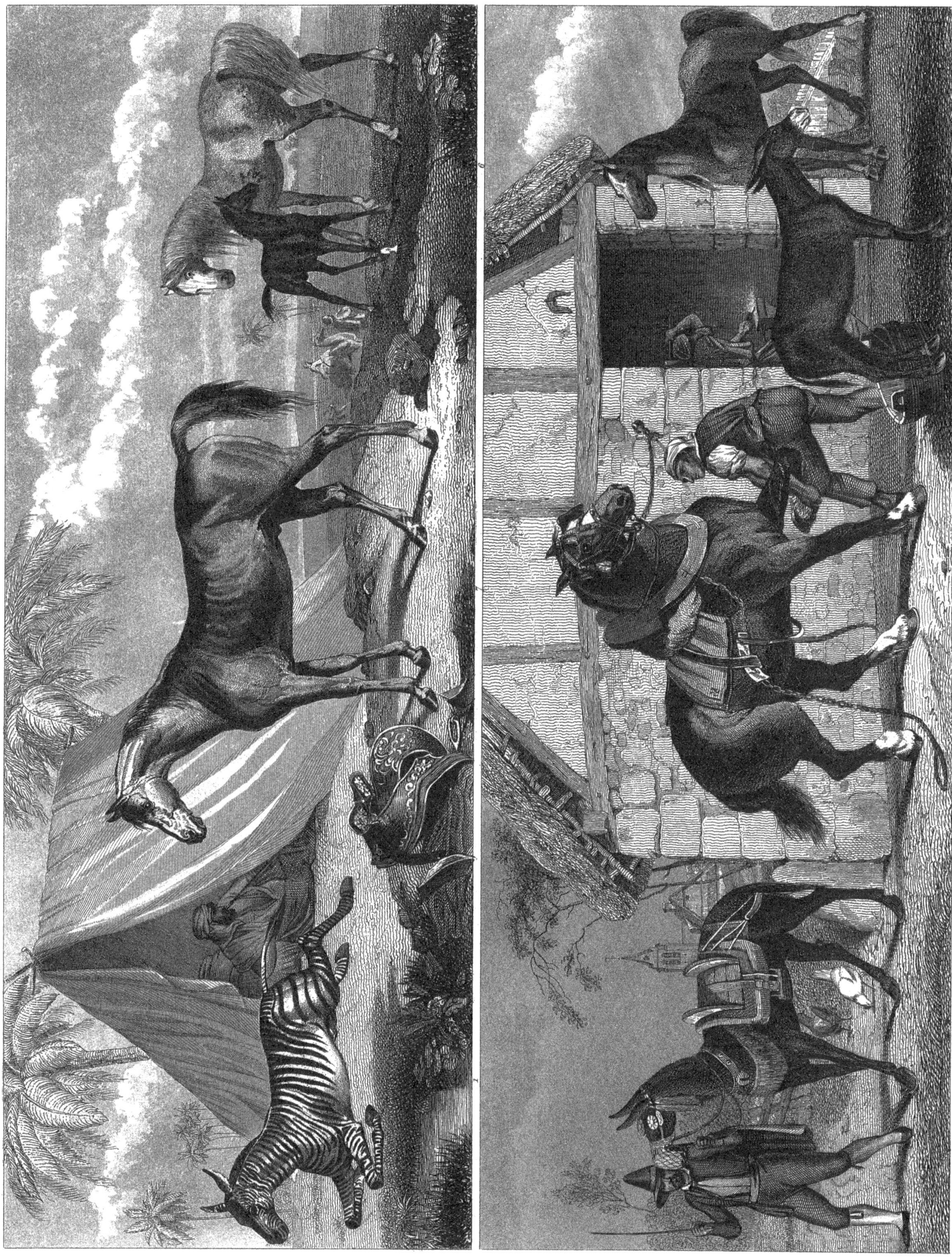

Plate 110 (Zoology, Plate 38)

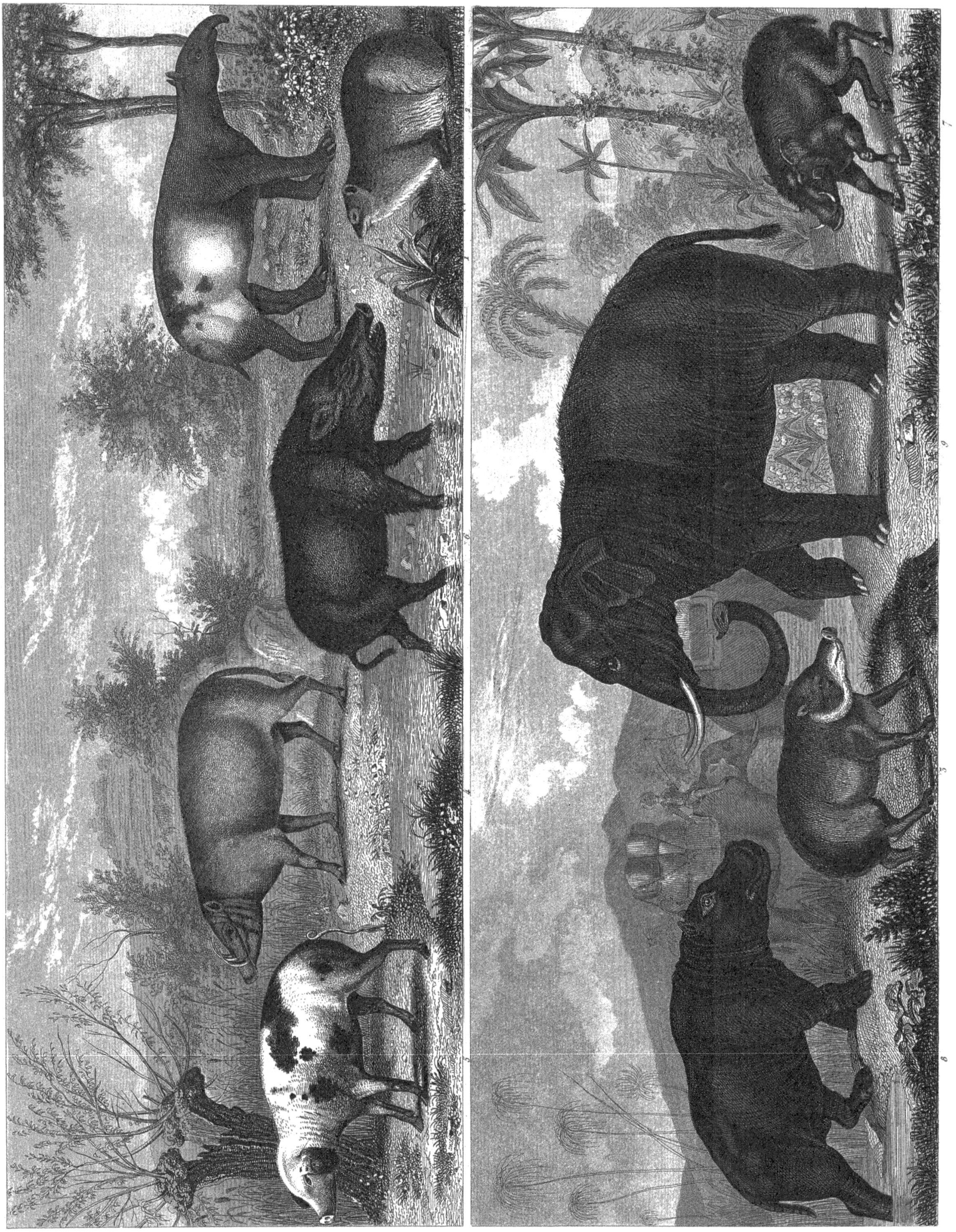

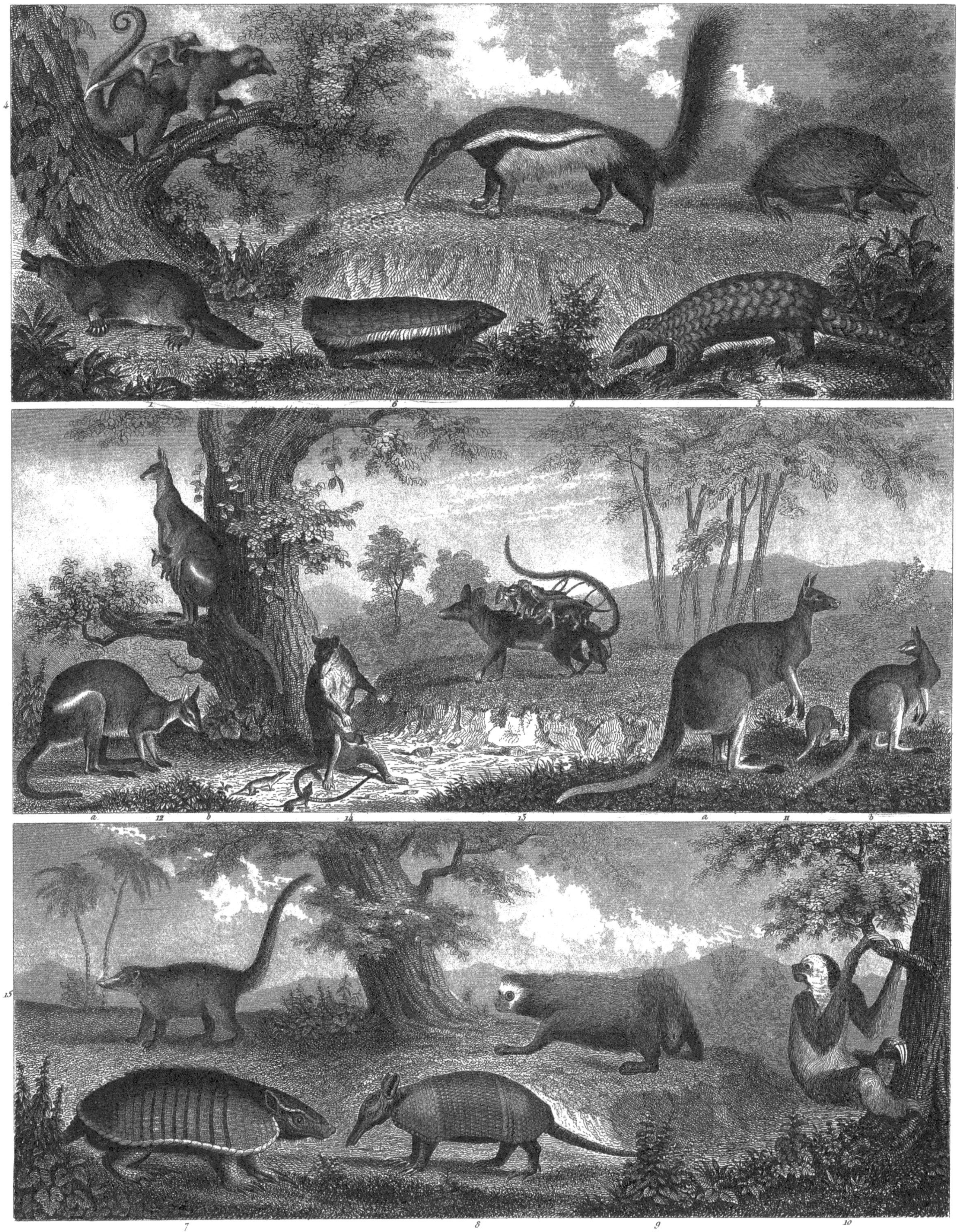

Plate 112 (Zoology, Plate 40)

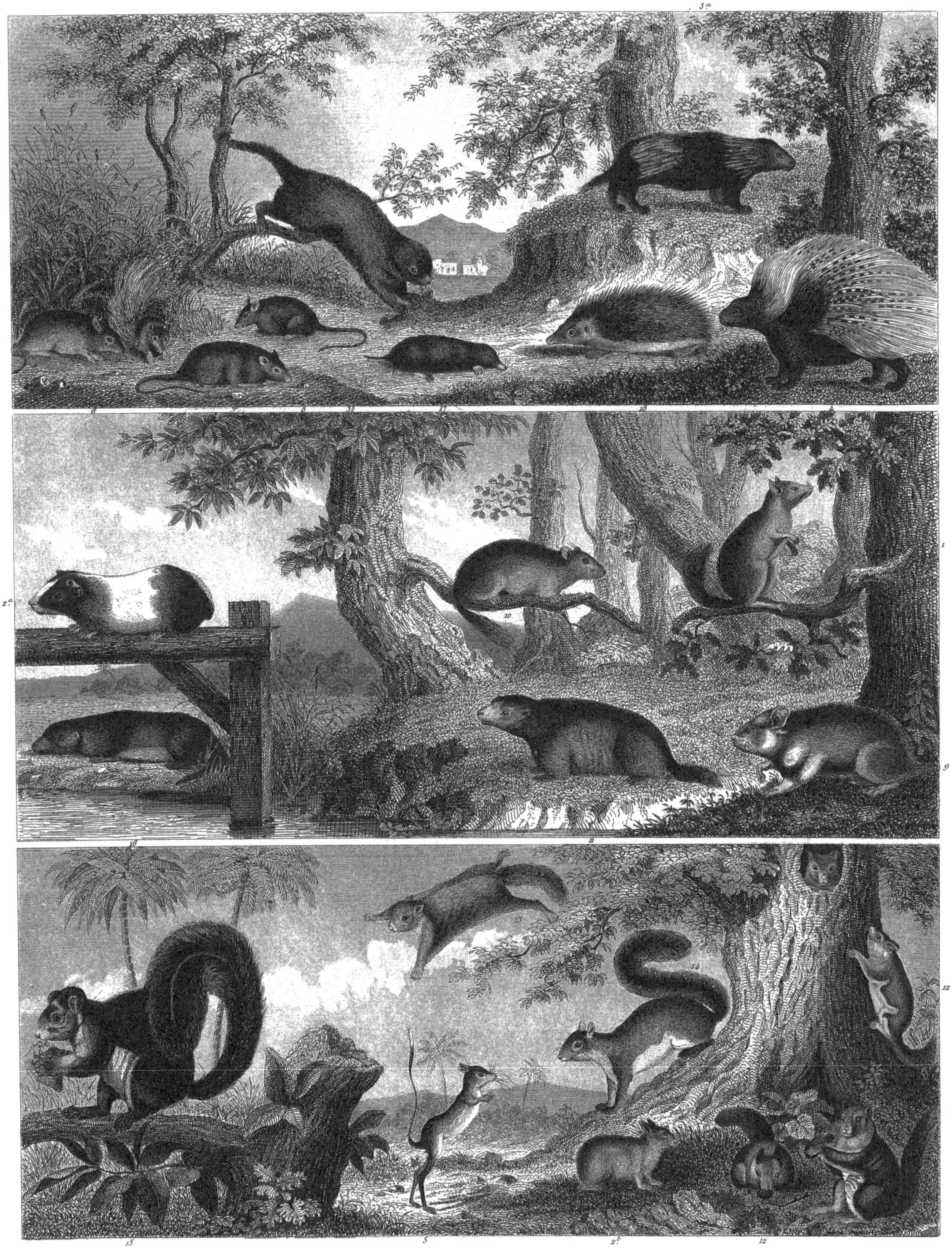

Plate 113 (Zoology, Plate 41)

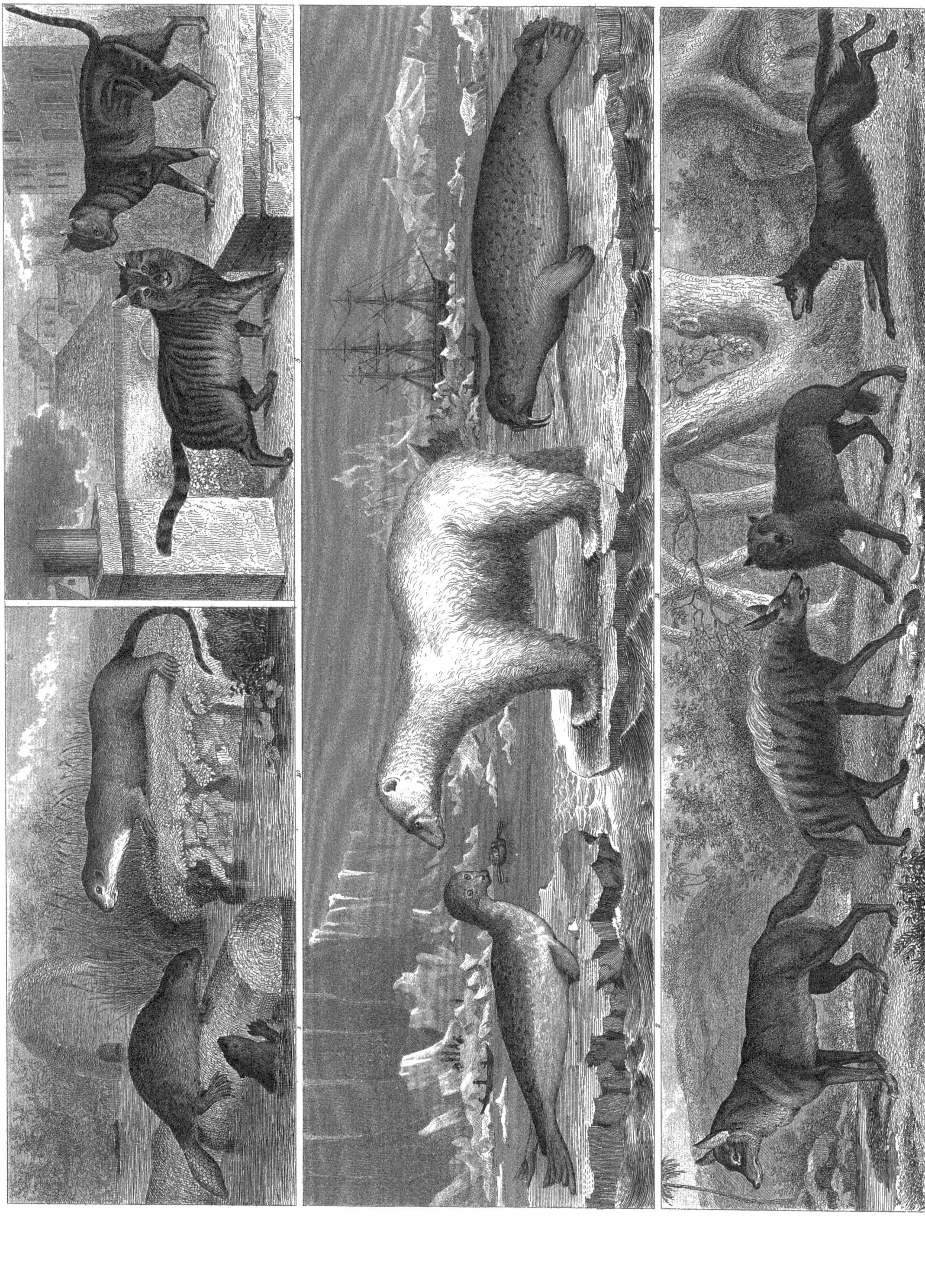

Plate 114 (Zoology, Plate 42)

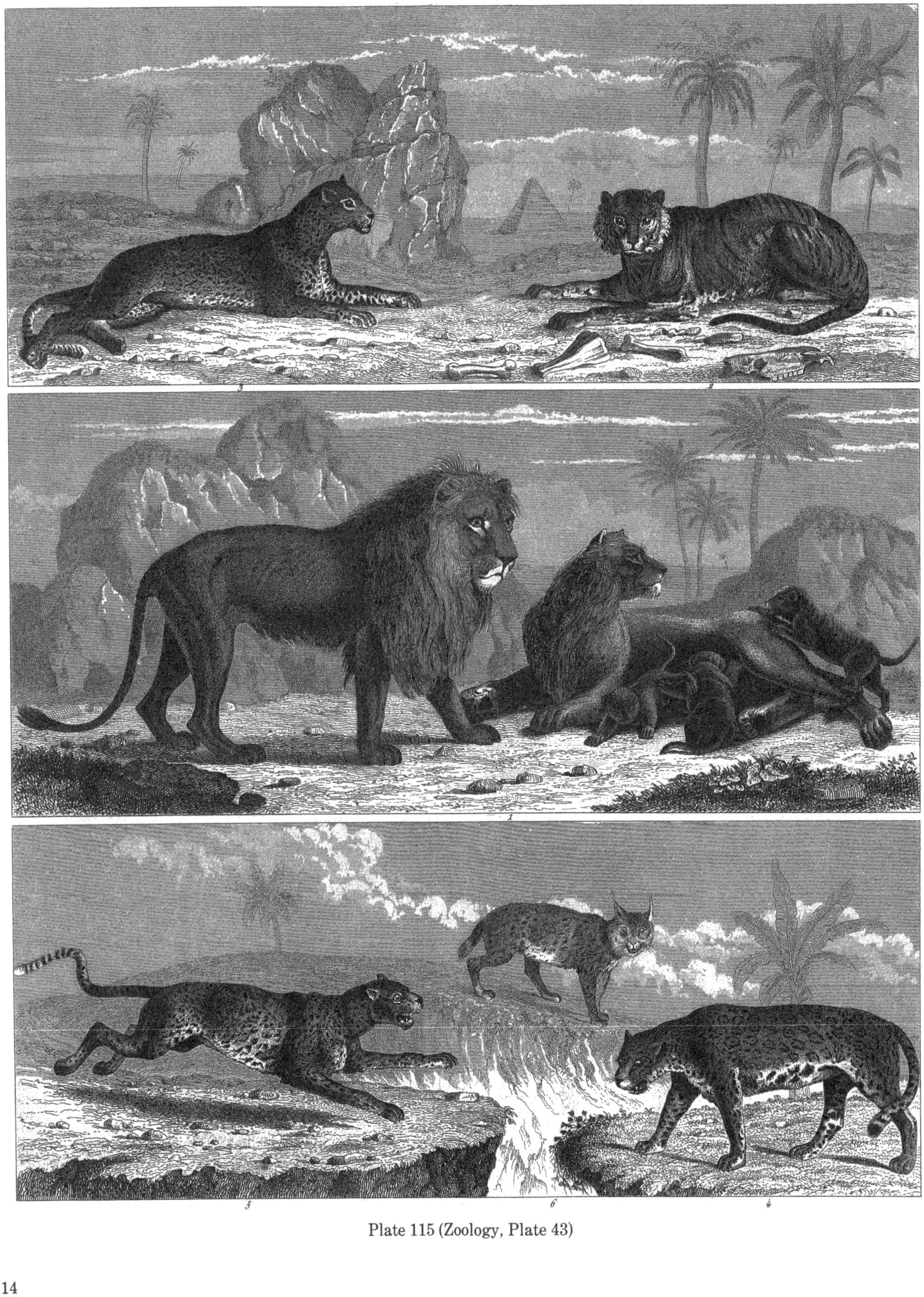

Plate 115 (Zoology, Plate 43)

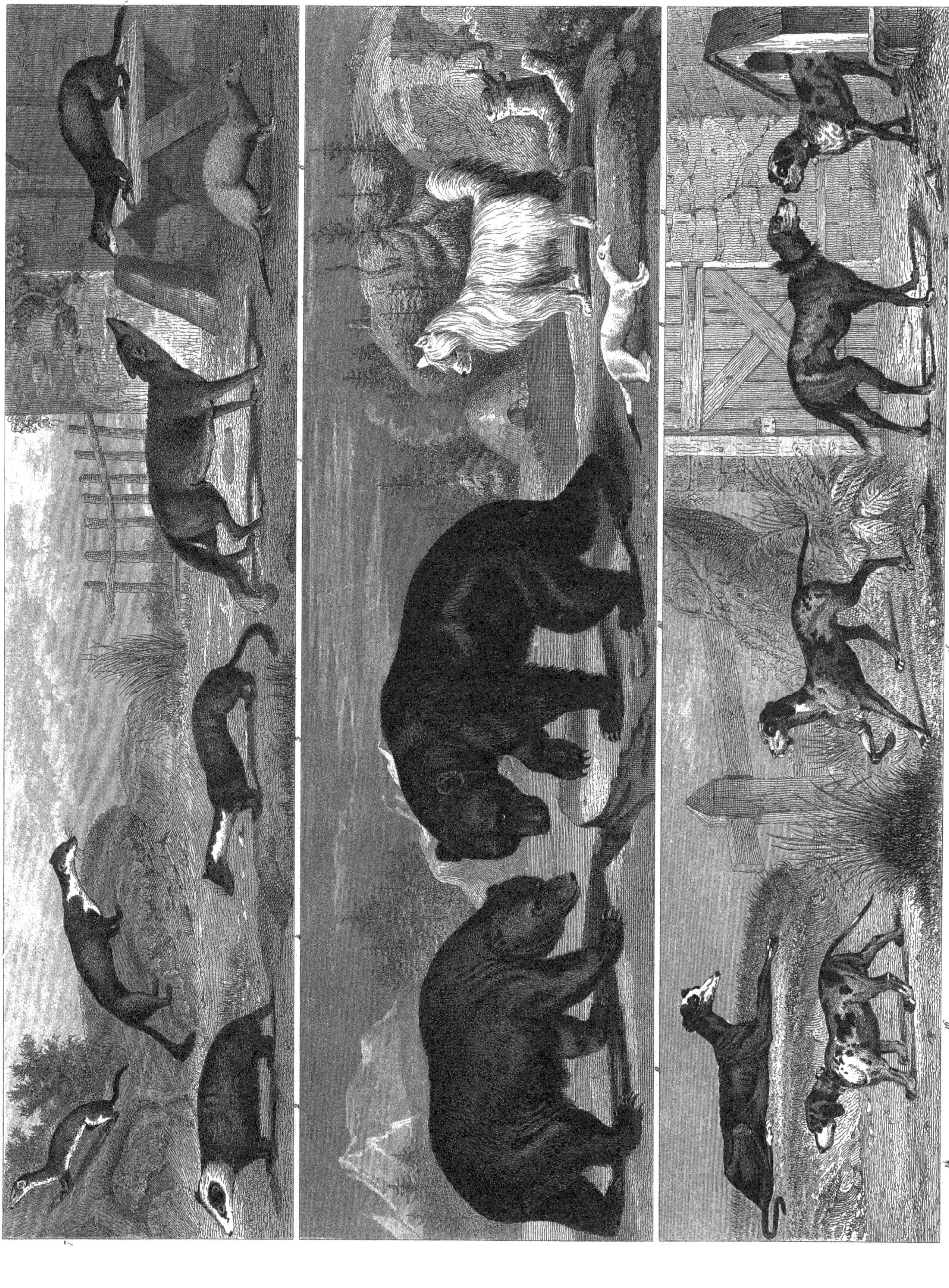

Plate 116 (Zoology, Plate 44)

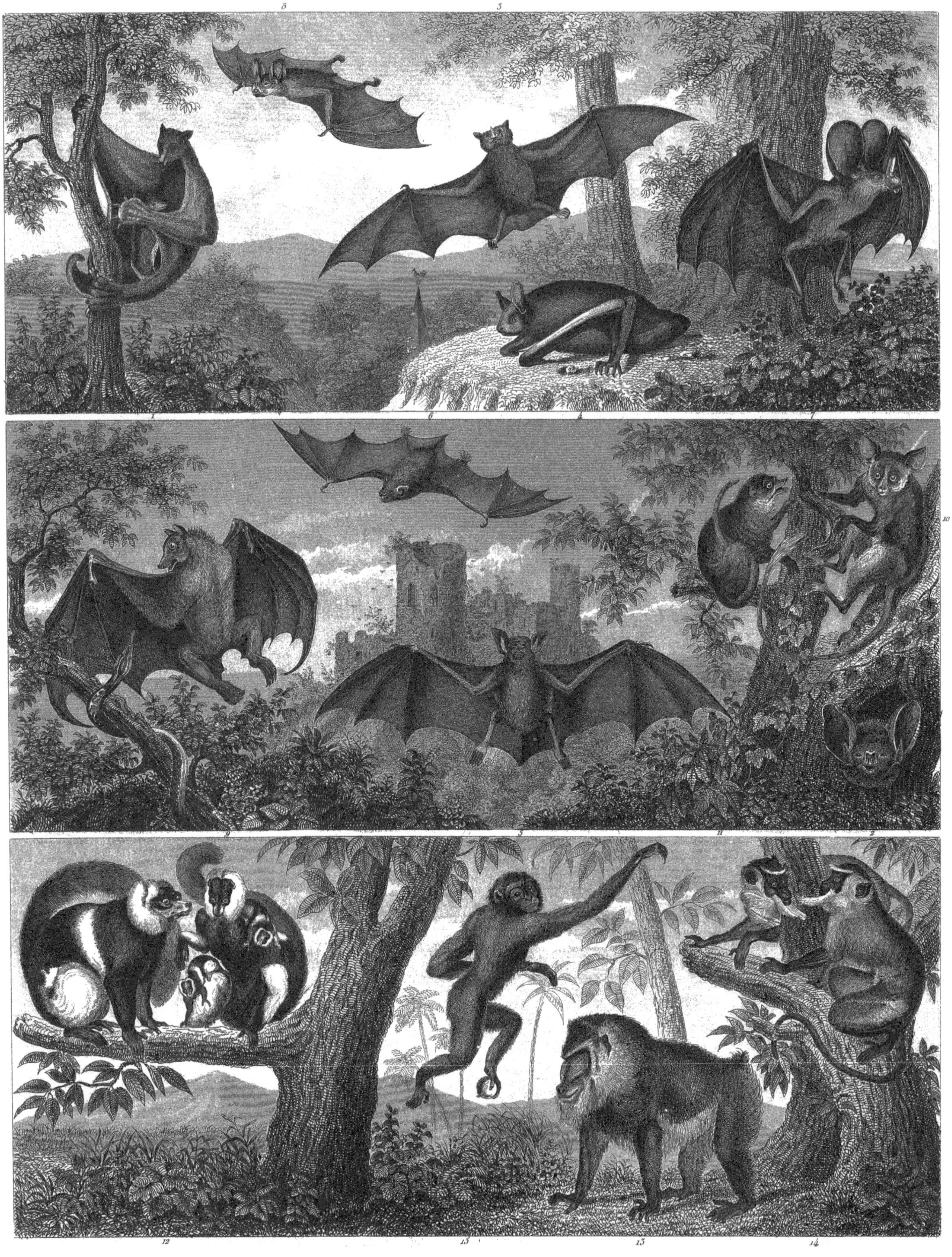

Plate 117 (Zoology, Plate 45)

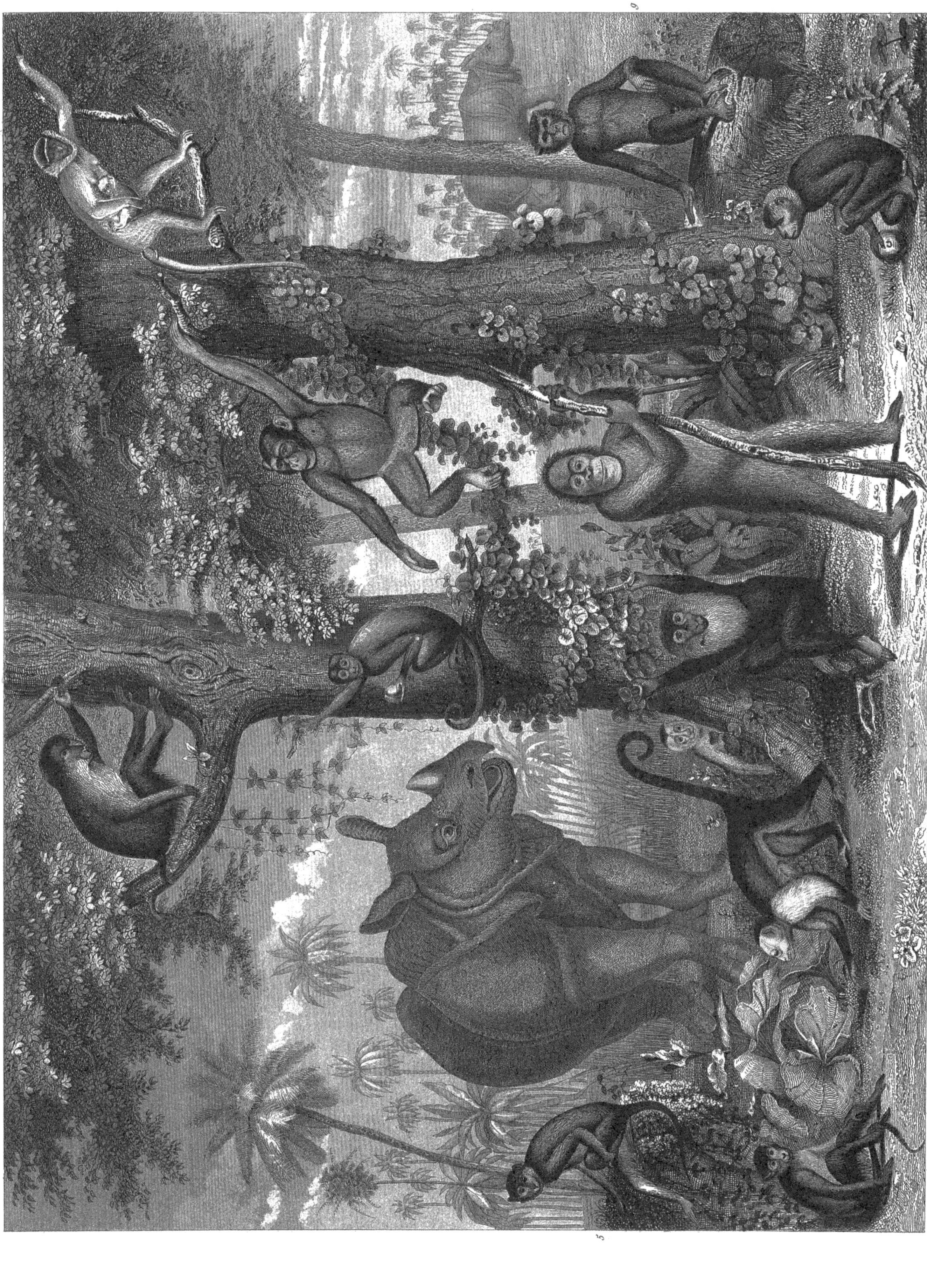

Plate 118 (Zoology, Plate 46)

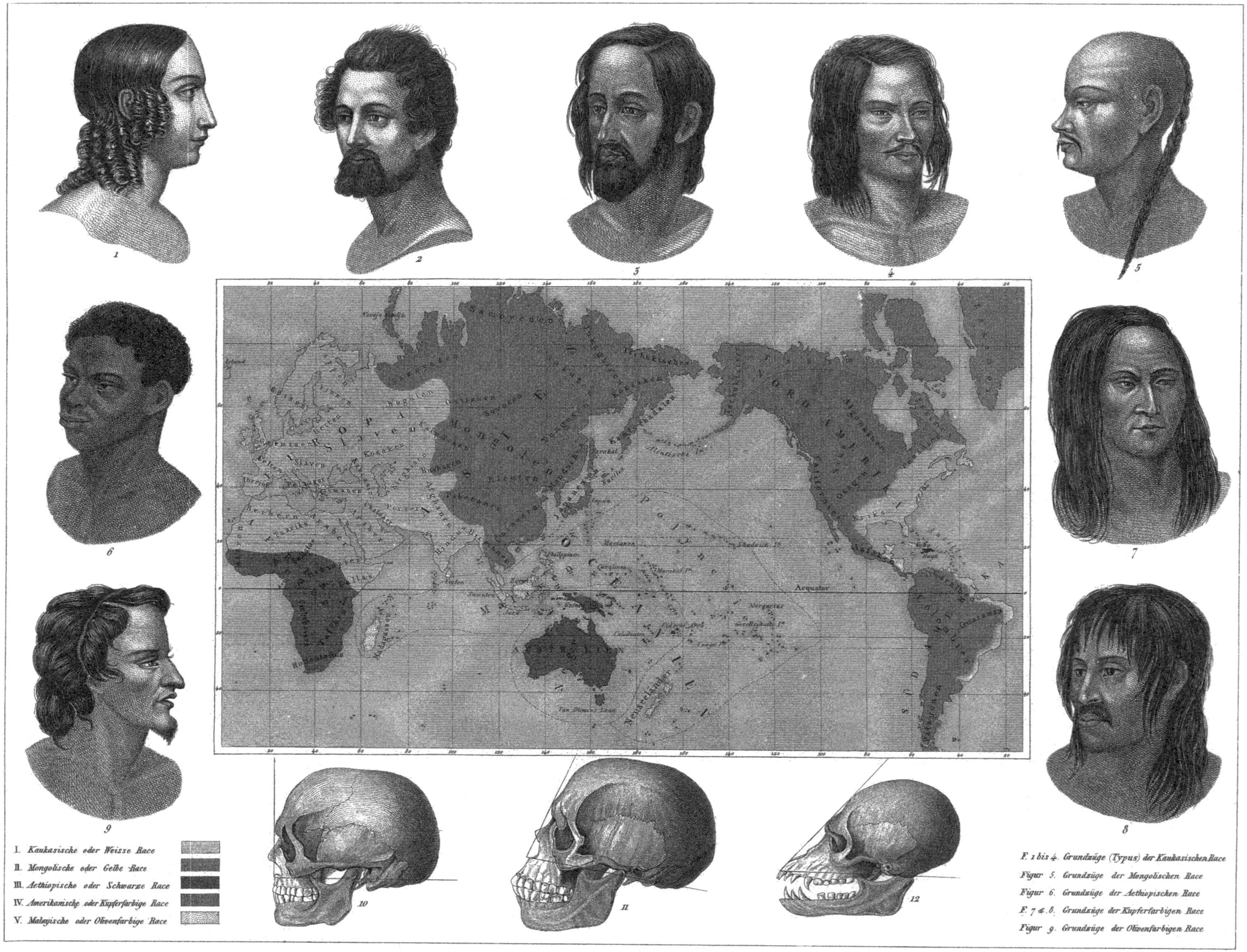
I. Kaukasische oder Weisse Race
II. Mongolische oder Gelbe Race
III. Aethiopische oder Schwarze Race
IV. Amerikanische oder Kupferfarbige Race
V. Malayische oder Olivenfarbige Race
F. 1 bis 4. Grundzüge (Typus) der Kaukasischen Race
Figur 5. Grundzüge der Mongolischen Race
Figur 6. Grundzüge der Aethiopischen Race
F. 7 & 8. Grundzüge der Kupferfarbigen Race
Figur 9. Grundzüge der Olivenfarbigen Race

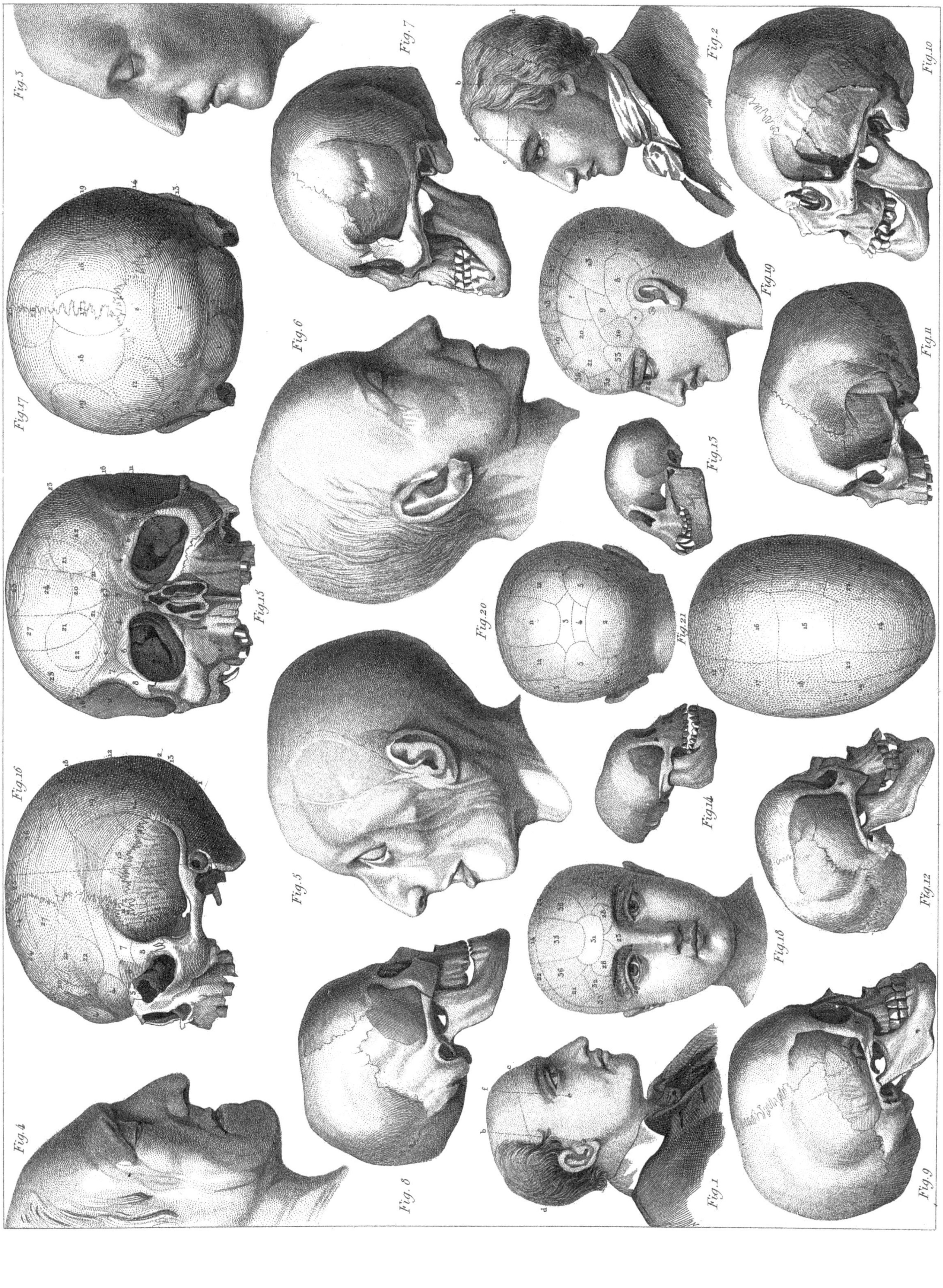

Plate 120 (Anthropology and Surgery, Plate 2)

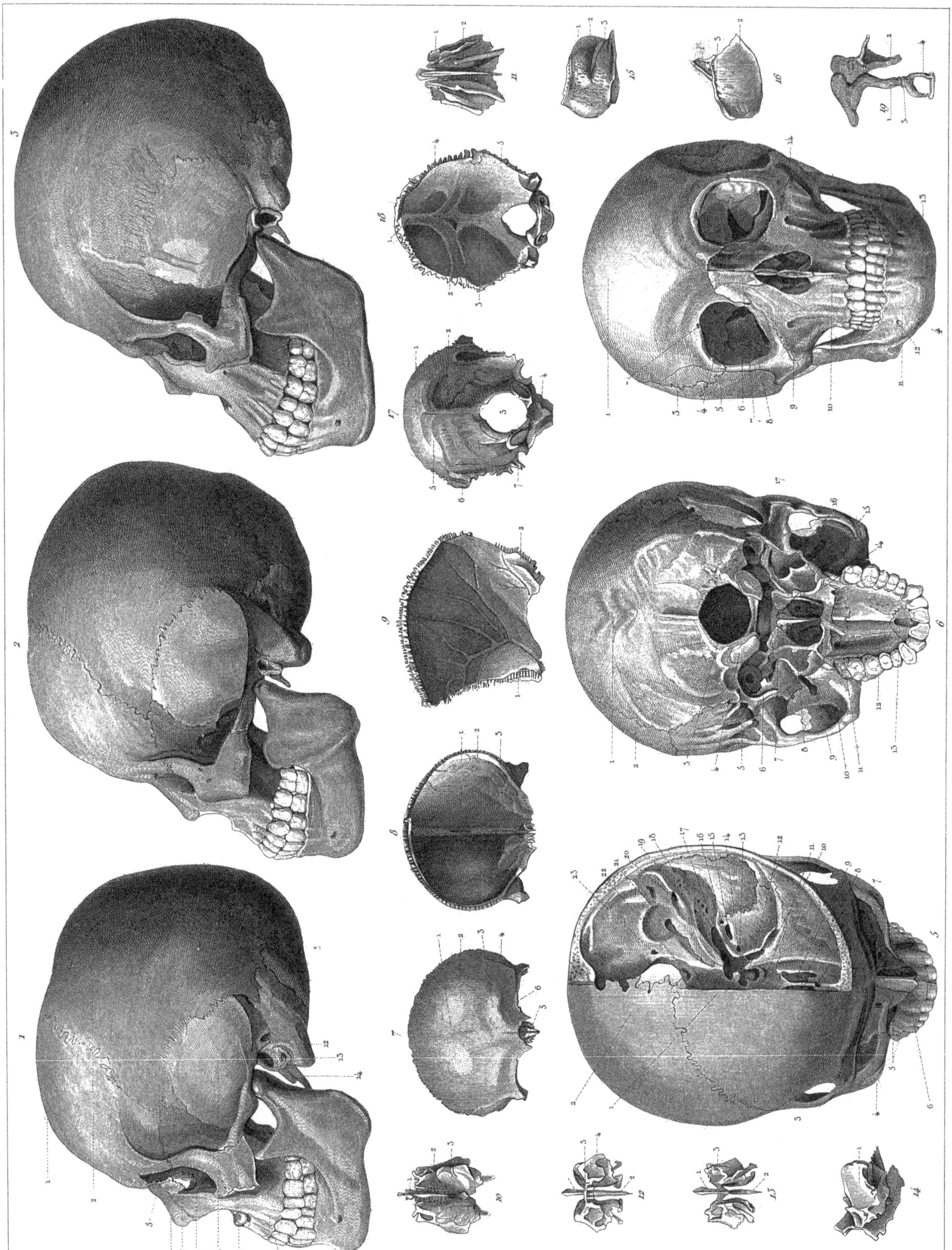

Plate 121 (Anthropology and Surgery, Plate 3)

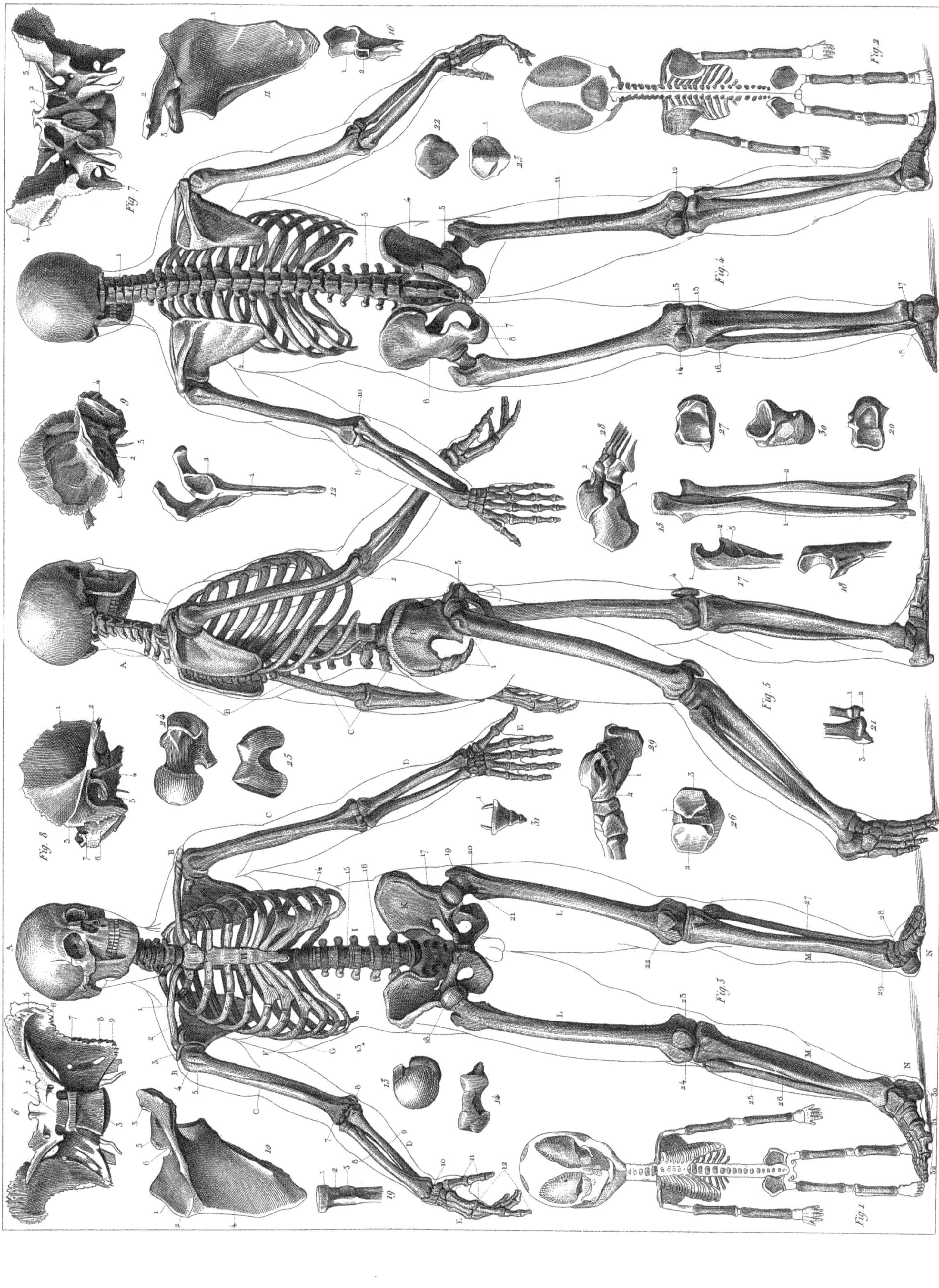

Plate 122 (Anthropology and Surgery, Plate 4)

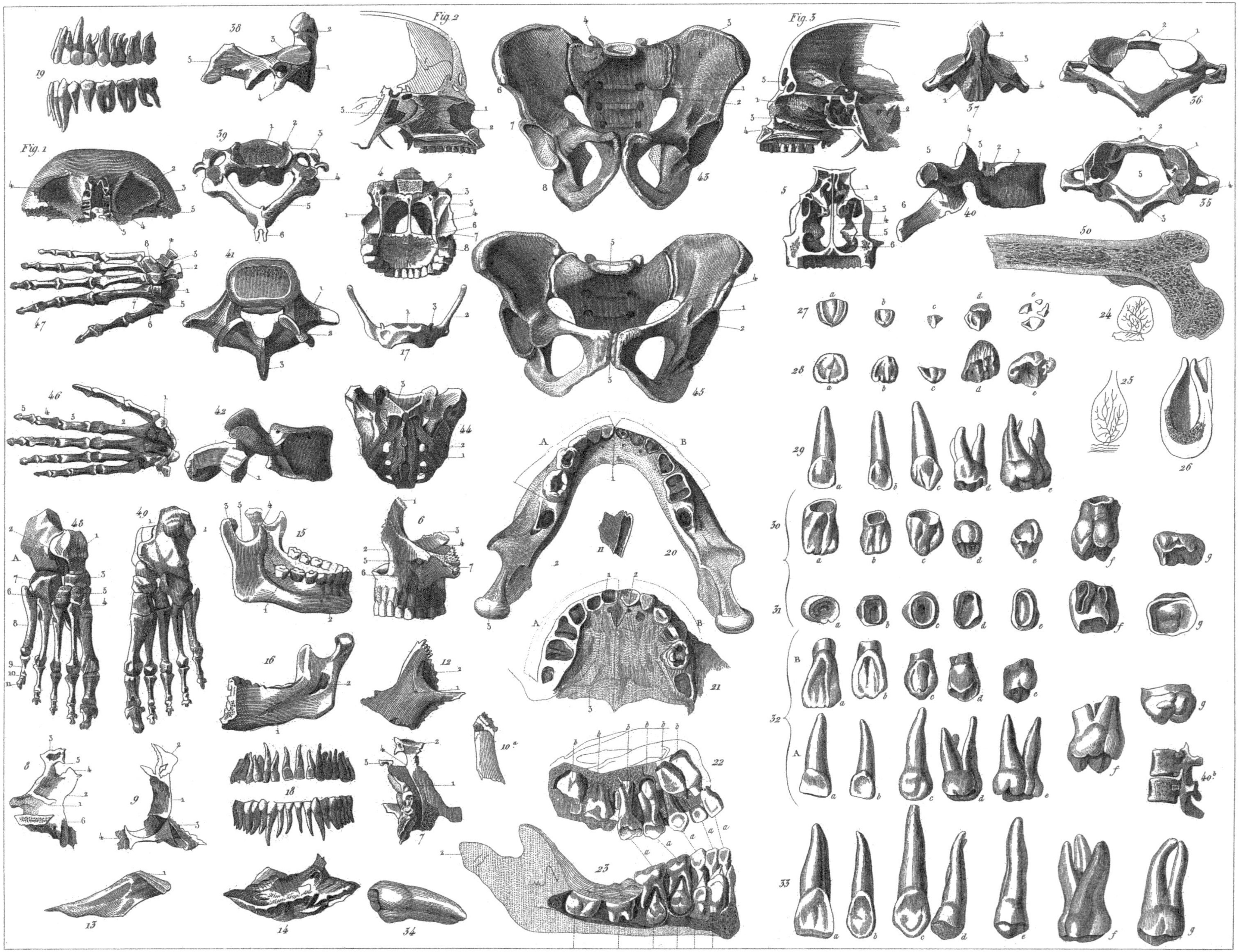

Plate 122 (Anthropology and Surgery, Plate 5)

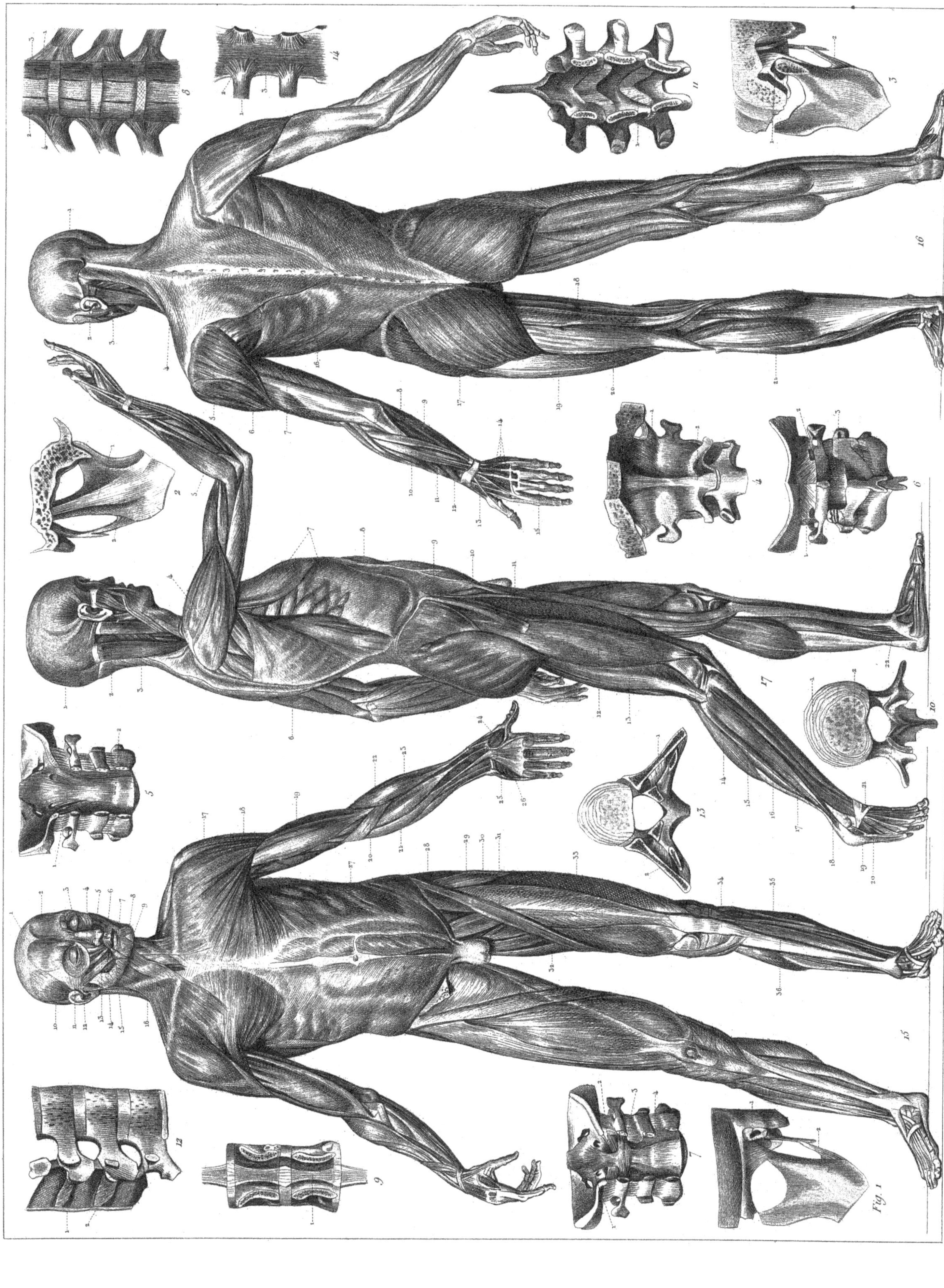

Plate 124 (Anthropology and Surgery, Plate 6)

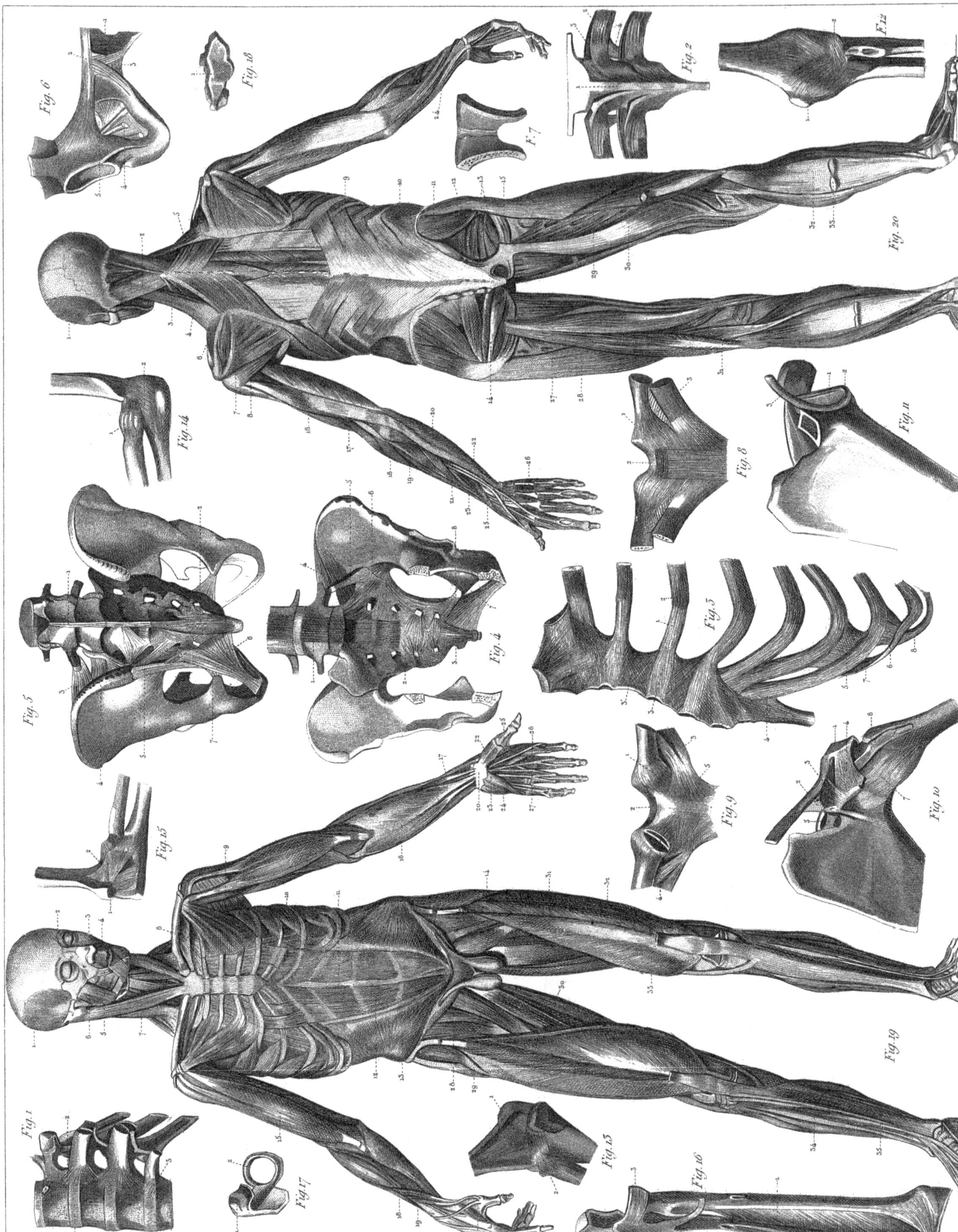

Plate 185 (Anatomy and Surgery, Plate 7)

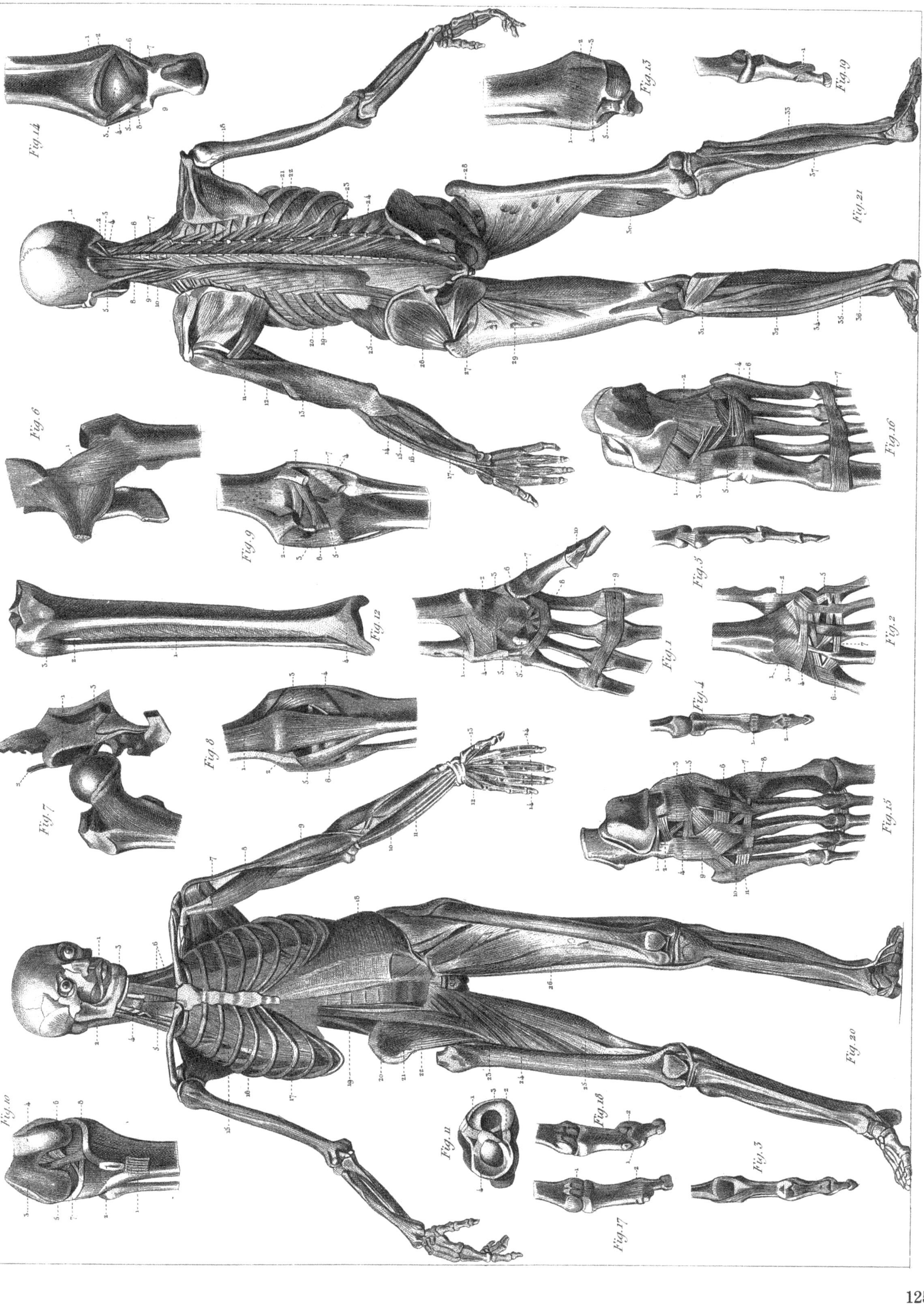

Plate 126 (Anthropology and Surgery, Plate 8)

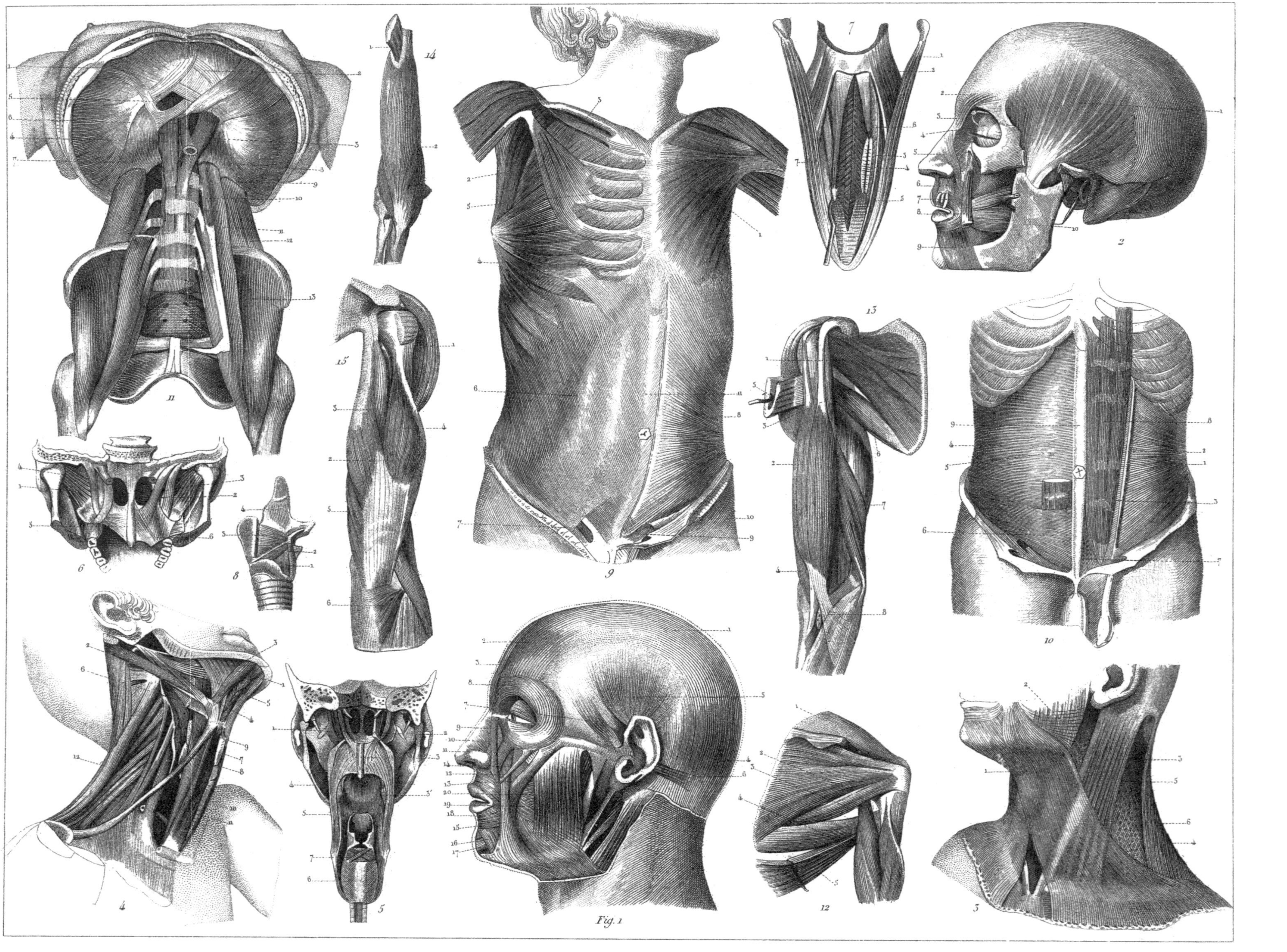

Plate 127 (Anthropology and Surgery, Plate 9)

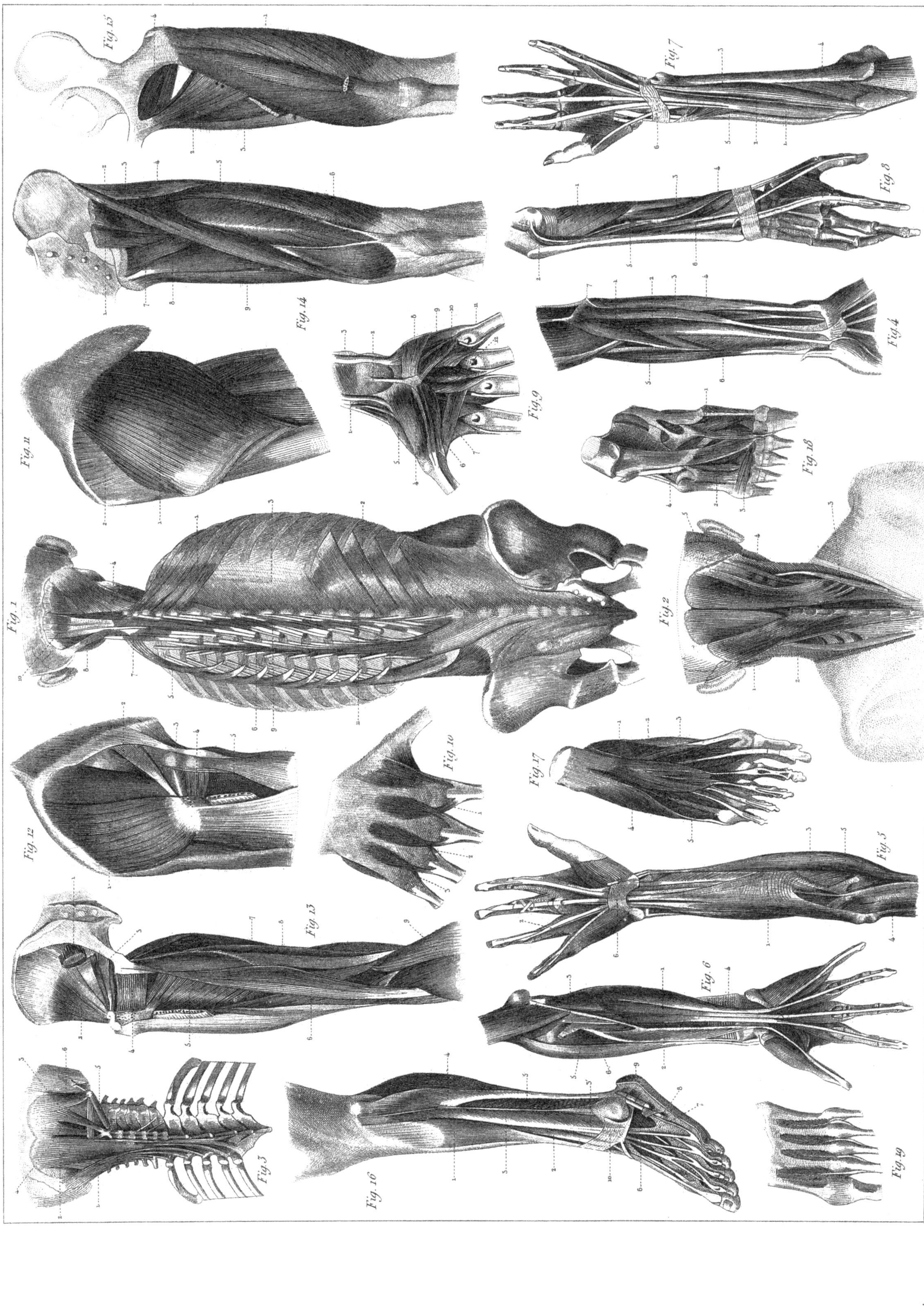

Plate 128 (Anthropology and Surgery, Plate 10)

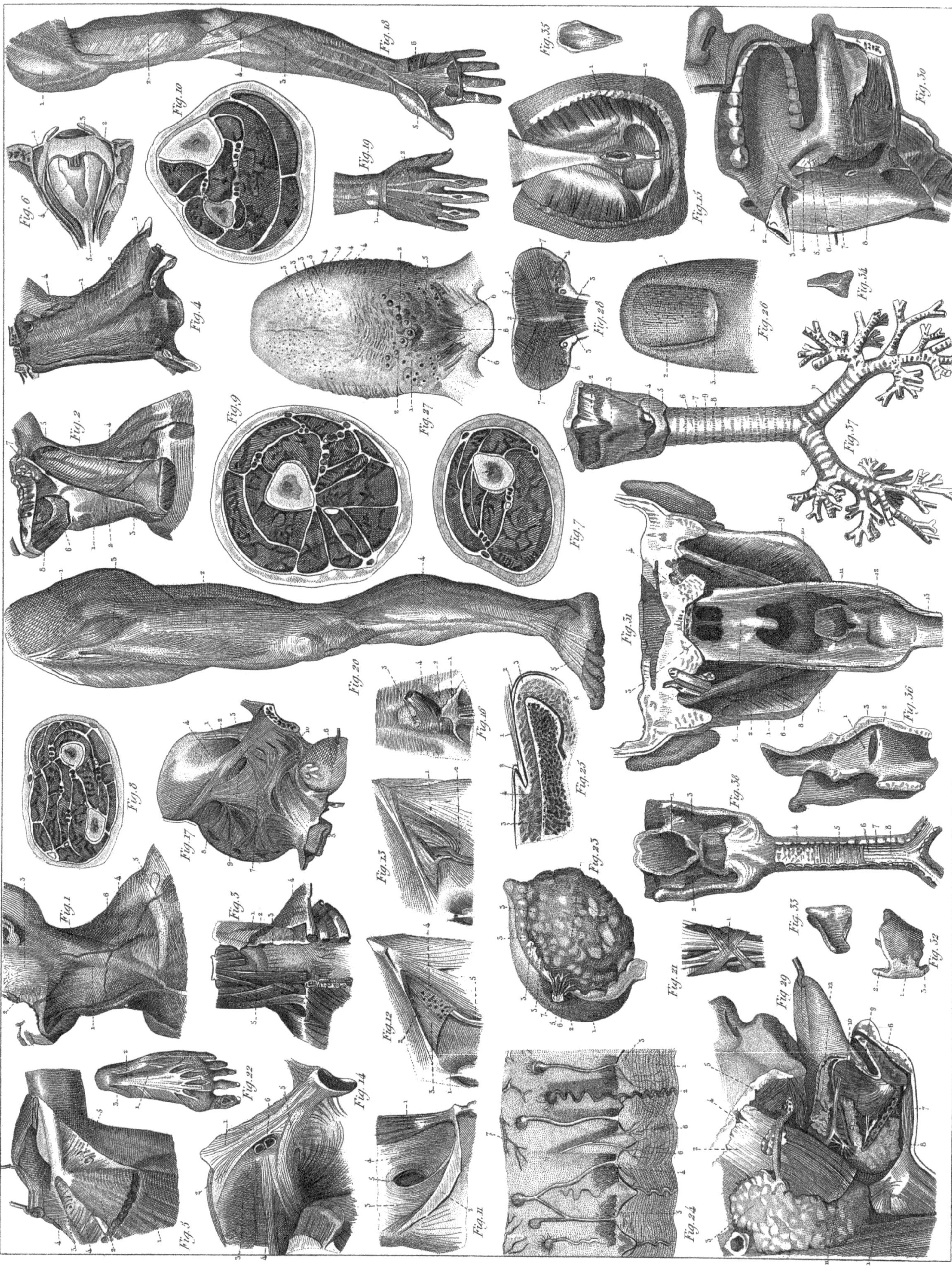

Plate 129 (Anthropology and Surgery, Plate 11)

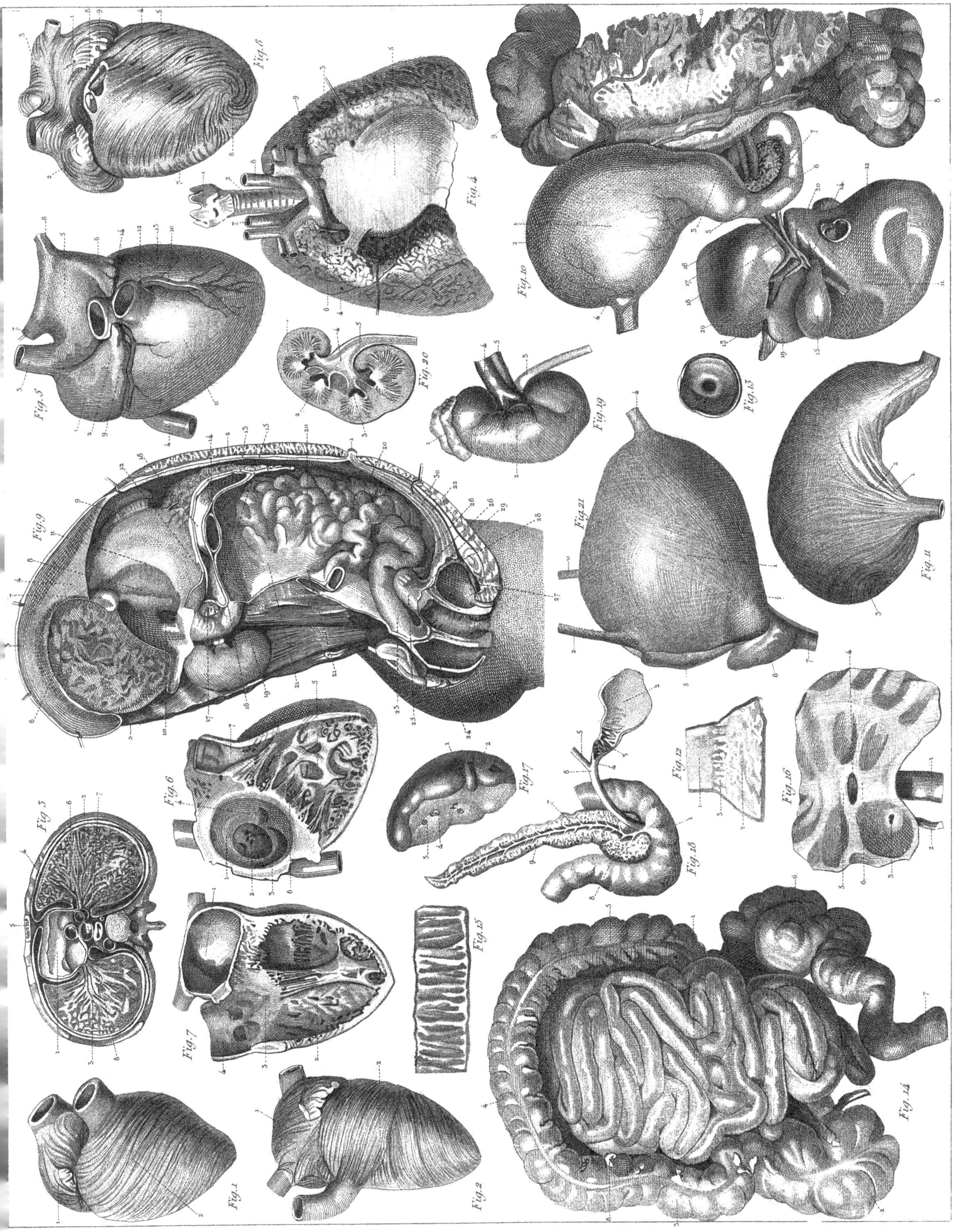

Plate 130 (Anthropology and Surgery, Plate 12)

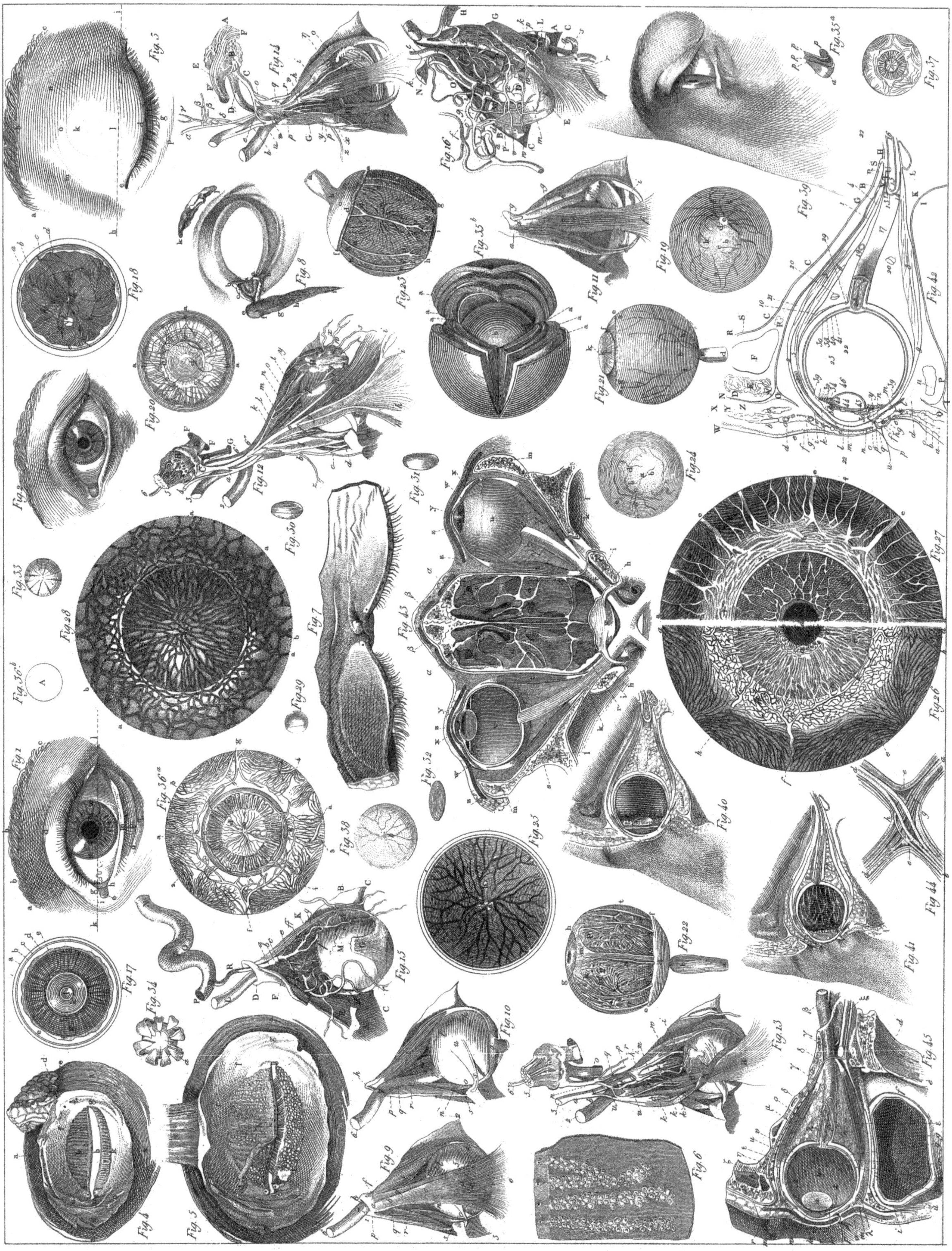

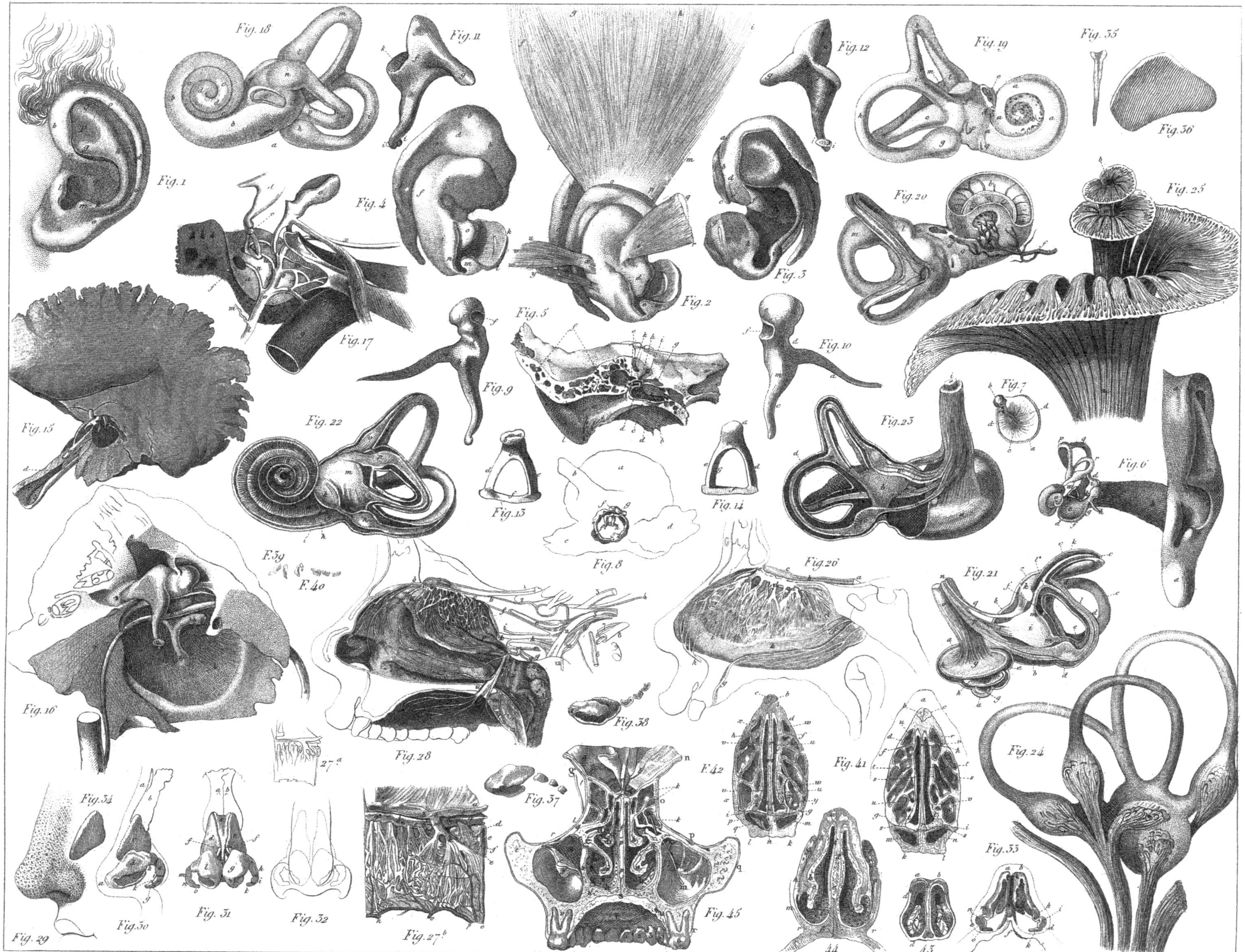

Plate 132 (Anthropology and Surgery, Plate 14)

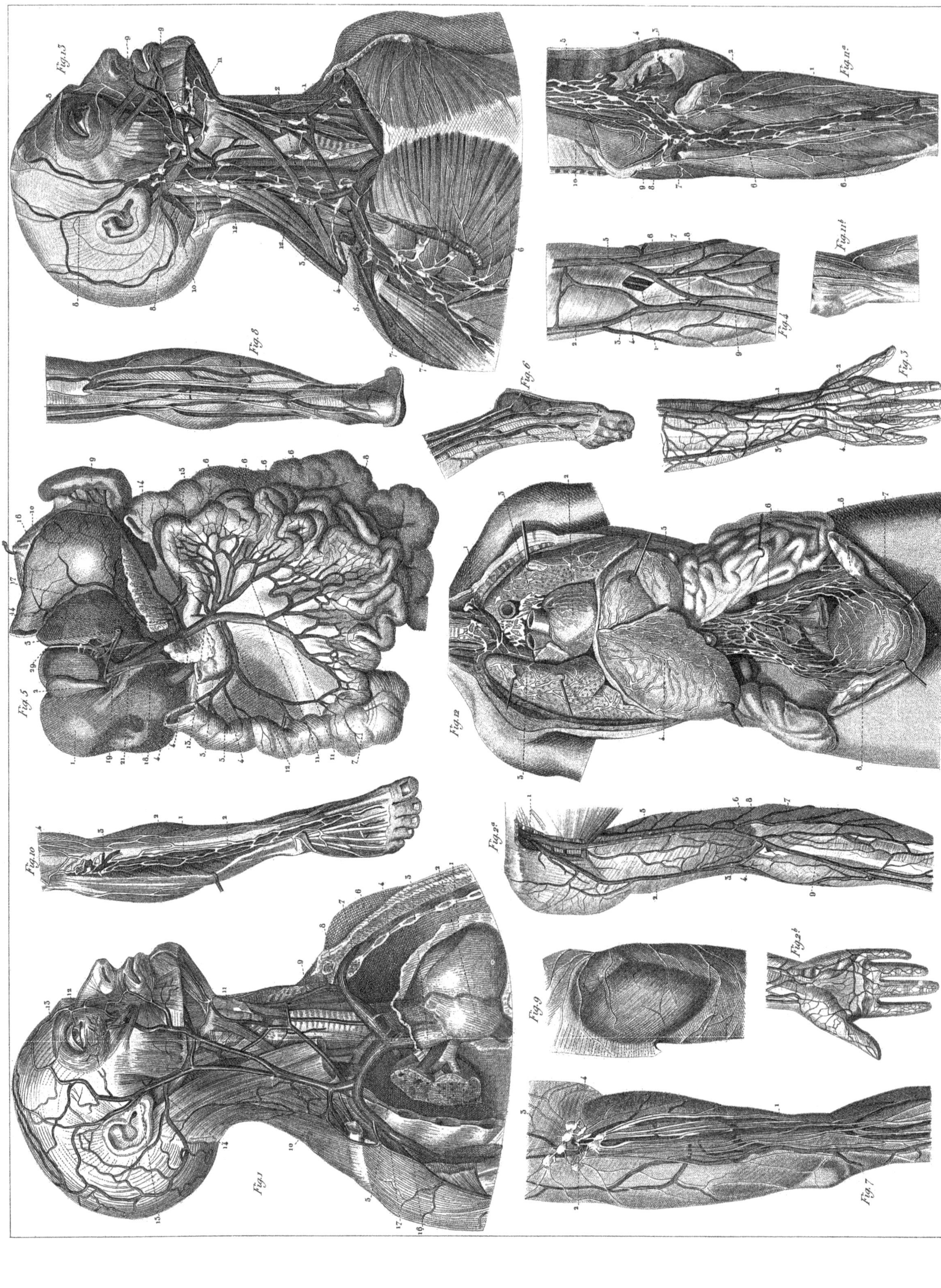

Plate 133 (Anthropology and Surgery, Plate 15)

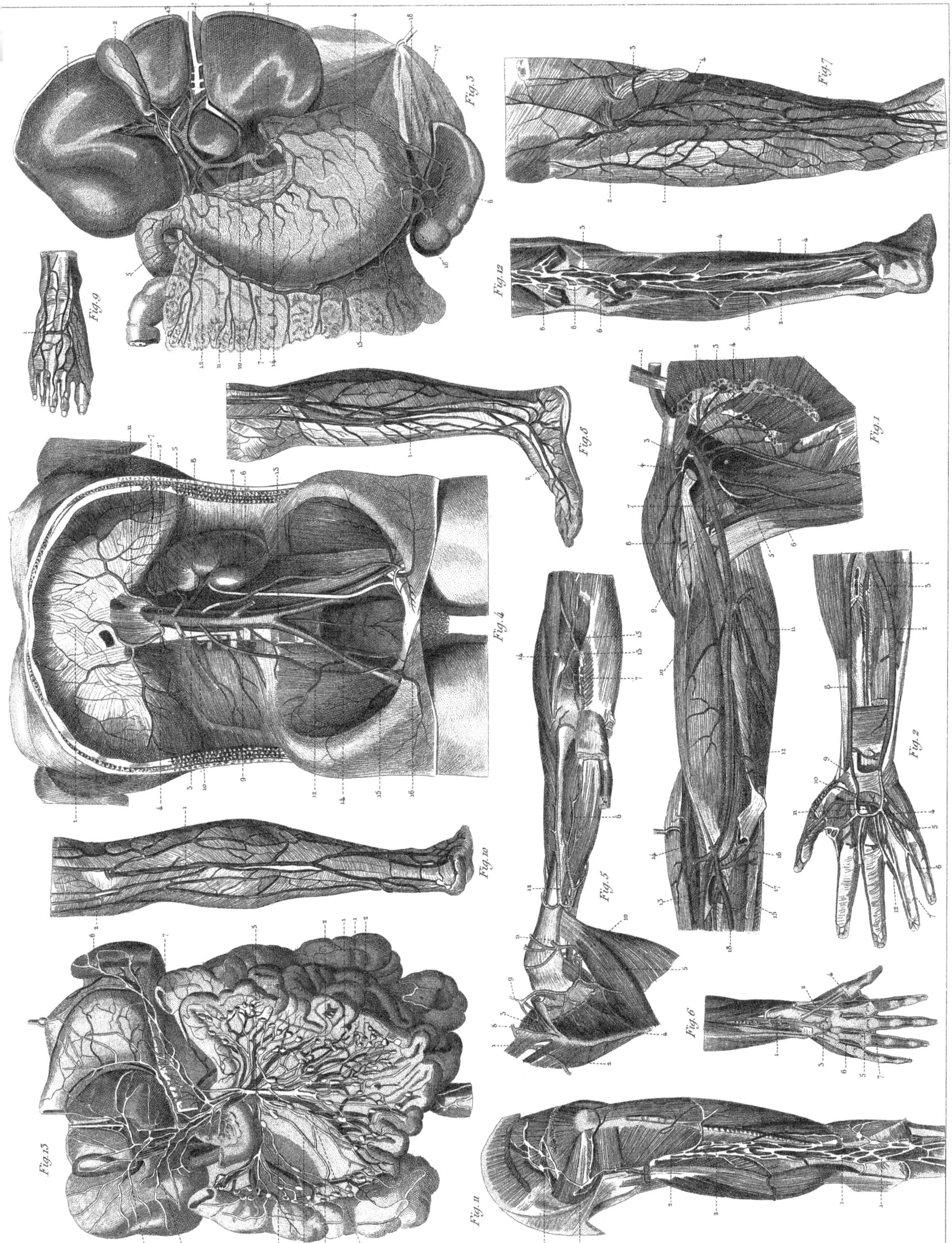

Plate 134 (Anthropology and Surgery, Plate 16)

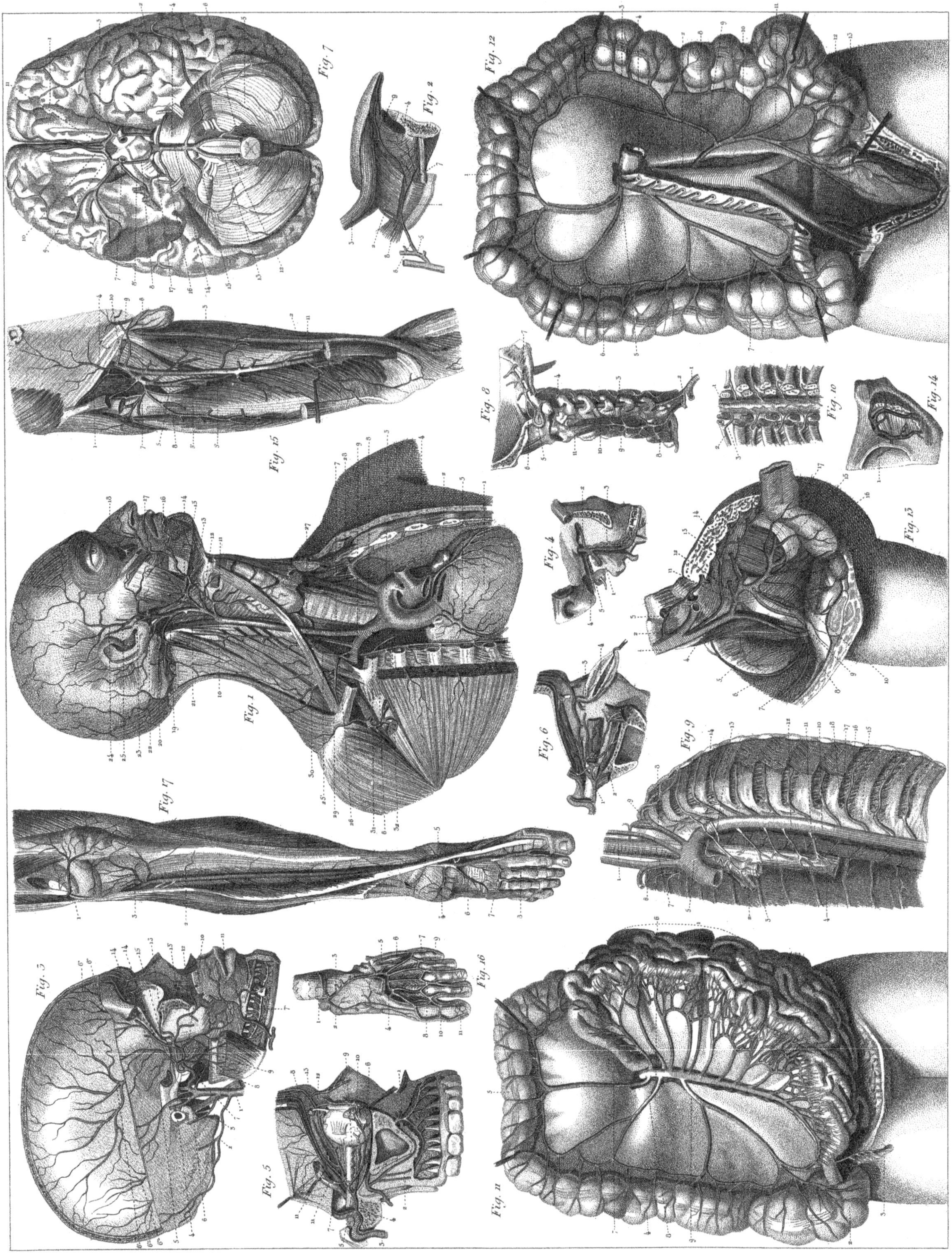

Plate 135 (Anthropology and Surgery, Plate 17)

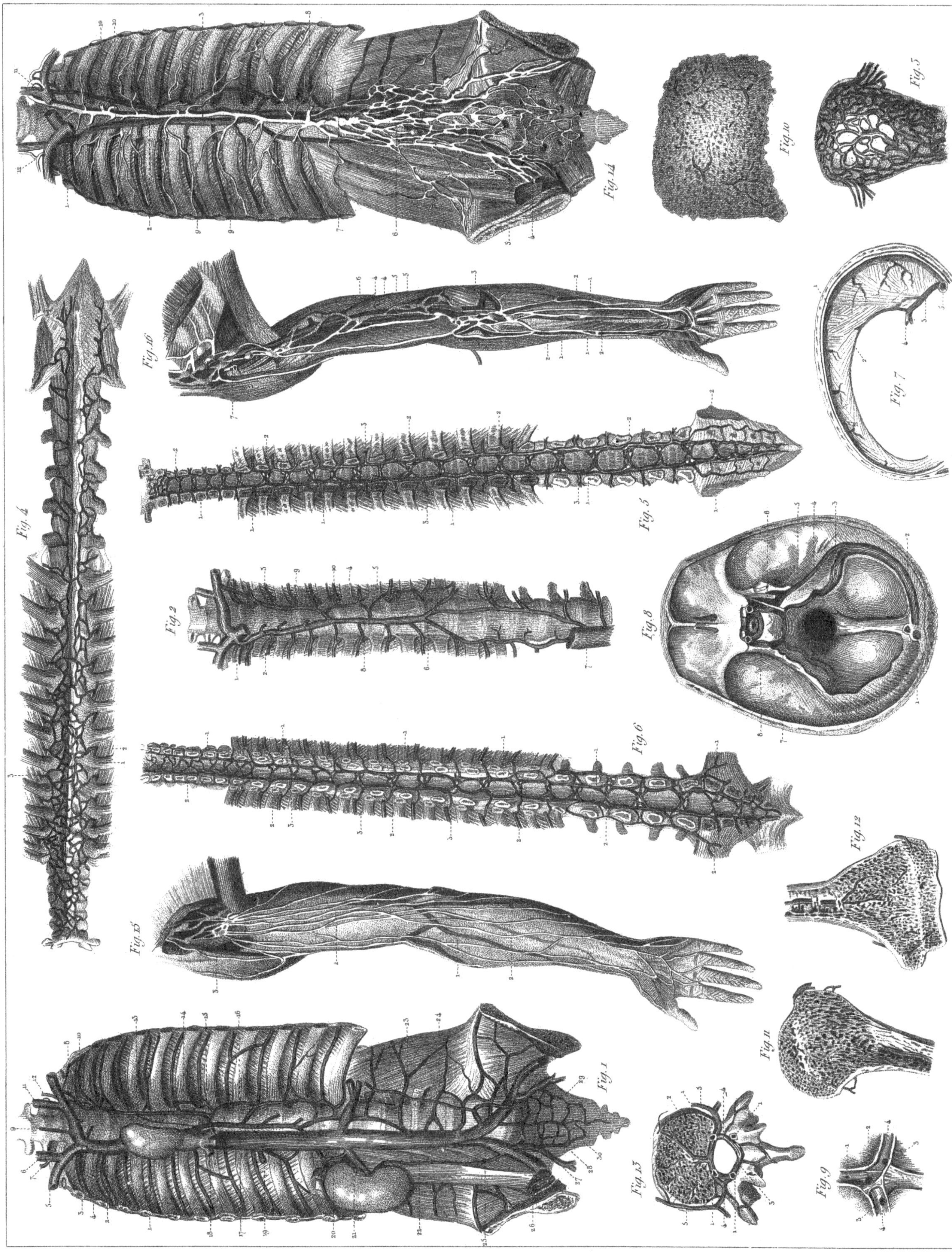

Plate 136 (Anthropology and Surgery, Plate 18)

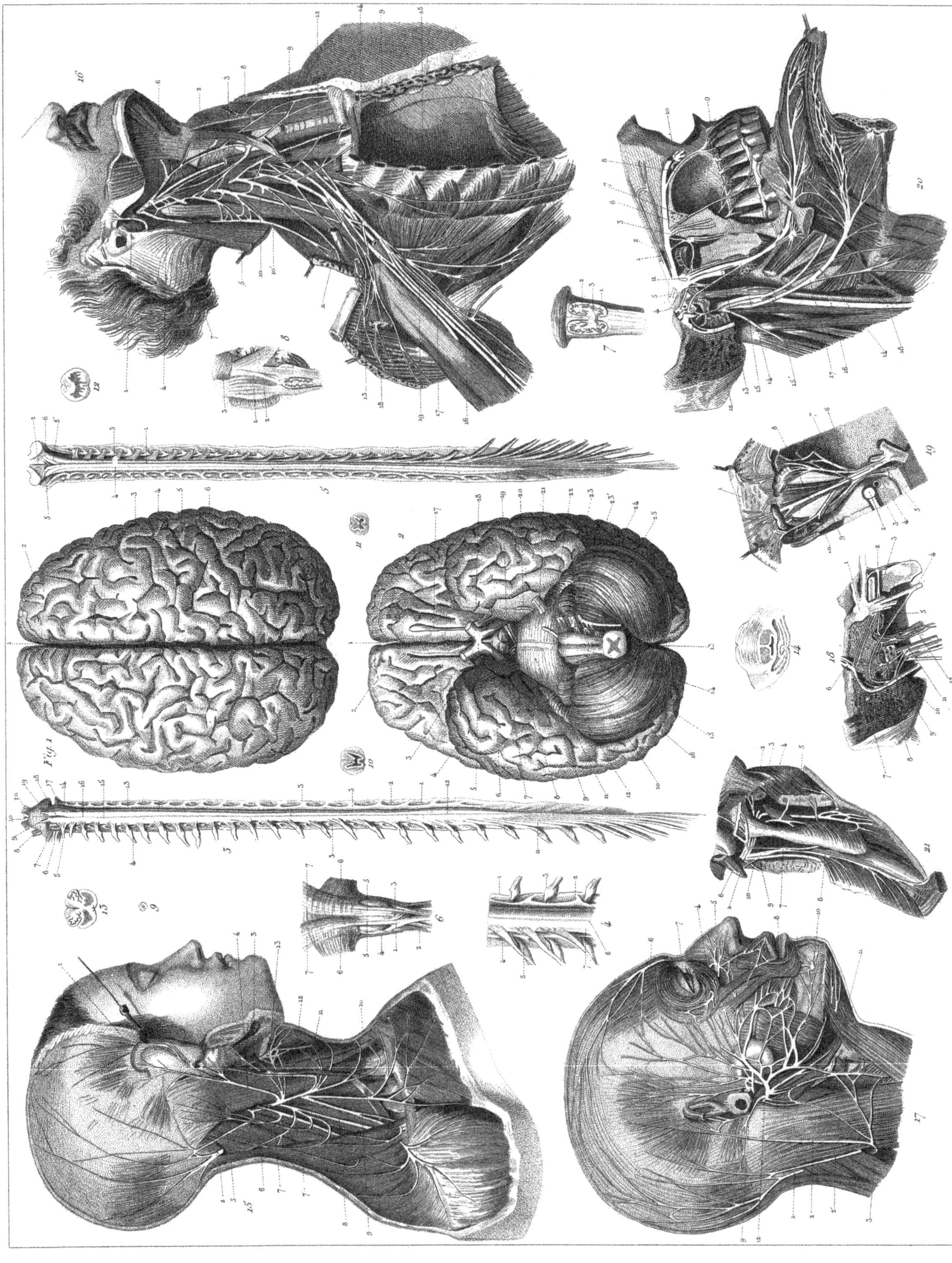

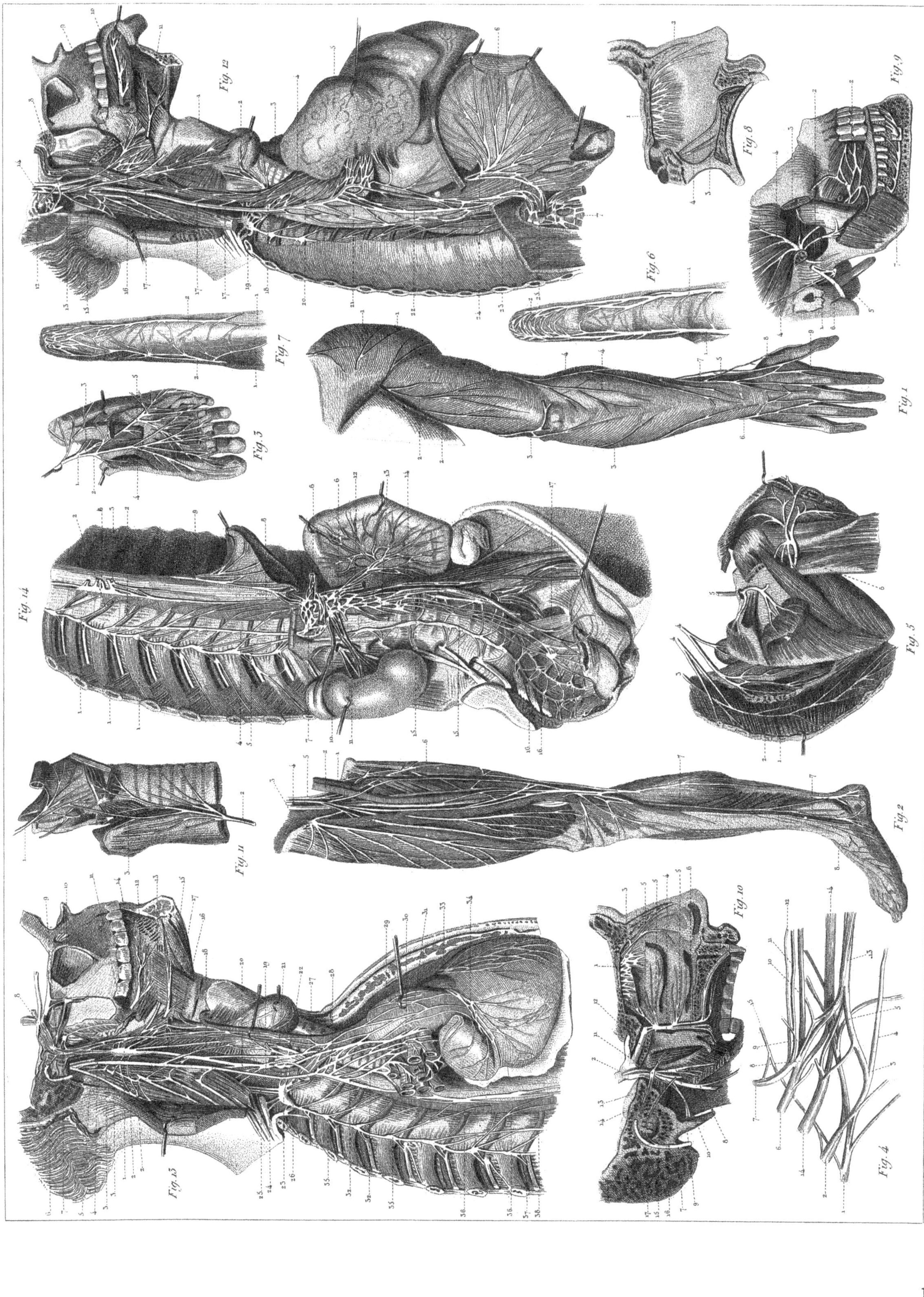

Plate 138 (Anthropology and Surgery, Plate 20)

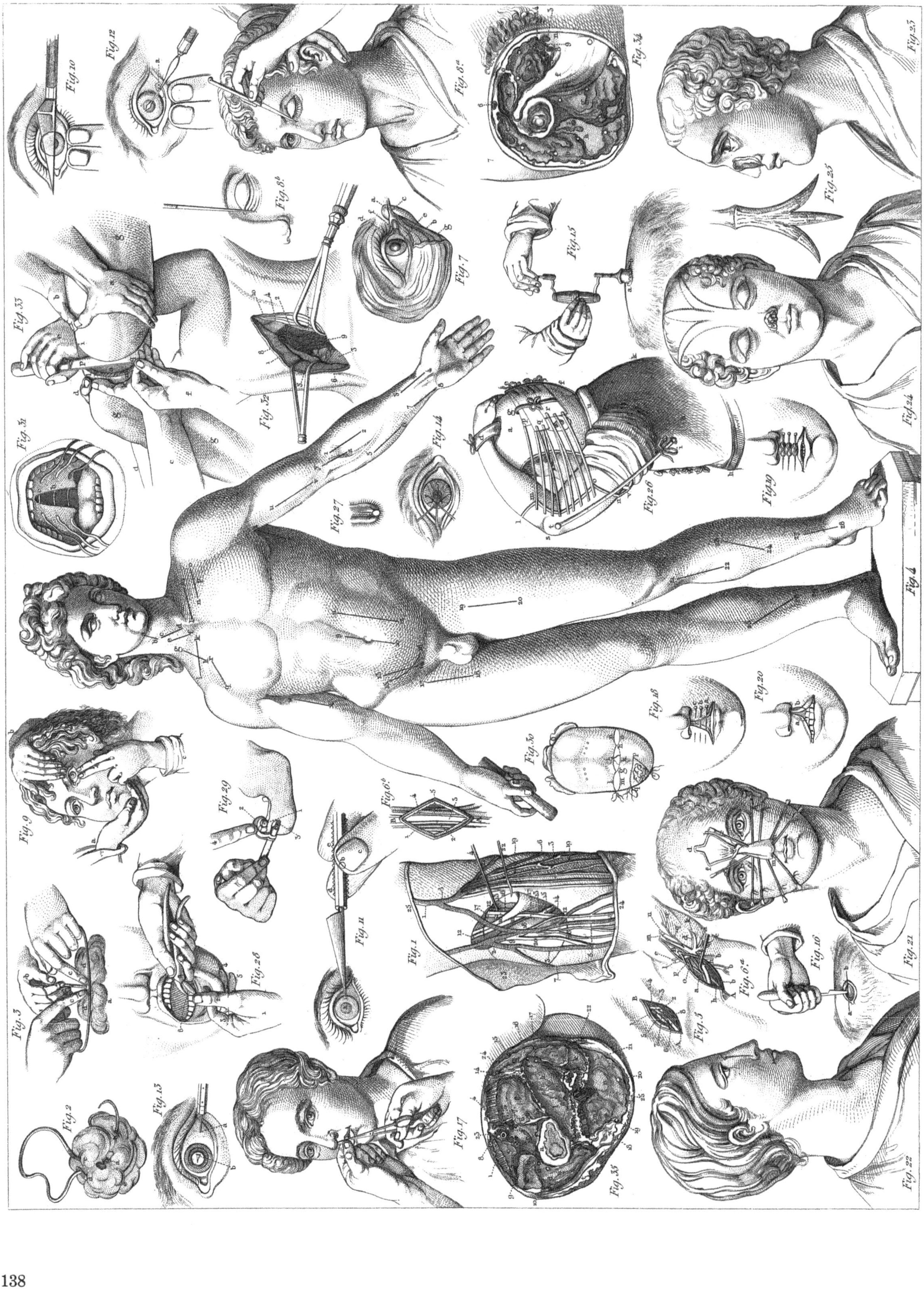

Plate 120 (Anthropology and Surgery, Plate 21)

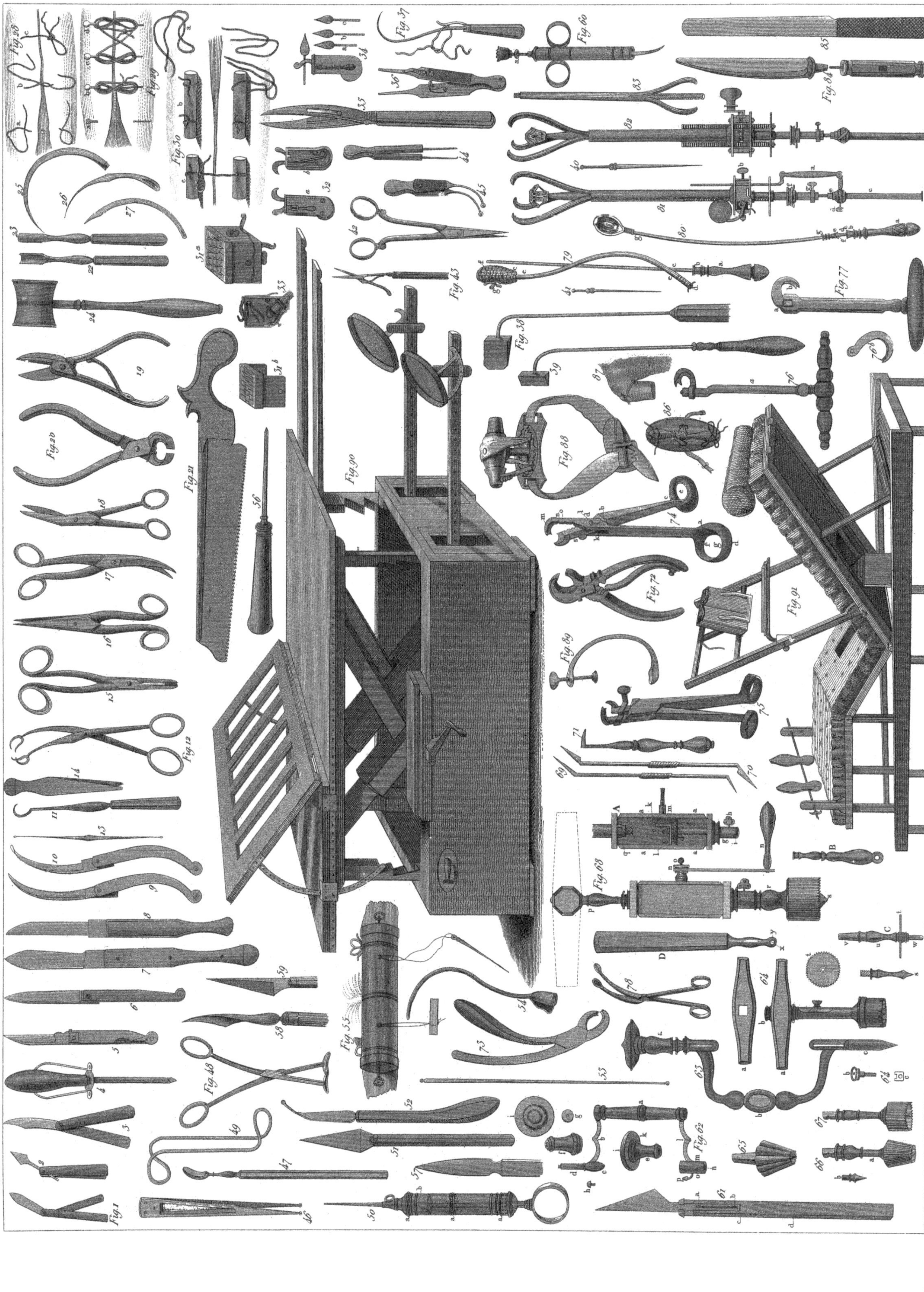

Plate 140 (Anthropology and Surgery, Plate 22)

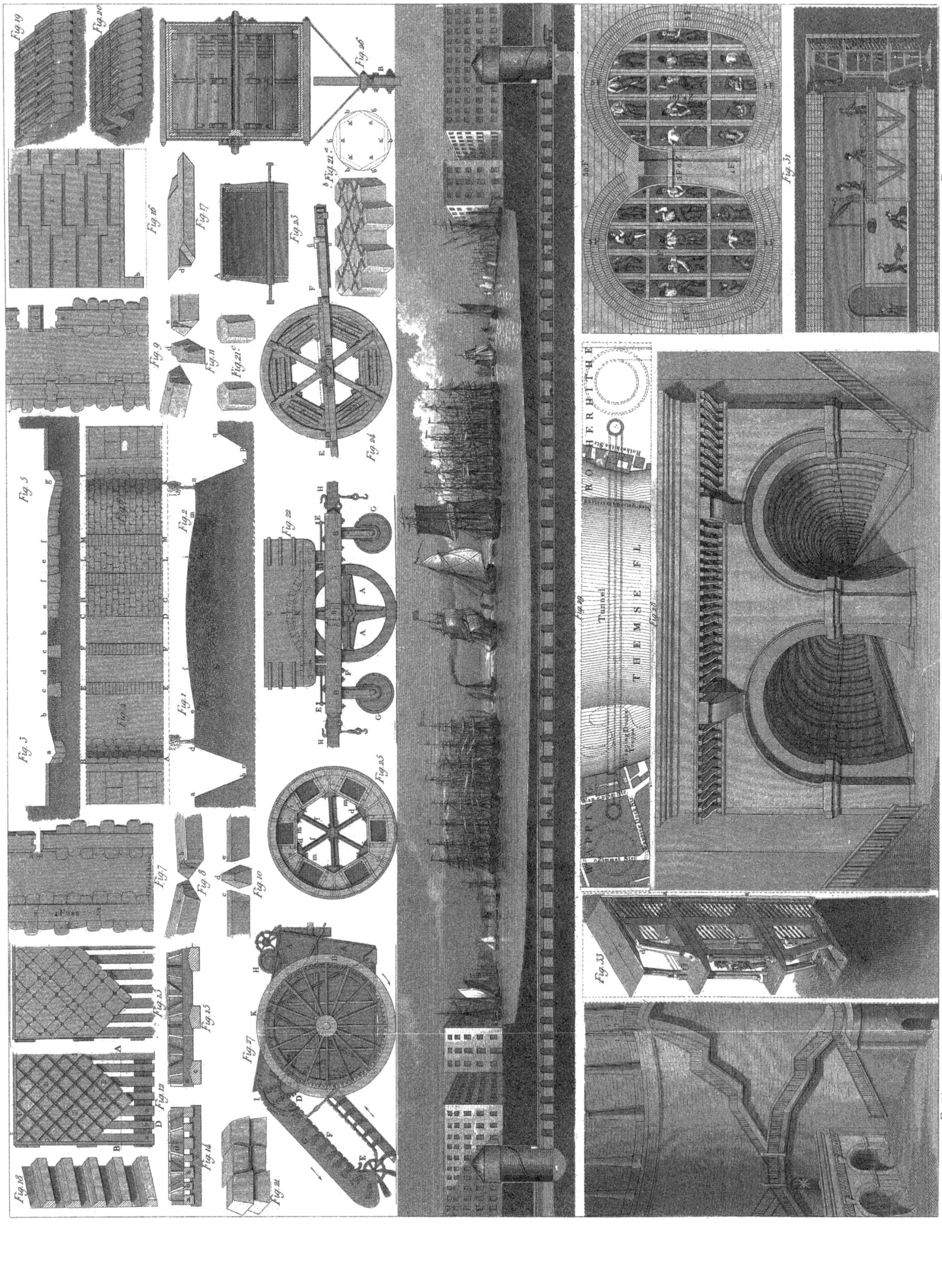

Plate 141 (Technology Plate 1)

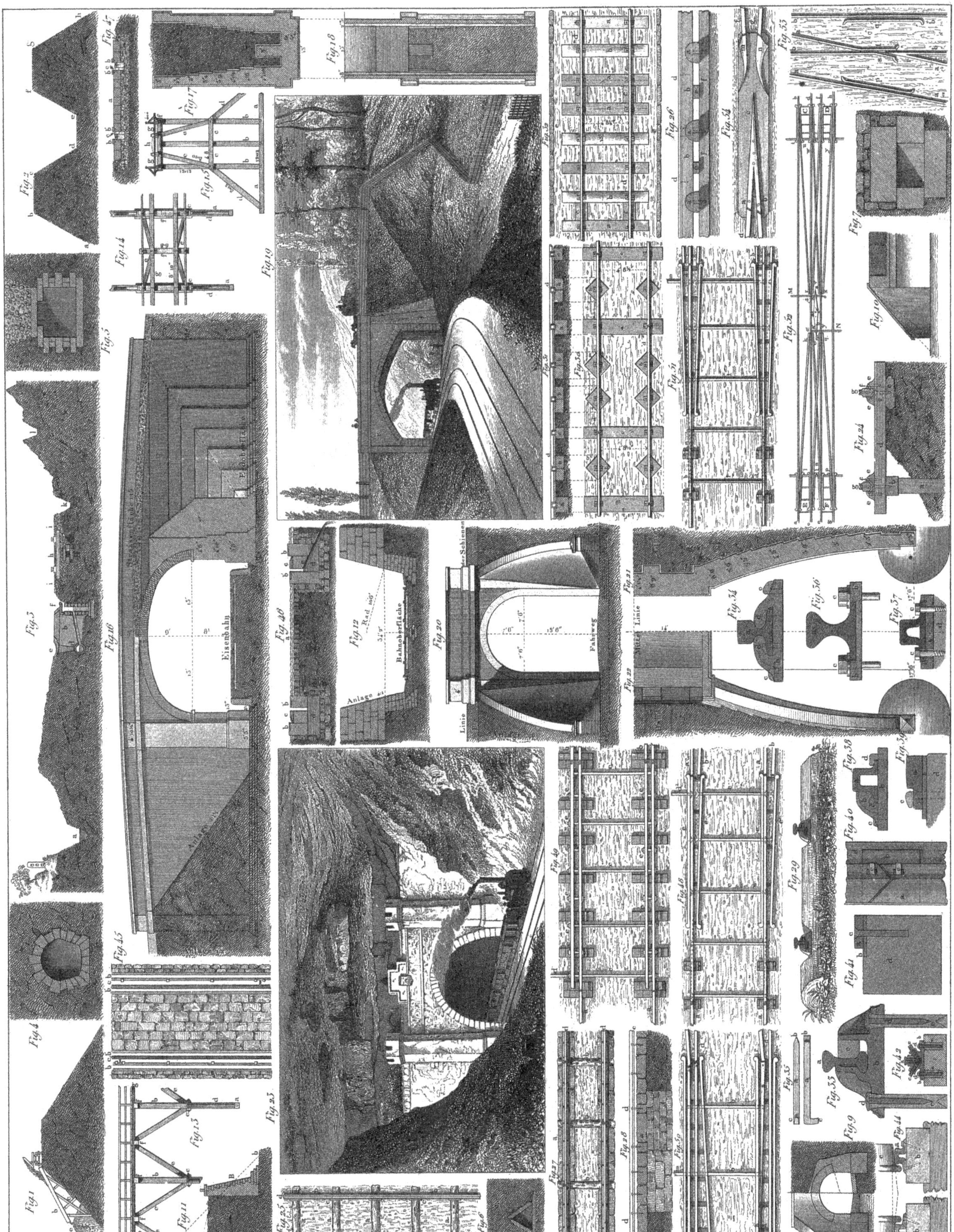

Plate 142 (Technology, Plate 2)

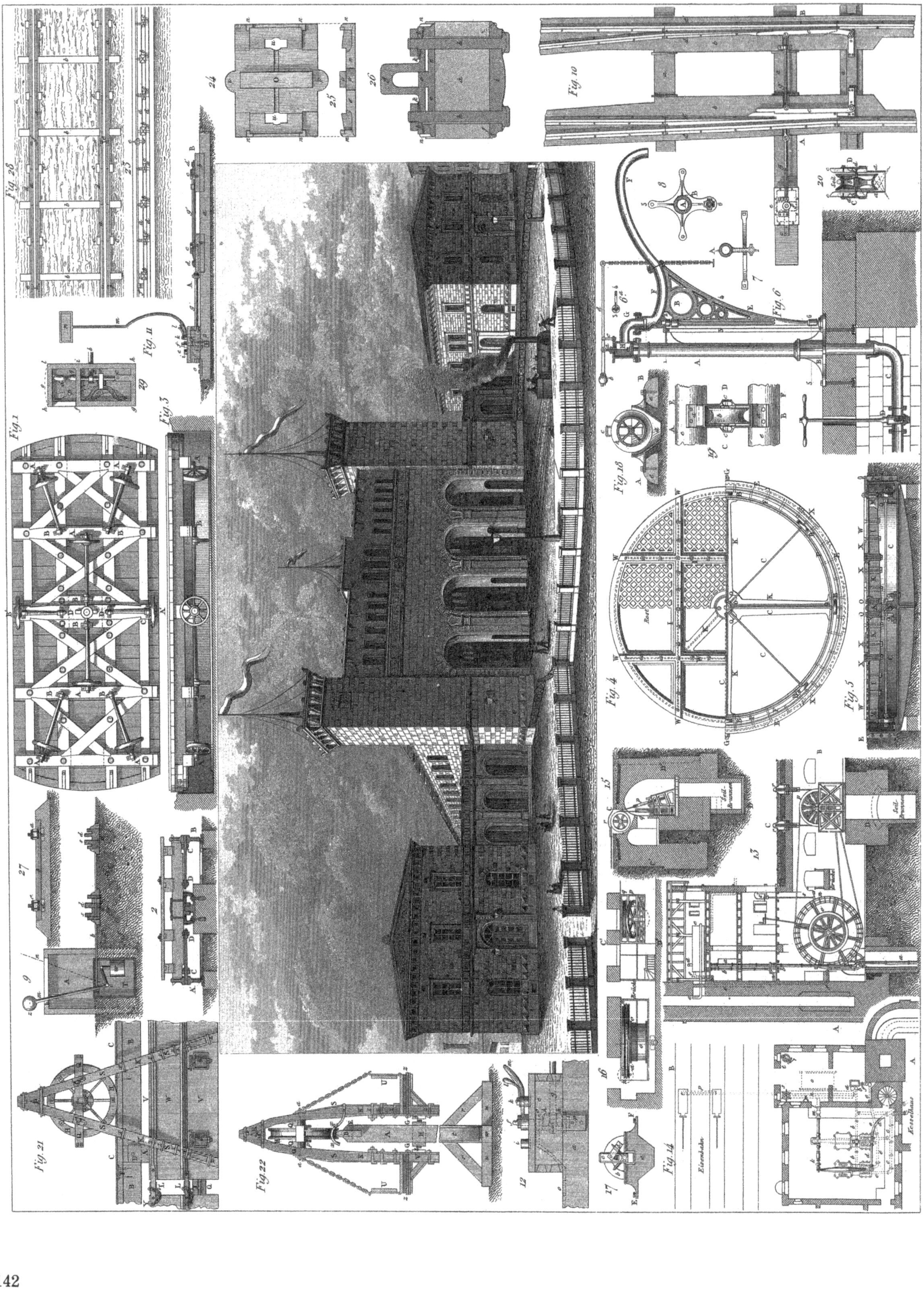

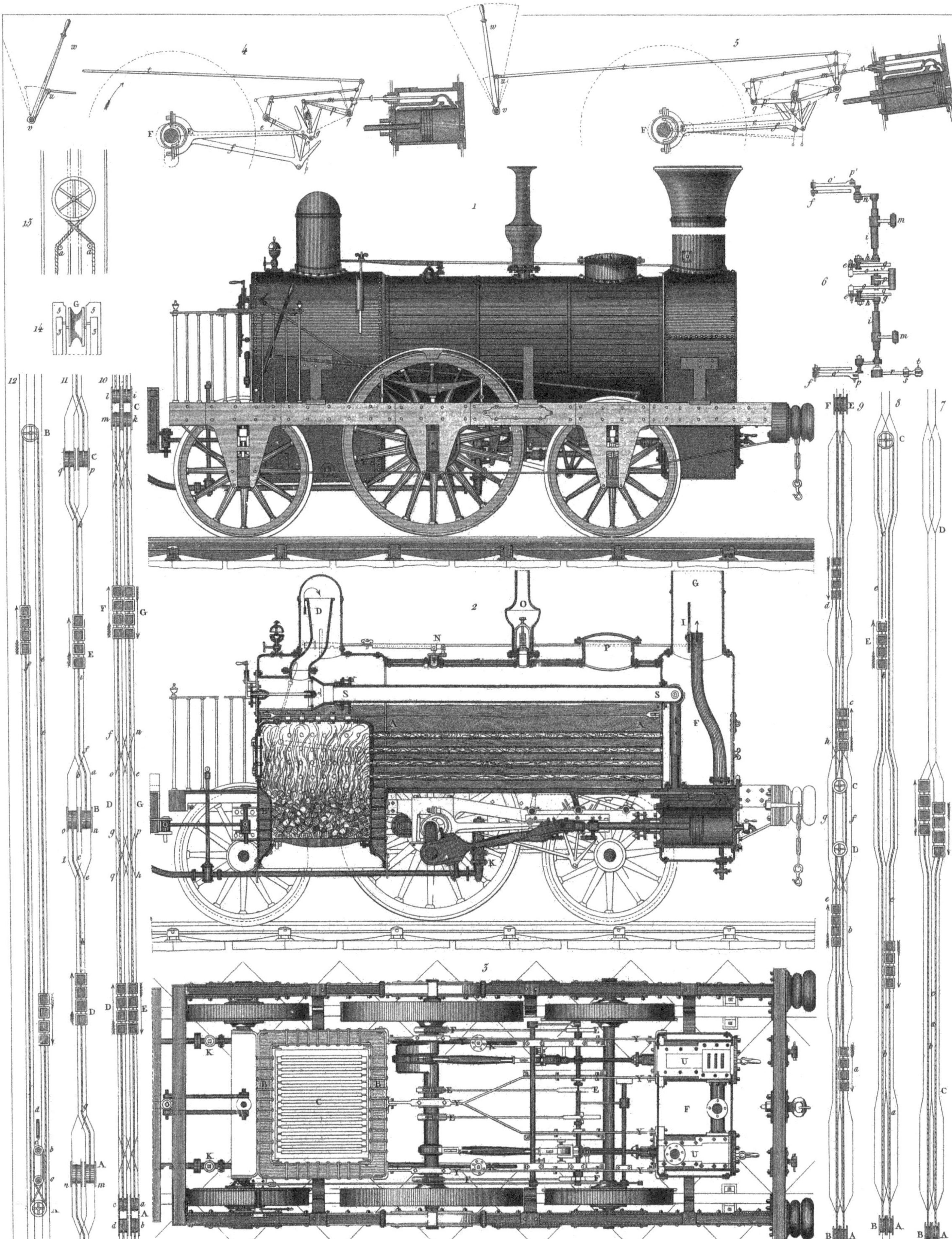

Plate 144 (Technology, Plate 4)

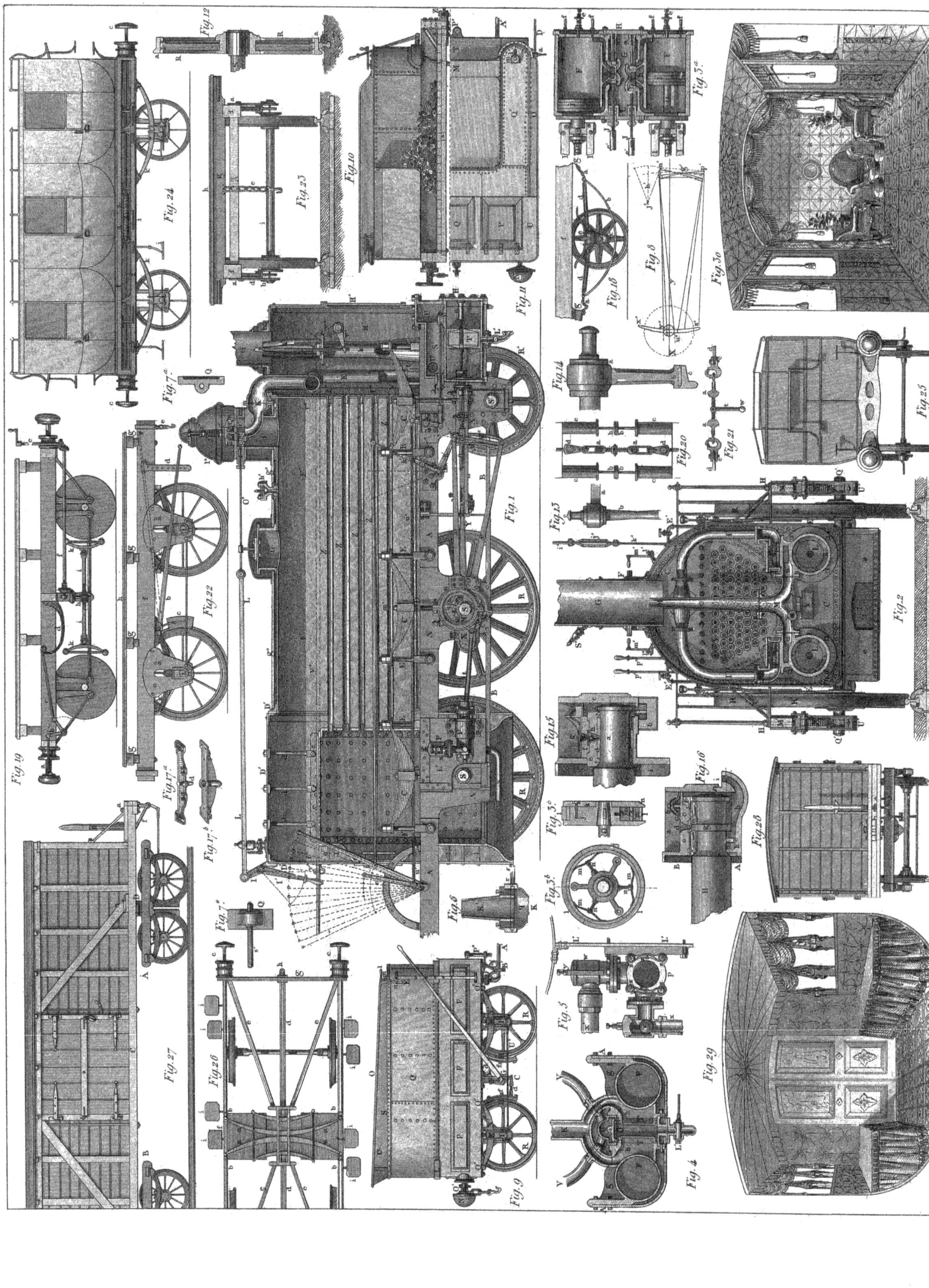

Plate 145 (Technology, Plate 5)

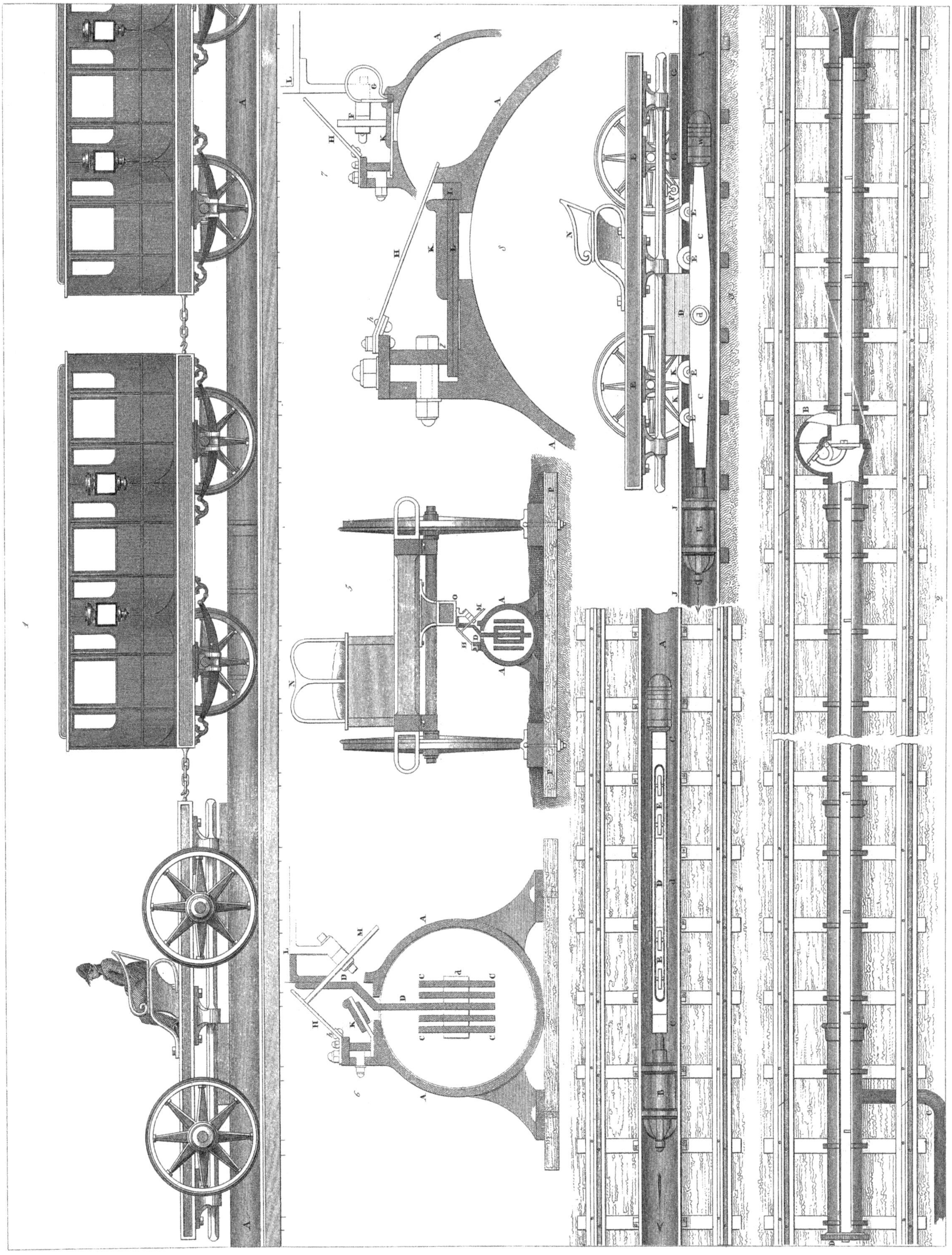

Plate 146 (Technology, Plate 6)

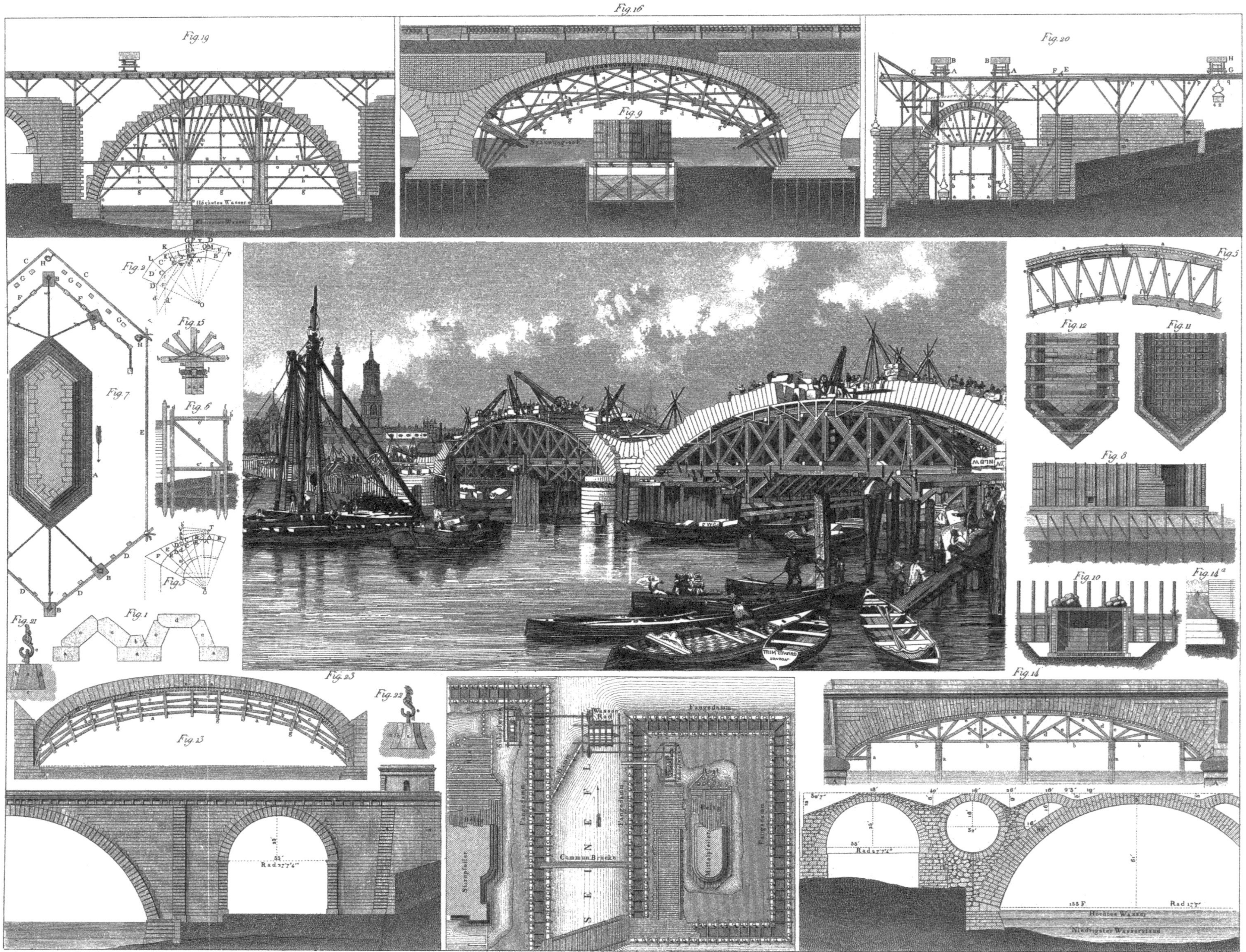

Plate 147 (Technology, Plate 7)

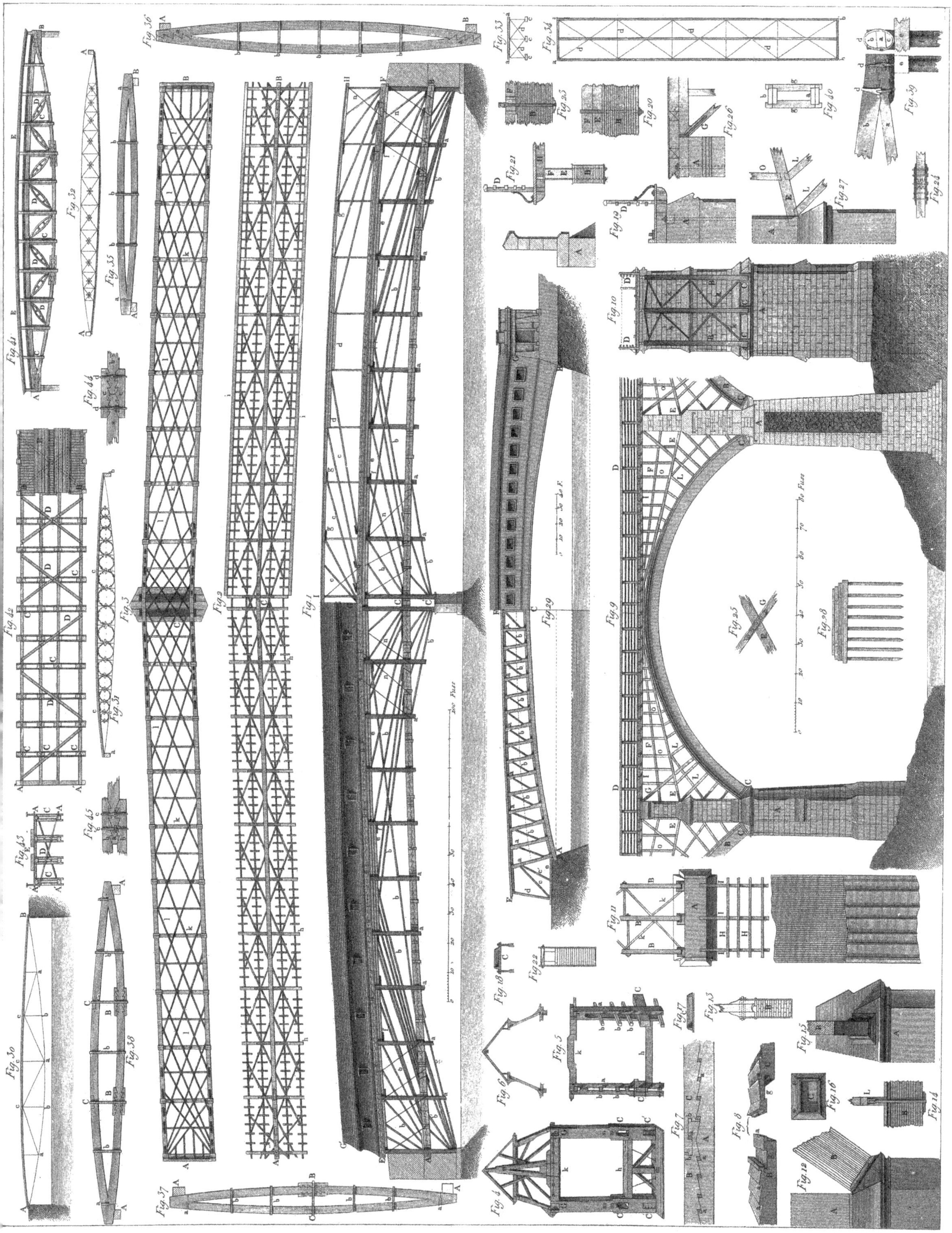

Plate 148 (Technology, Plate 8)

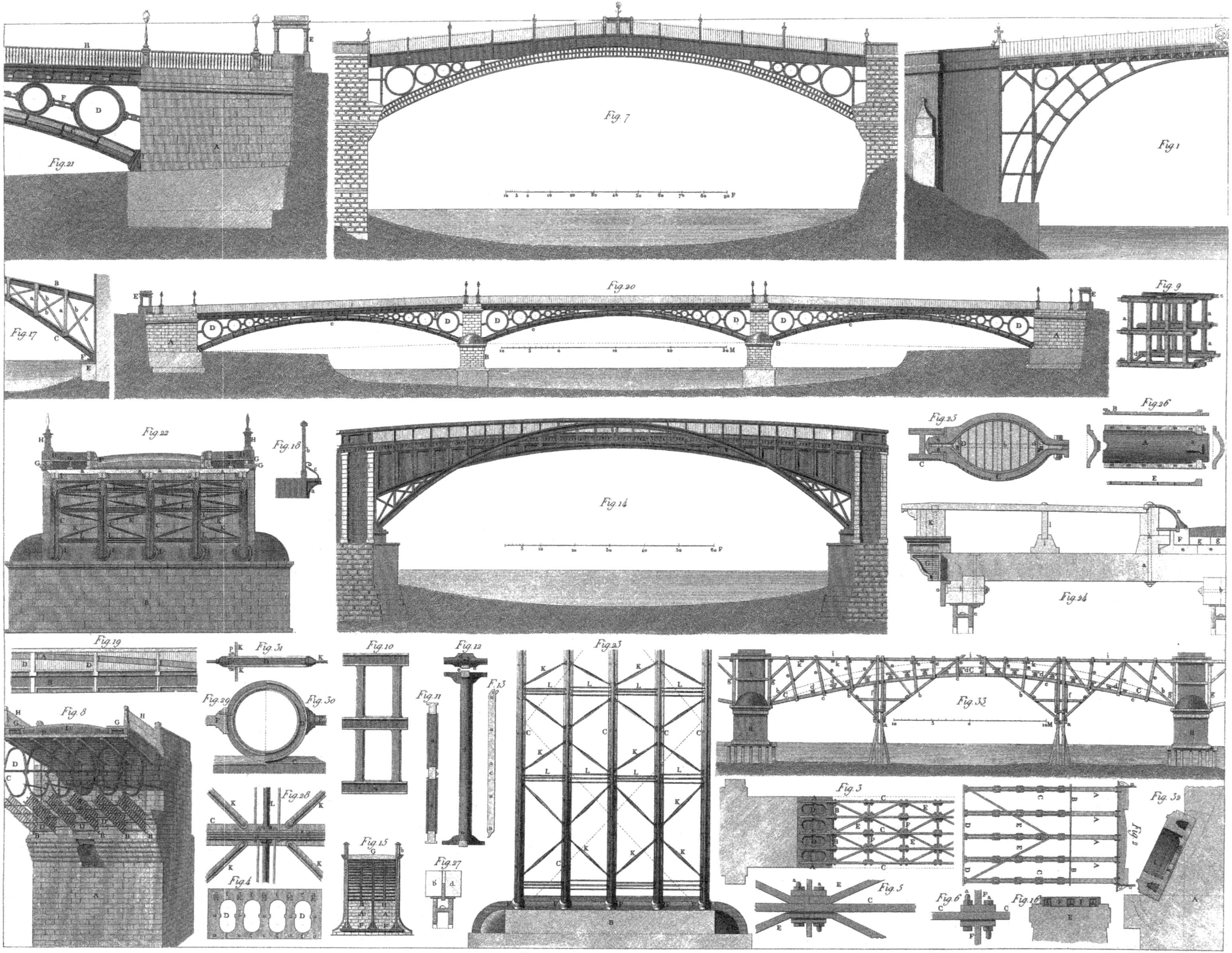

Plate 149 (Technology, Plate 9)

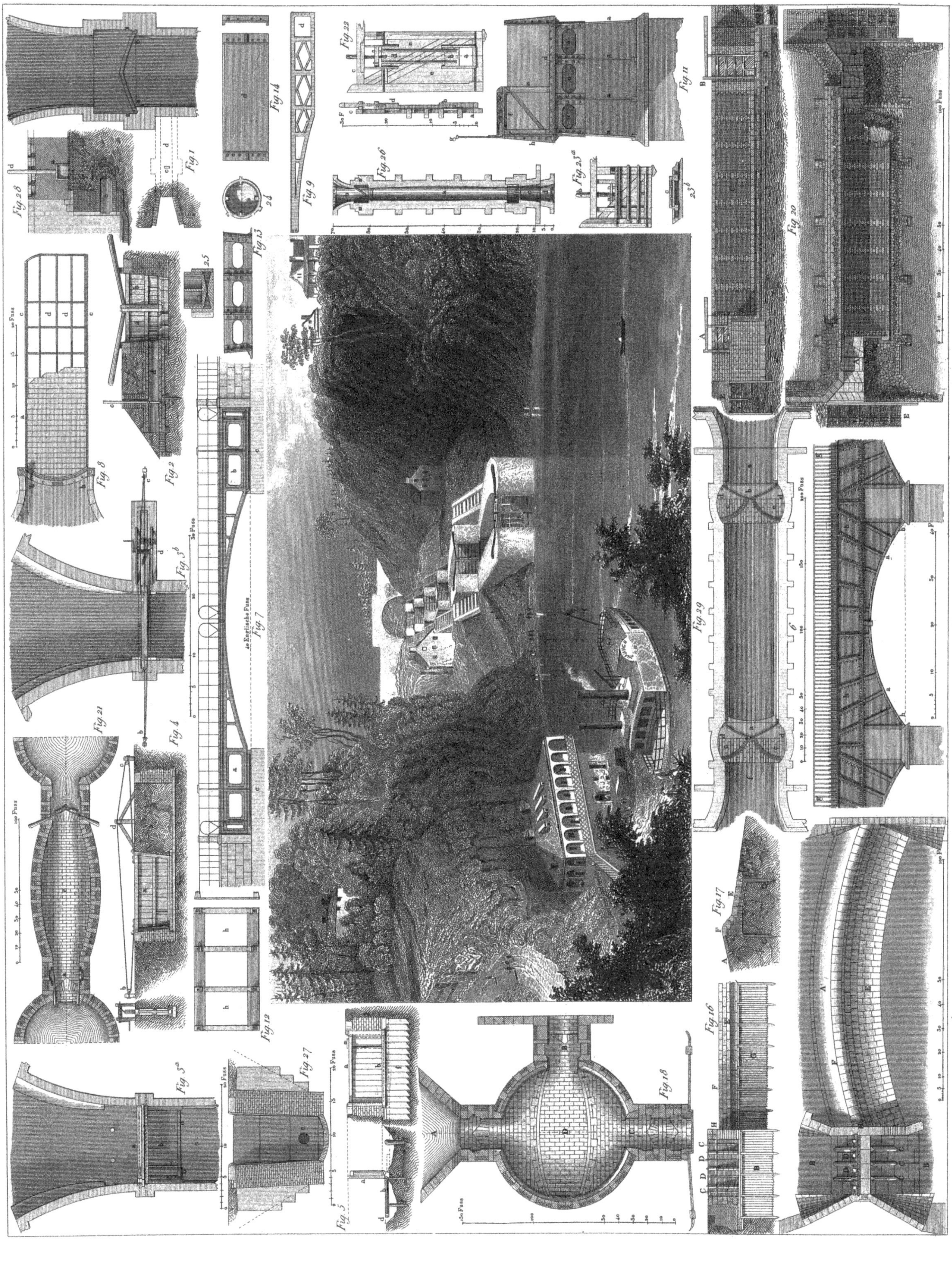

Plate 150 (Technology, Plate 10)

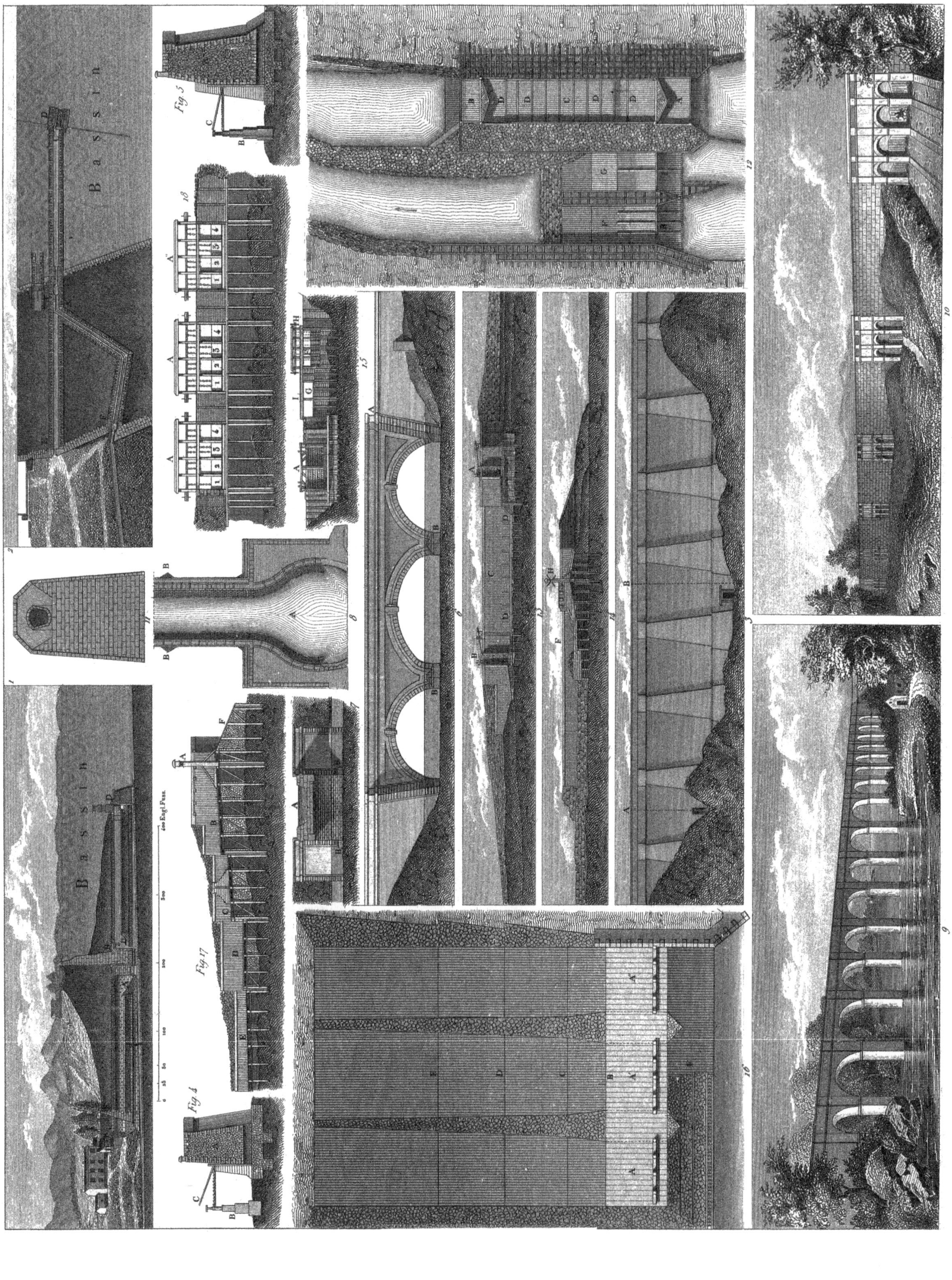

Plate 151 (Technology, Plate 11)

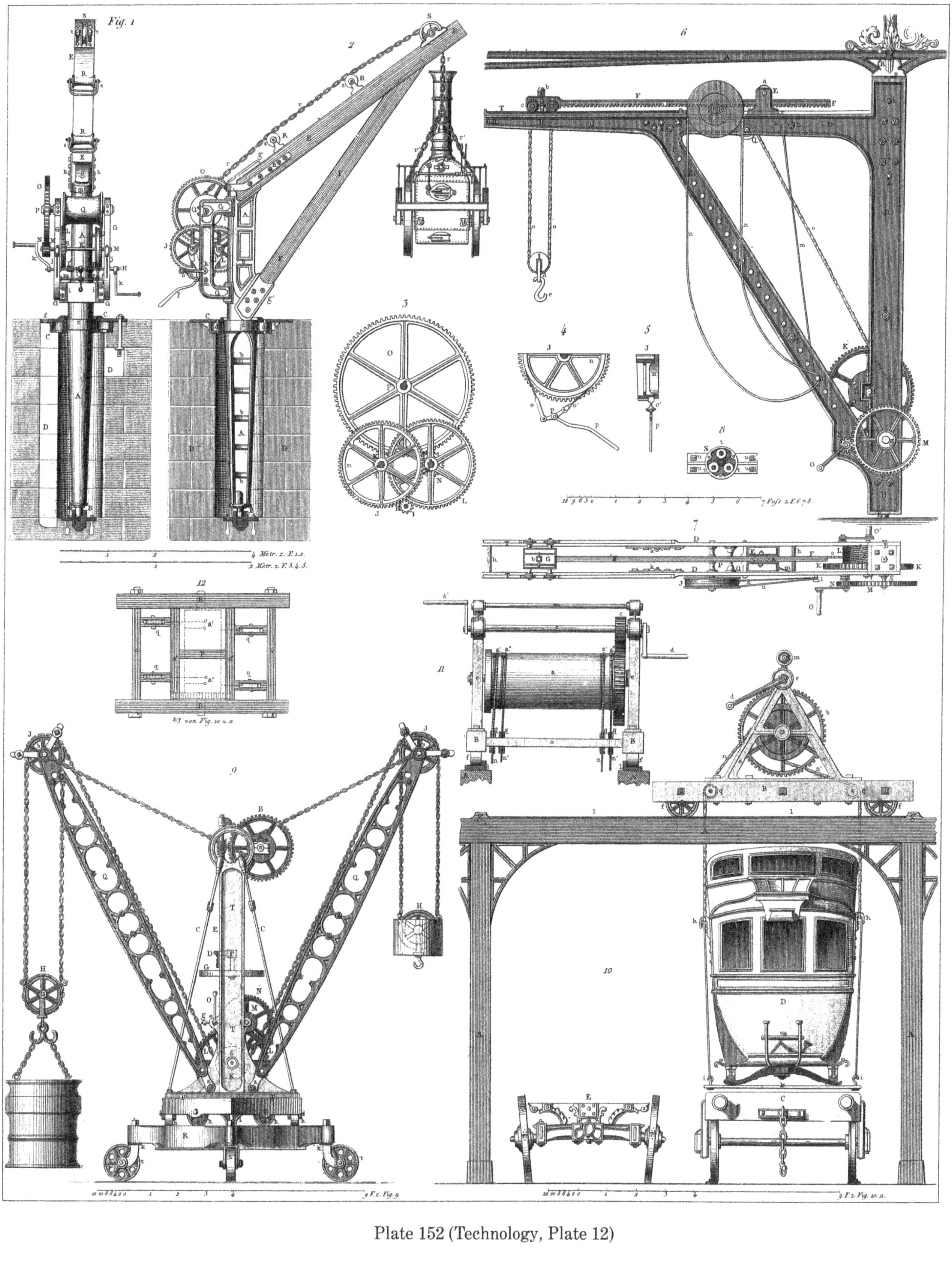

Plate 152 (Technology, Plate 12)

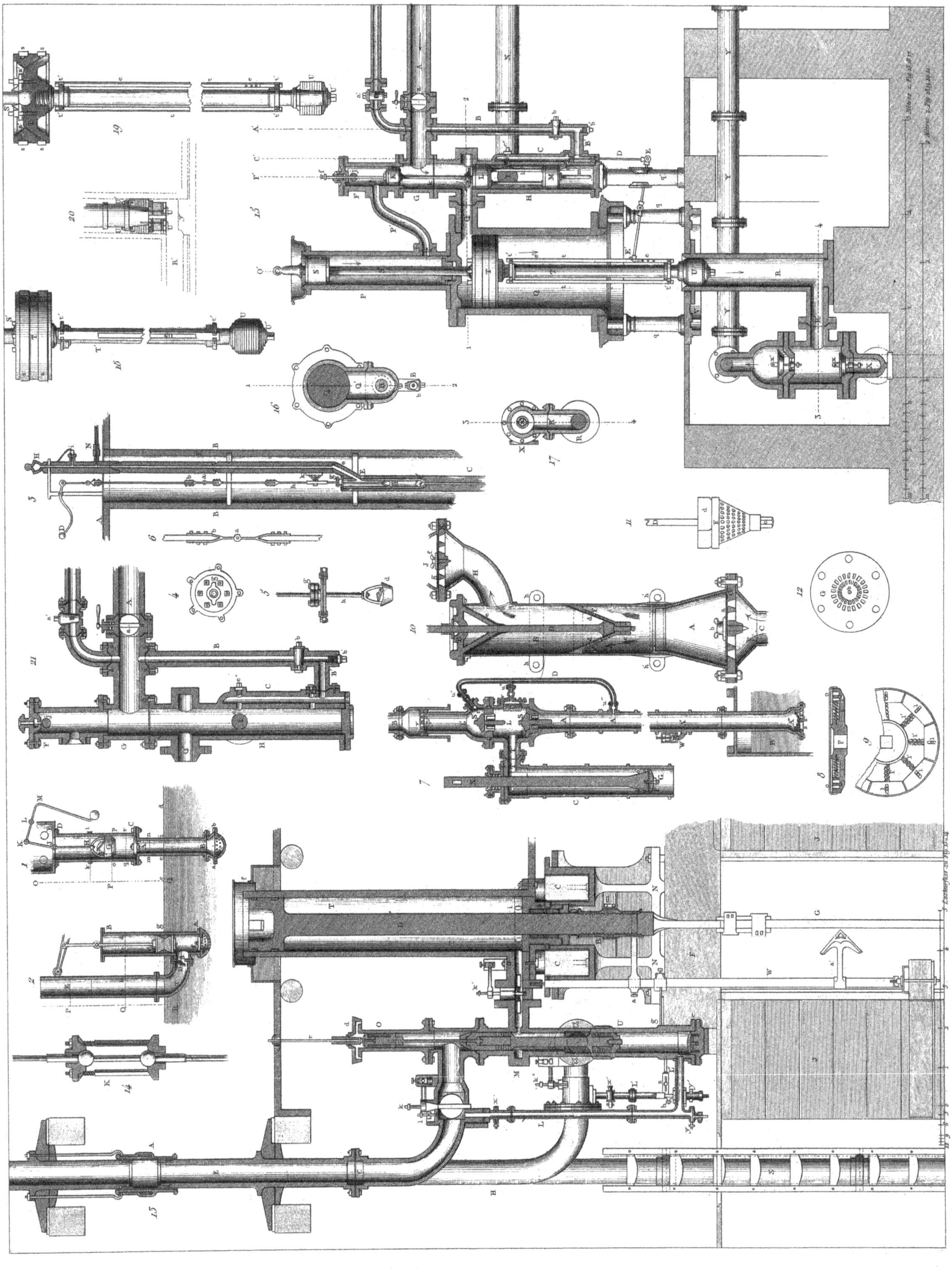

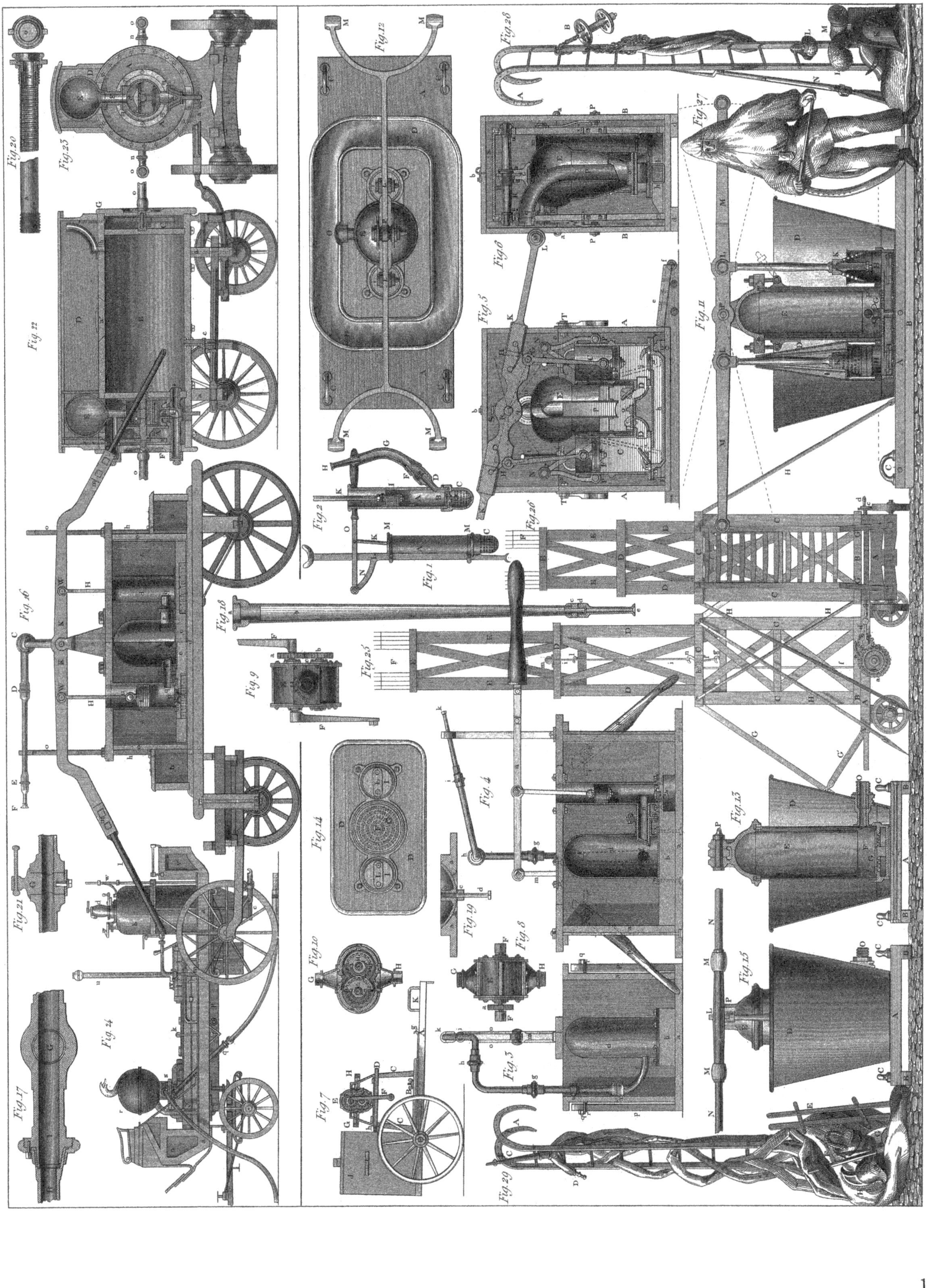

Plate 154 (Technology, Plate 14)

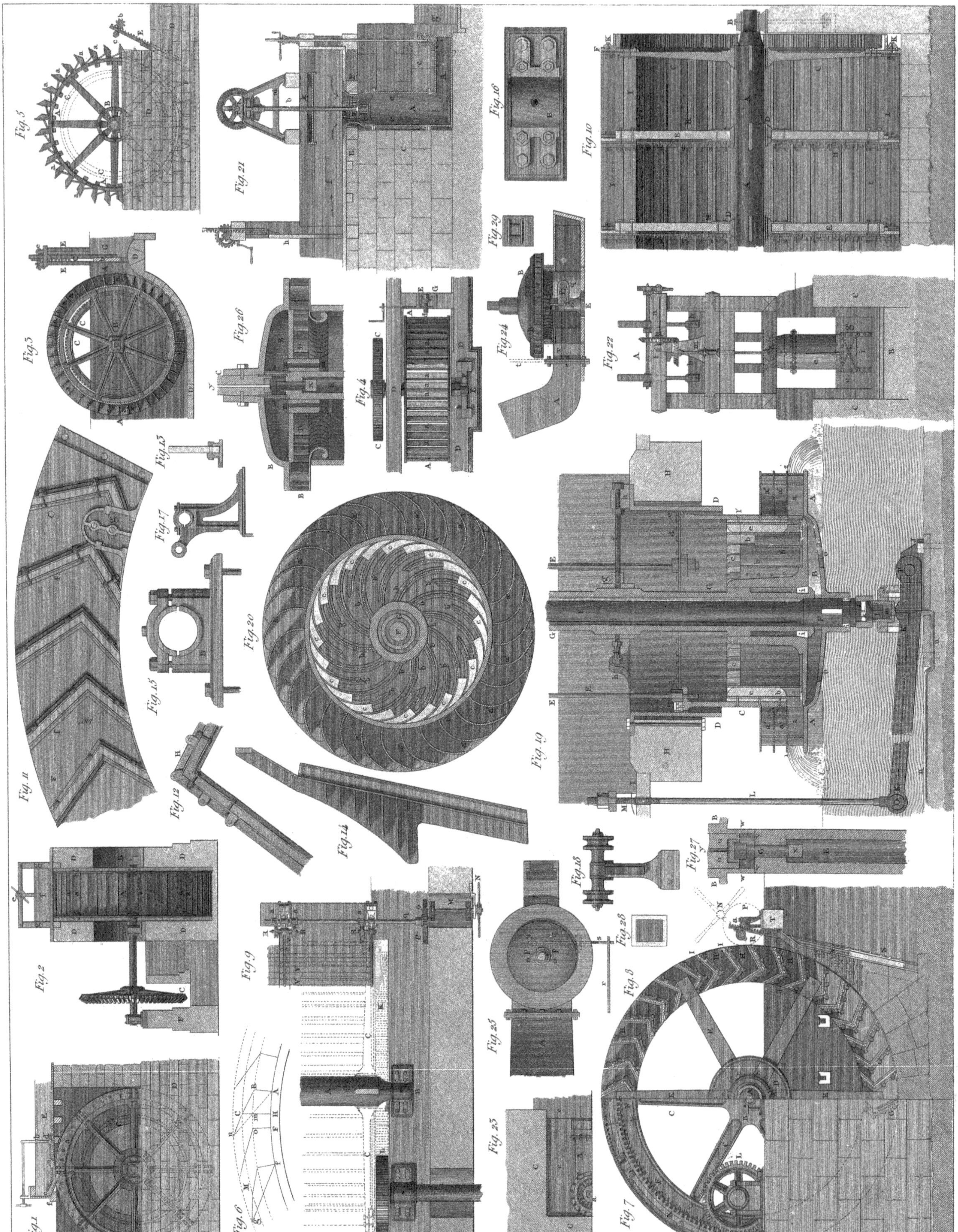

Plate 155 (Technology, Plate 15)

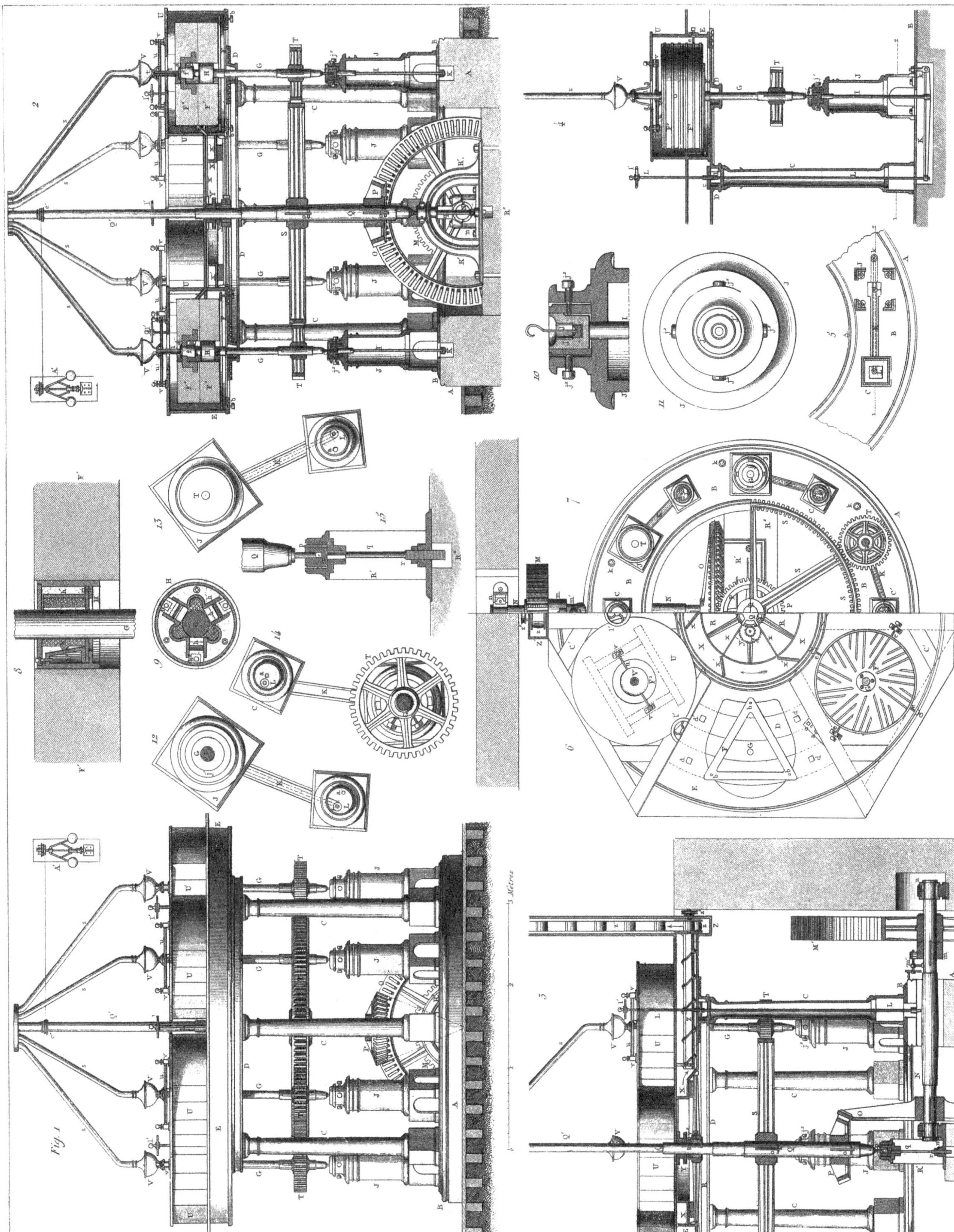

Plate 156 (Technology, Plate 16)

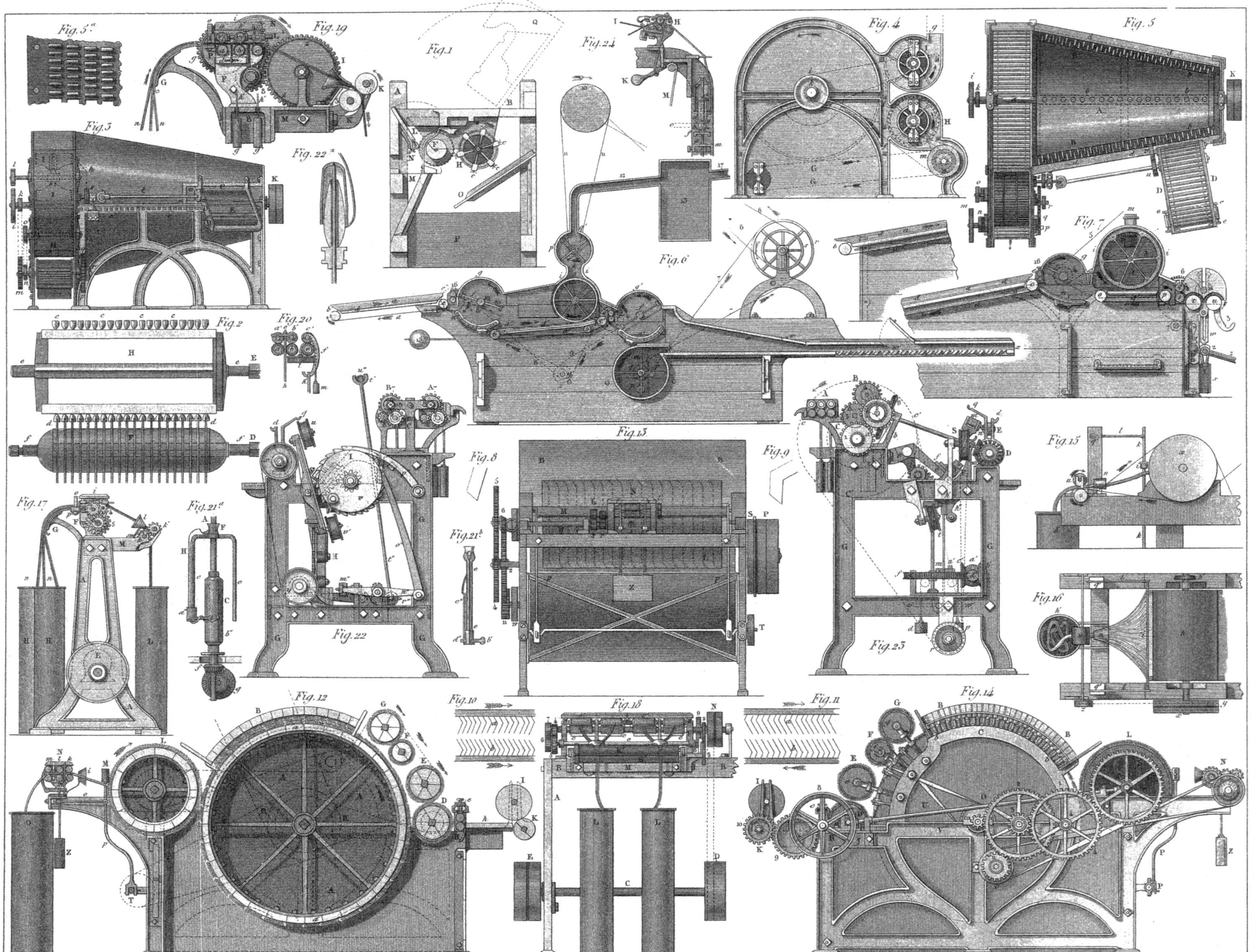

Plate 157 (Technology, Plat. 17)

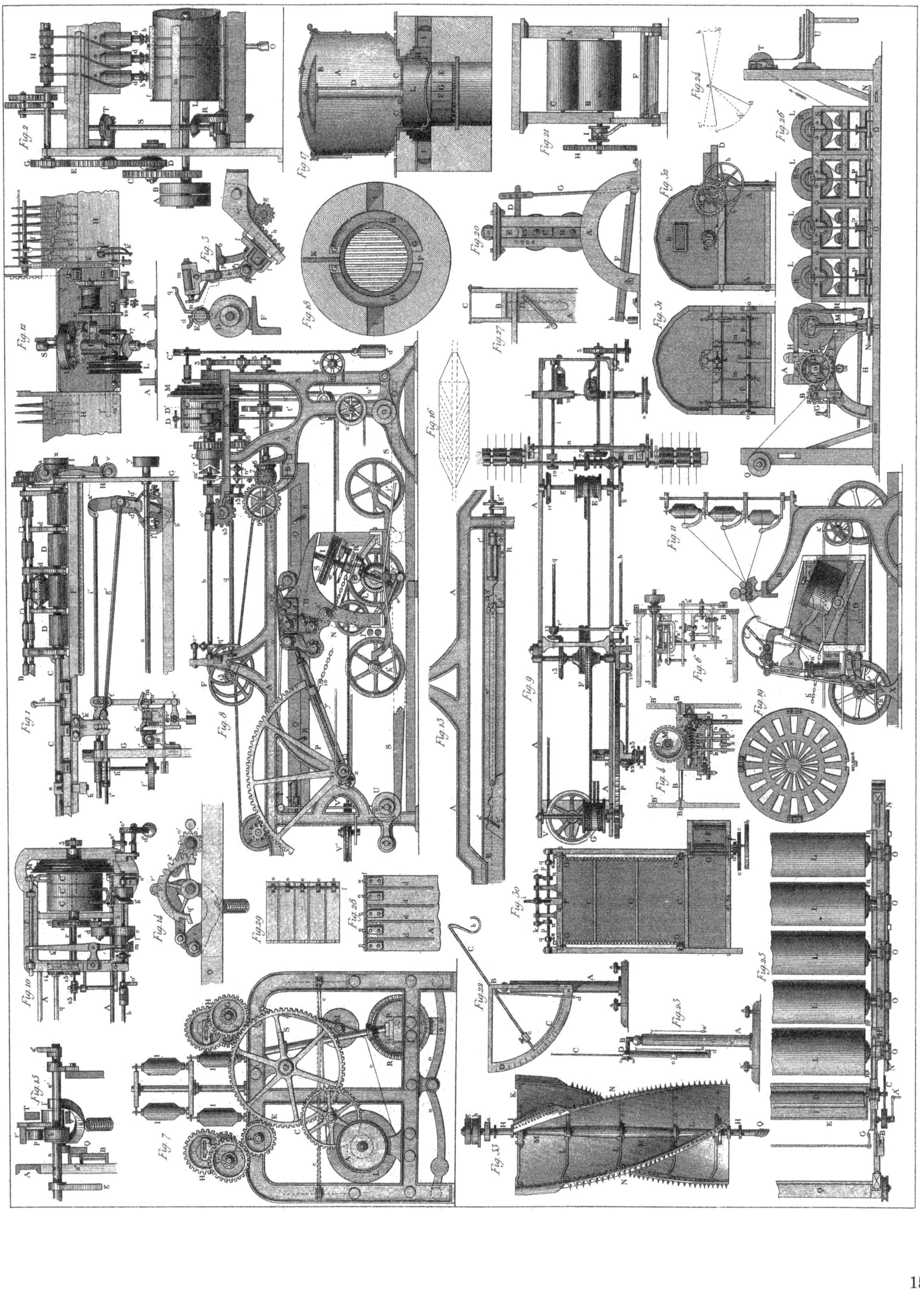

Plate 158 (Technology, Plate 18)

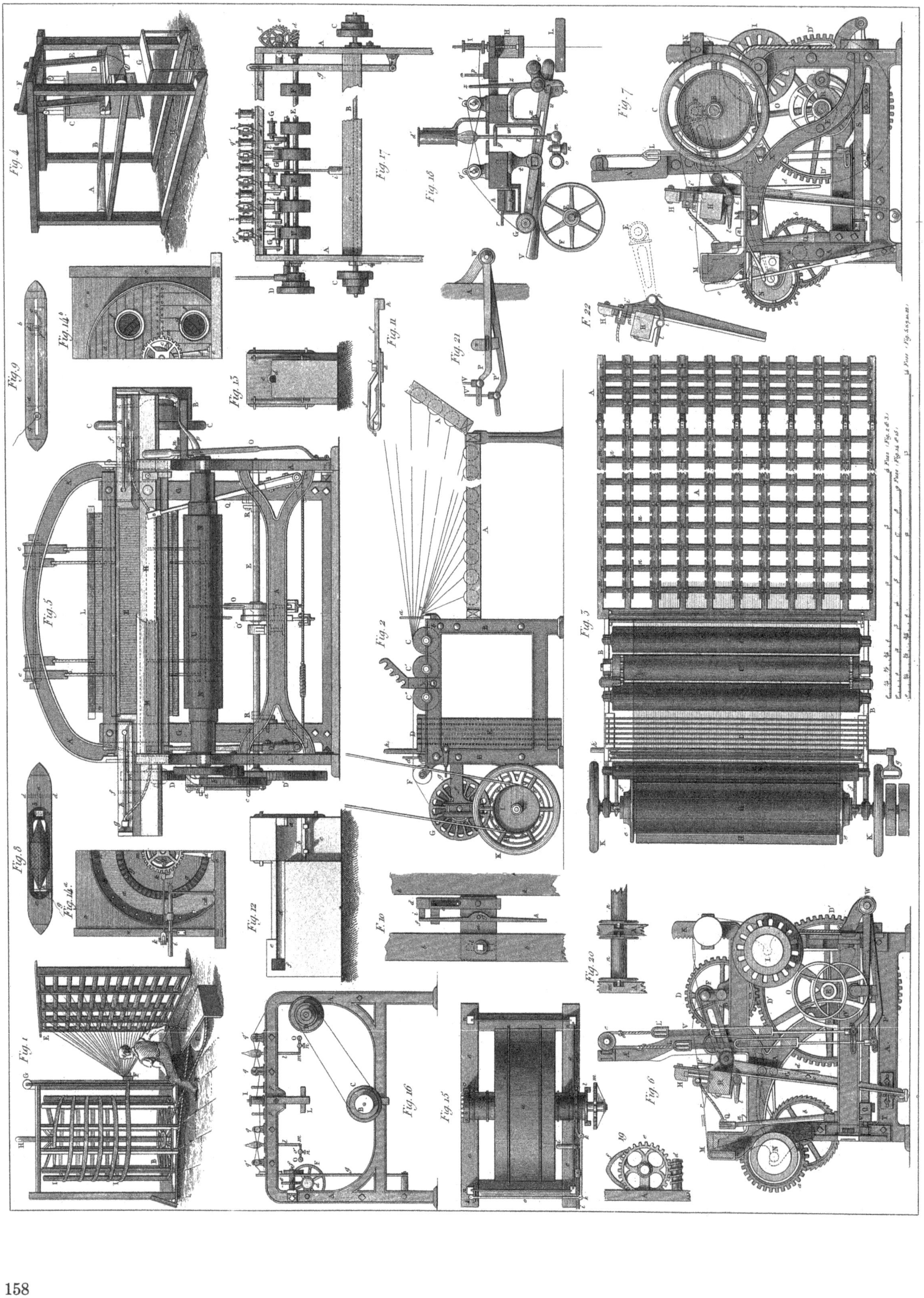

Plate 159 (Technology, Plate 19)

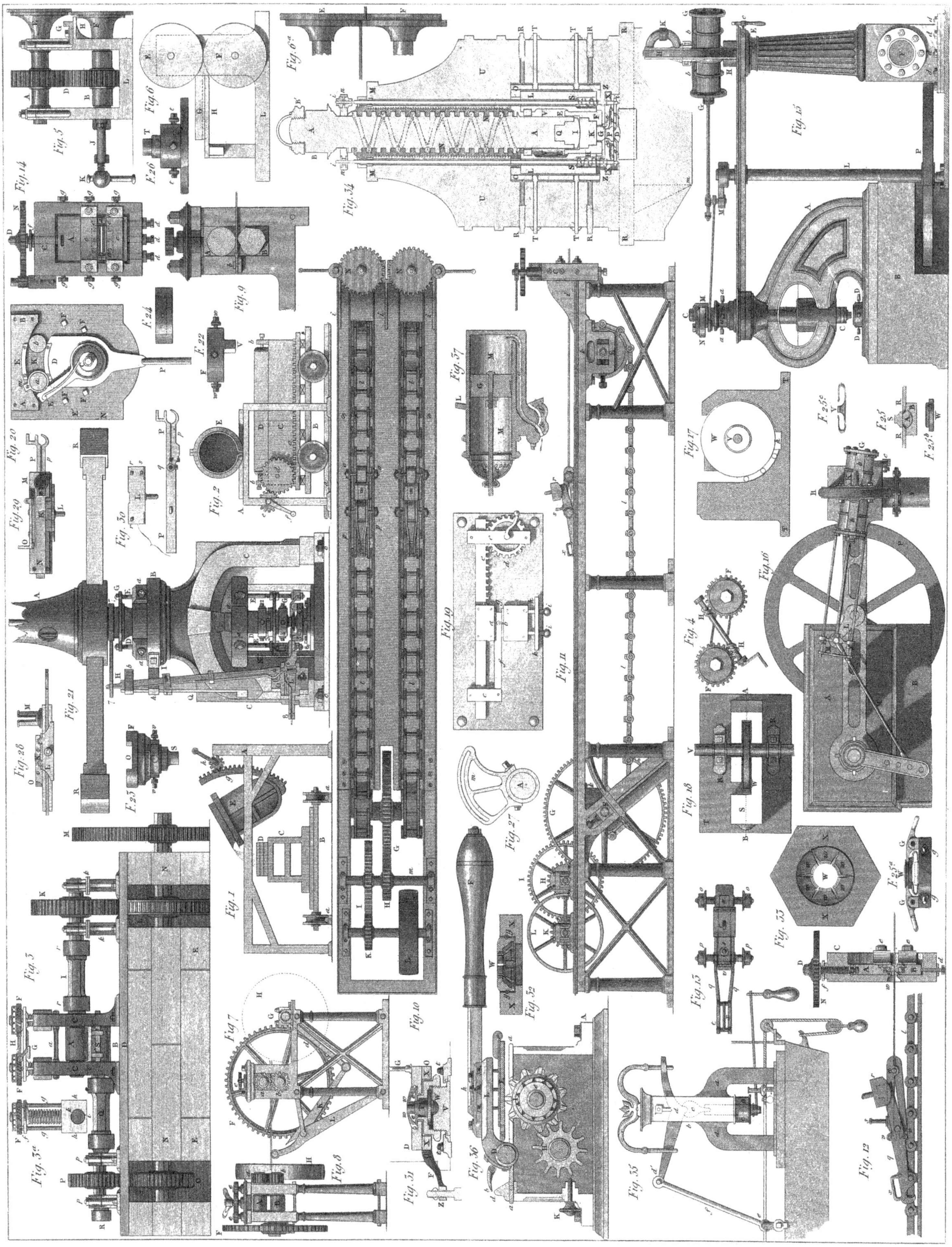

Plate 160 (Technology, Plate 20)

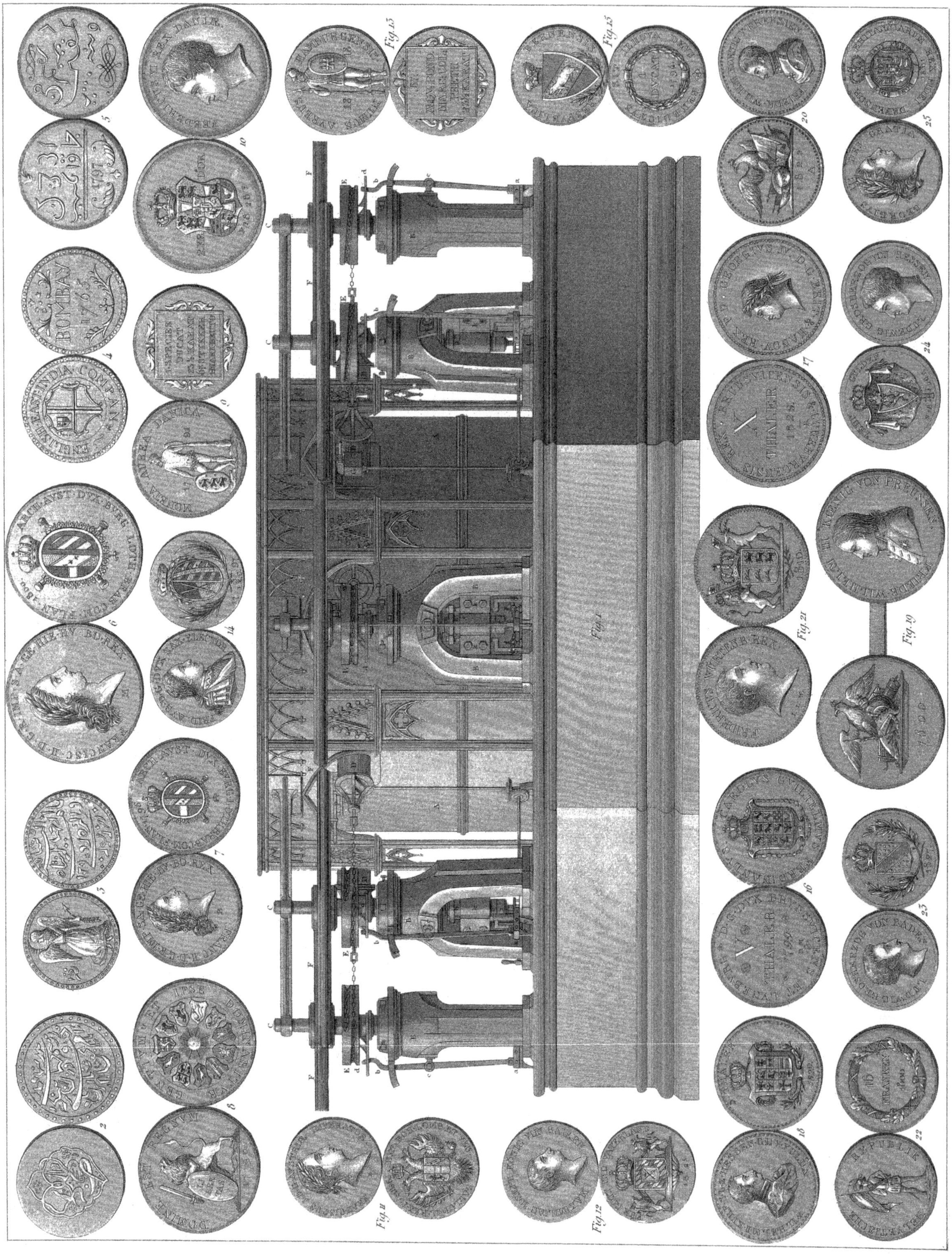

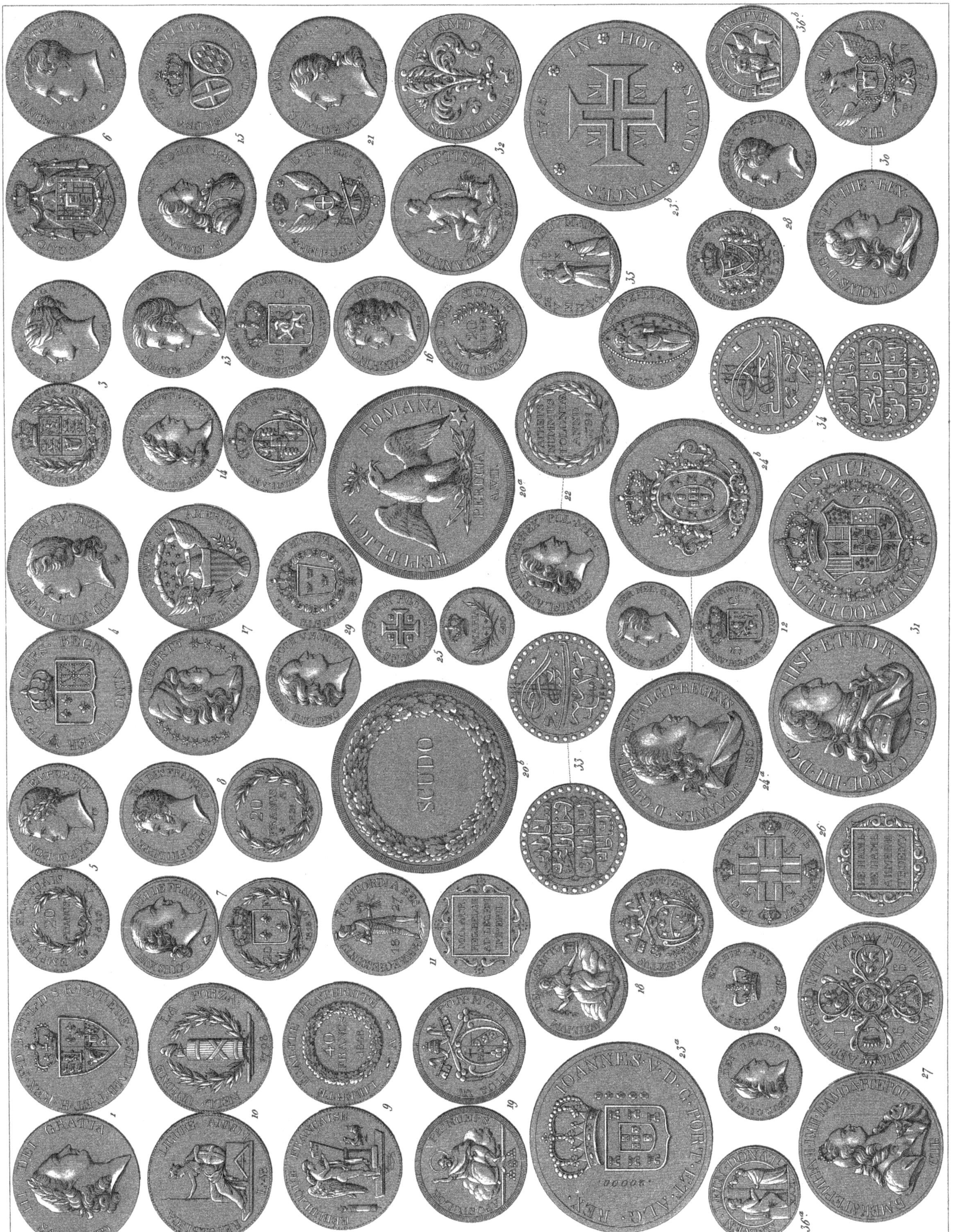

Plate 162 (Technology, Plate 22)

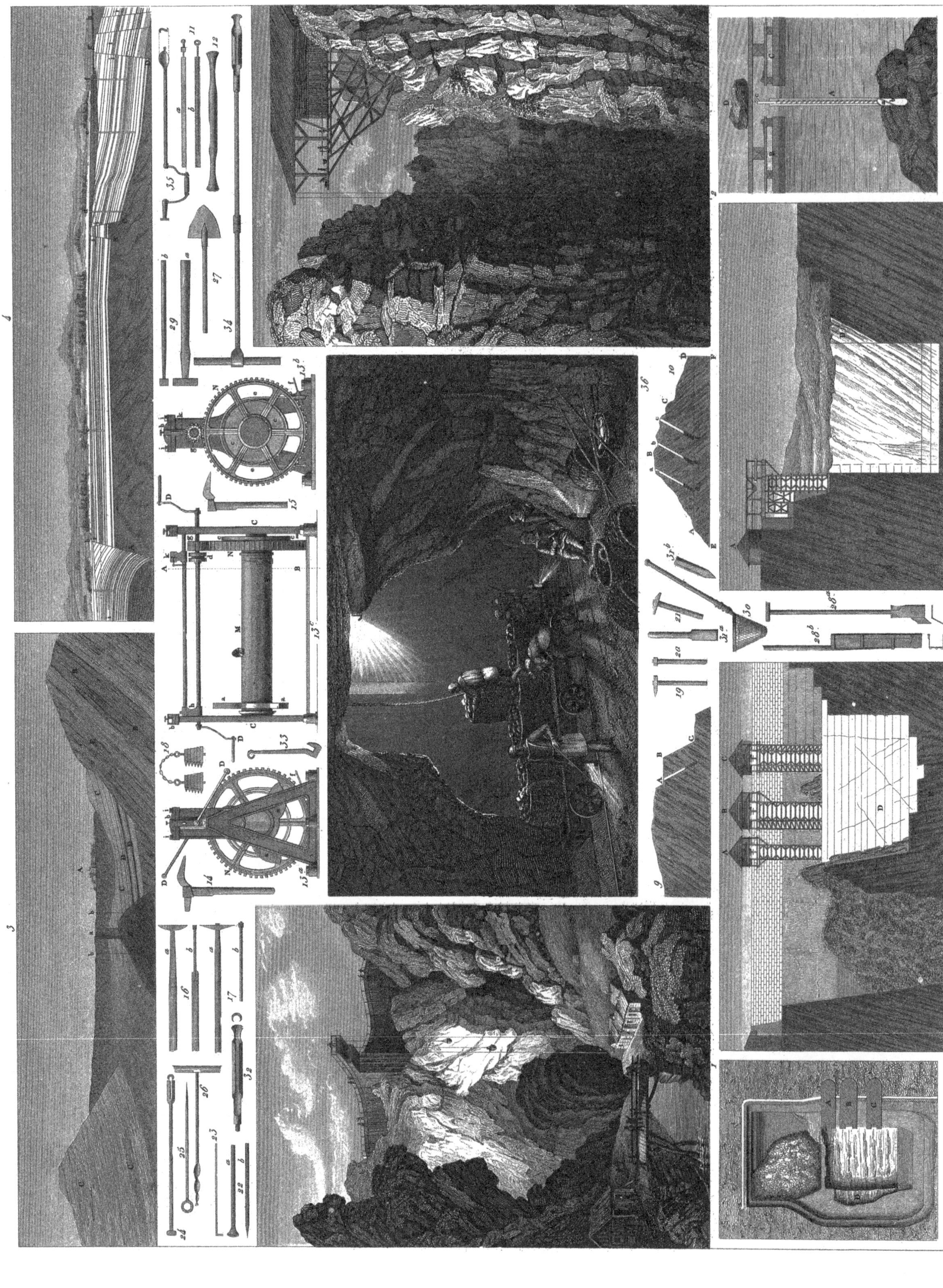

Plate 163 (Technology, Plate 23)

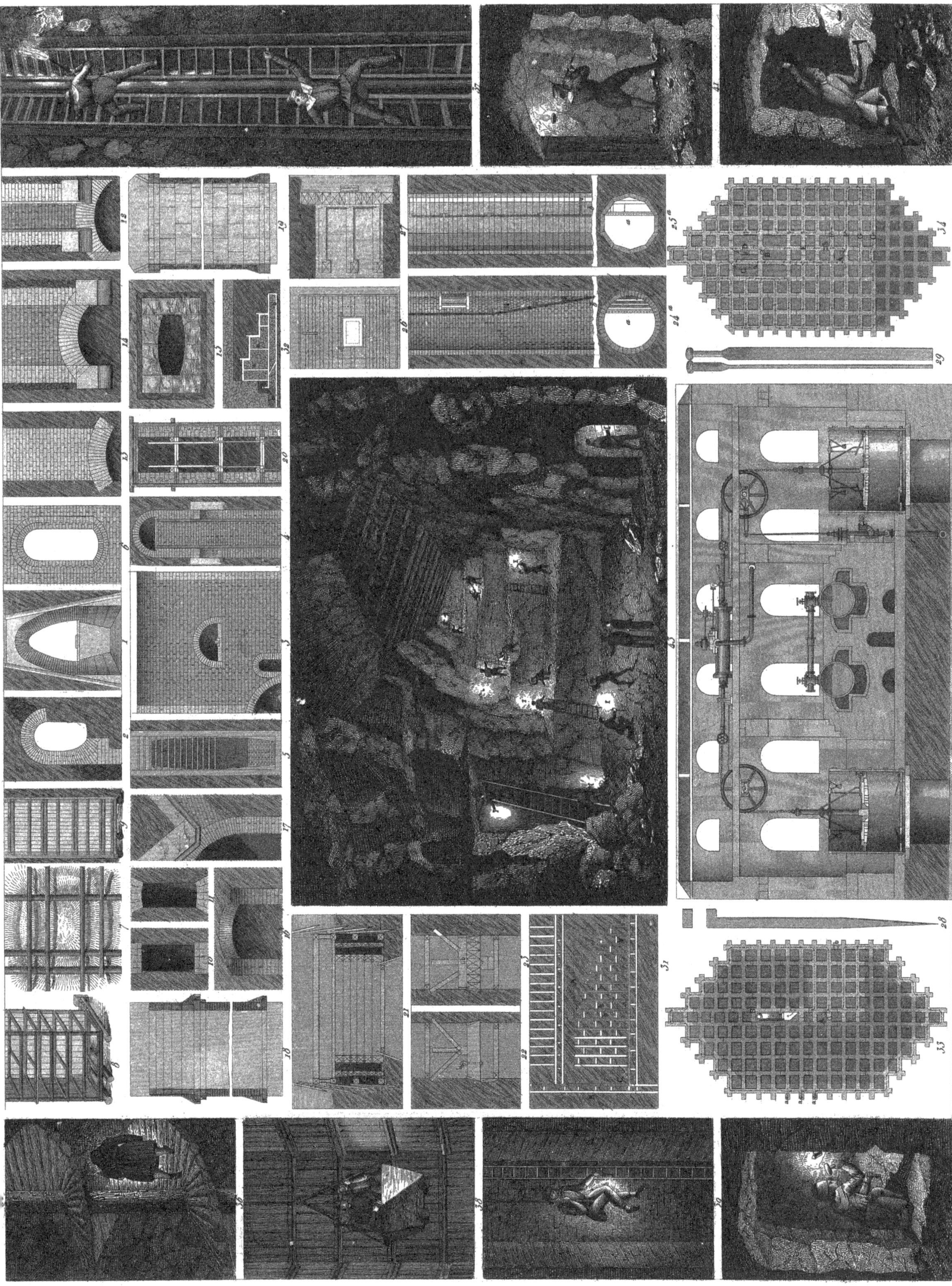

Plate 164 (Technology, Plate 24)

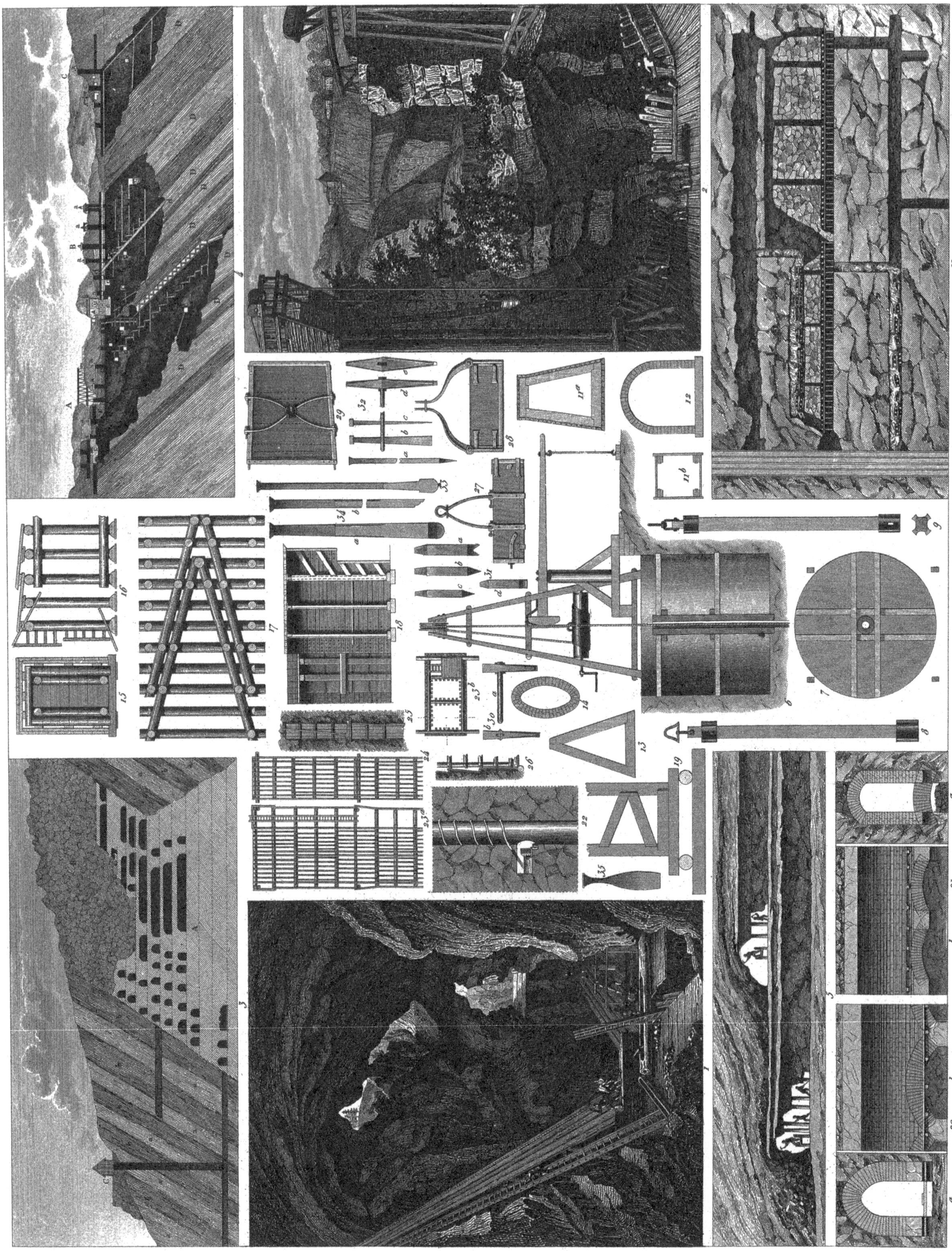

Plate 165 (Technology, Plate 25)

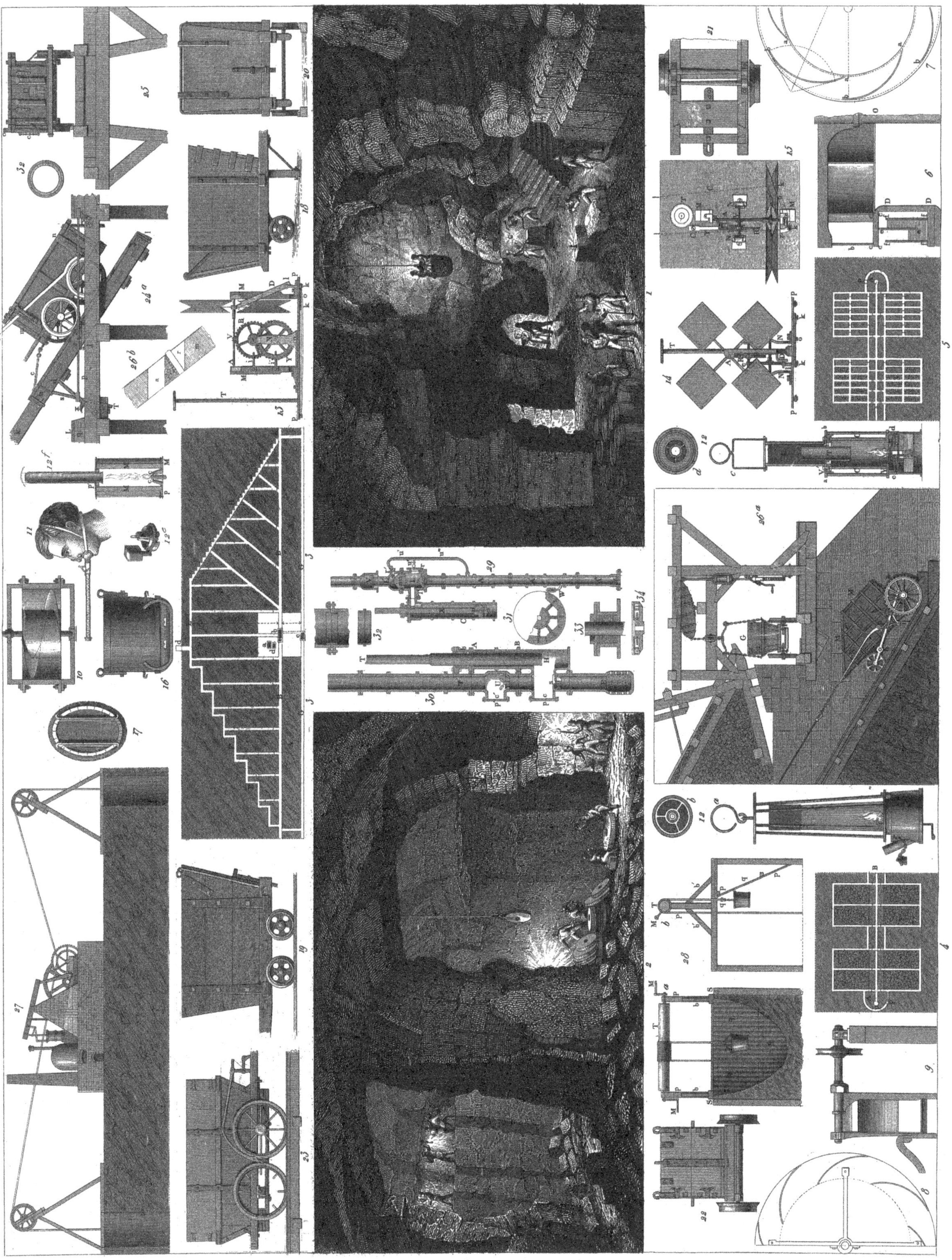

Plate 166 (Technology, Plate 26)

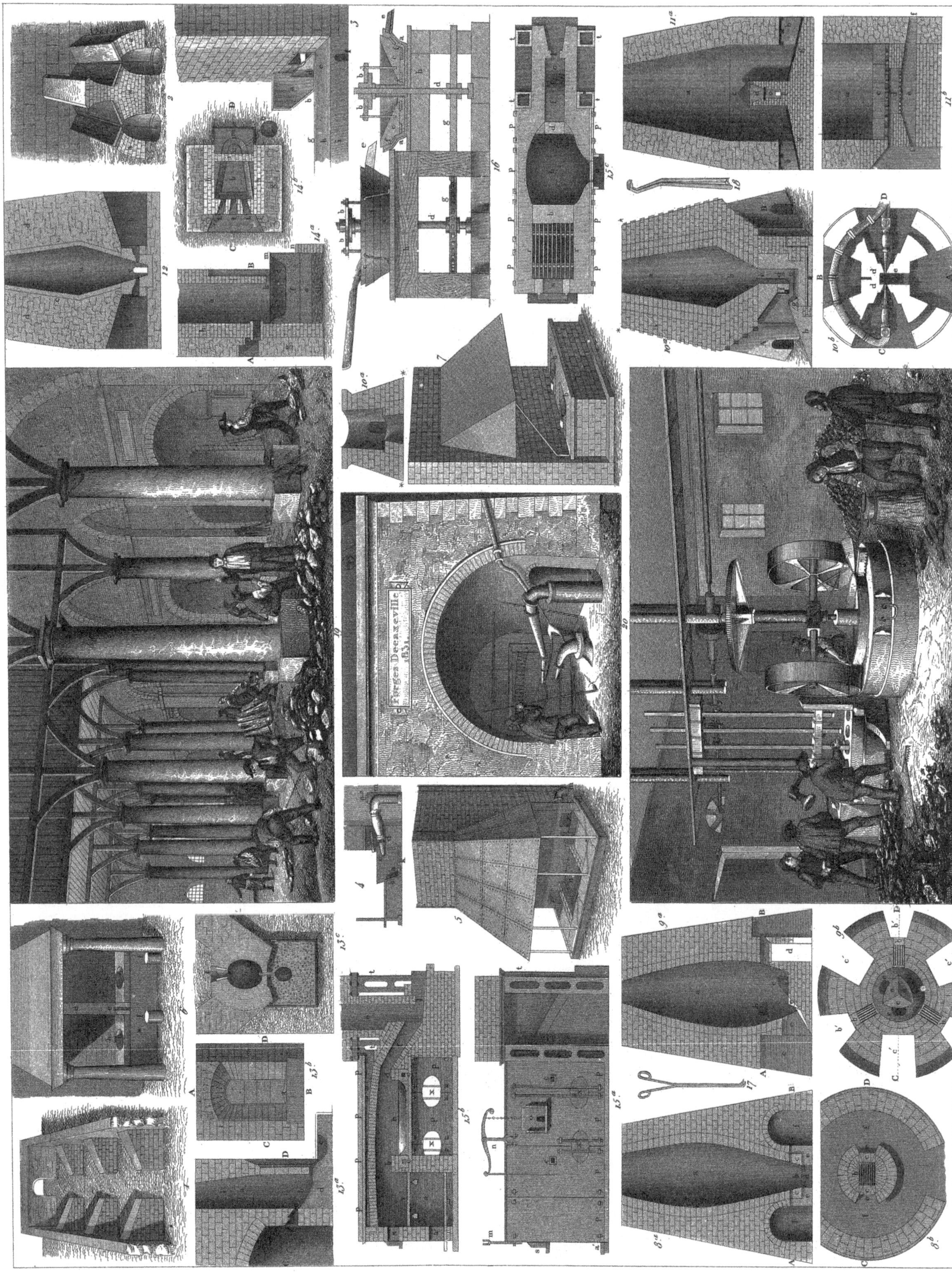

Plate 167 (Technology, Plate 27)

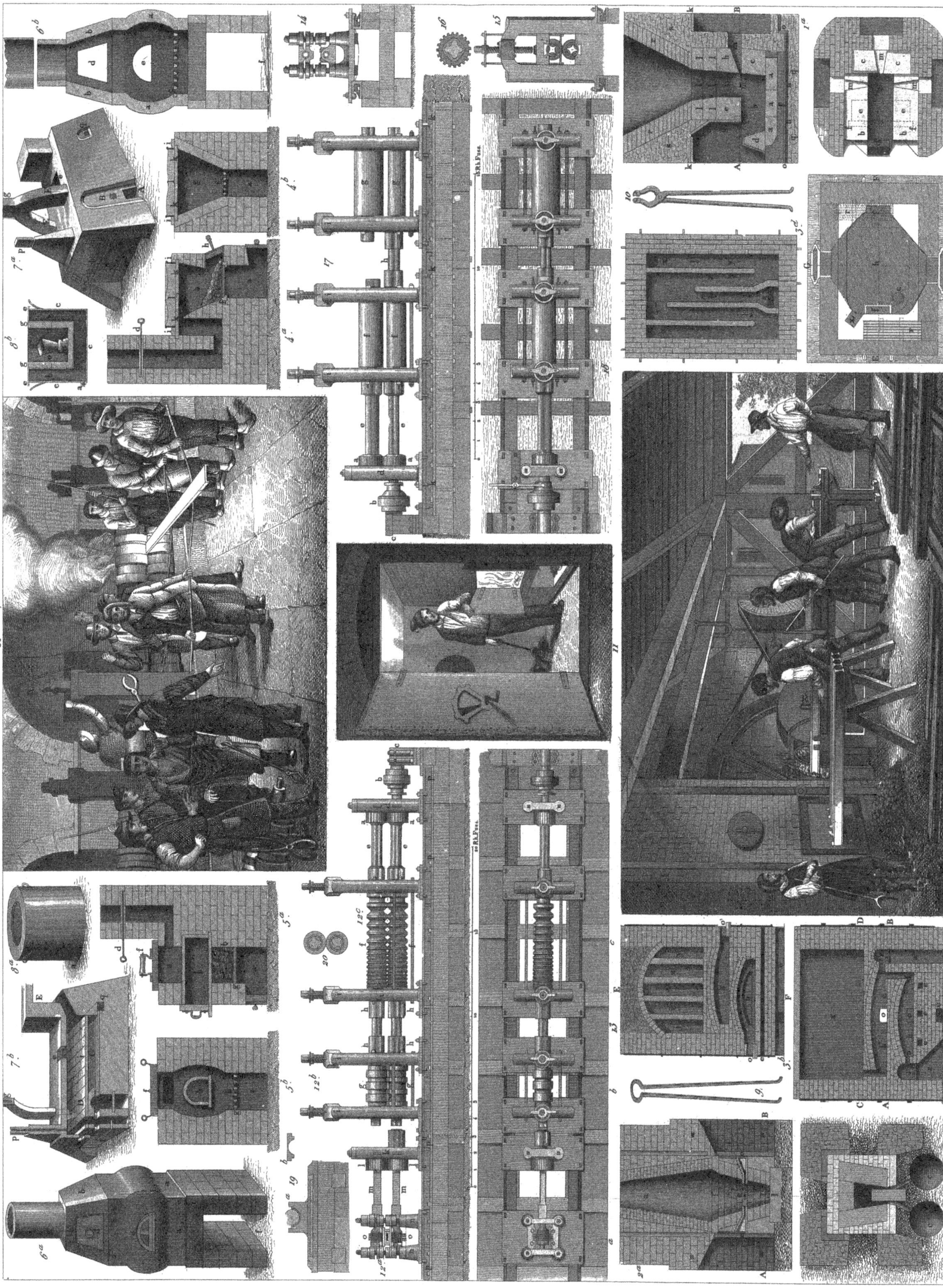

Plate 168 (Technology, Plate 28)

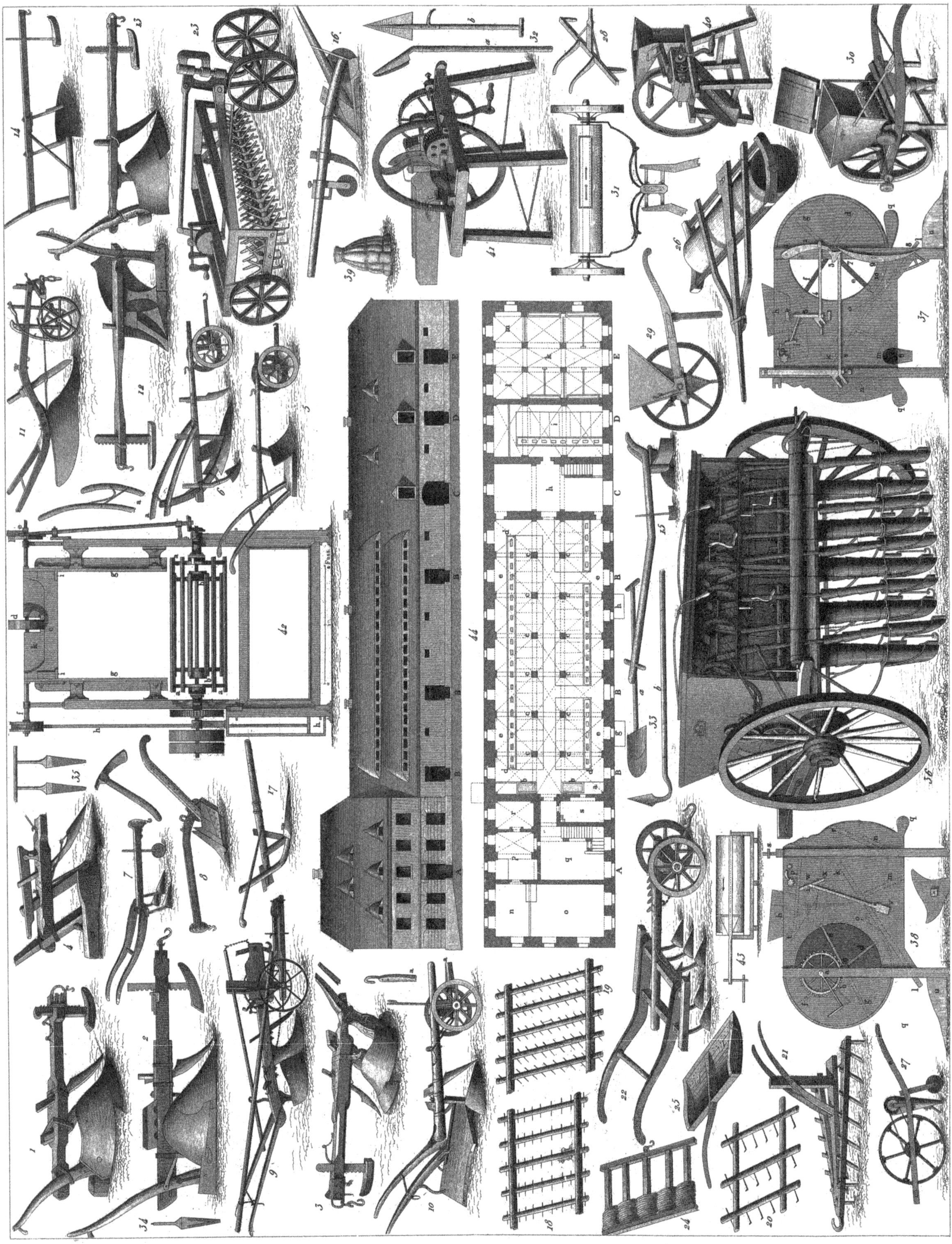

Plate 169 (Technology, Plate 29)

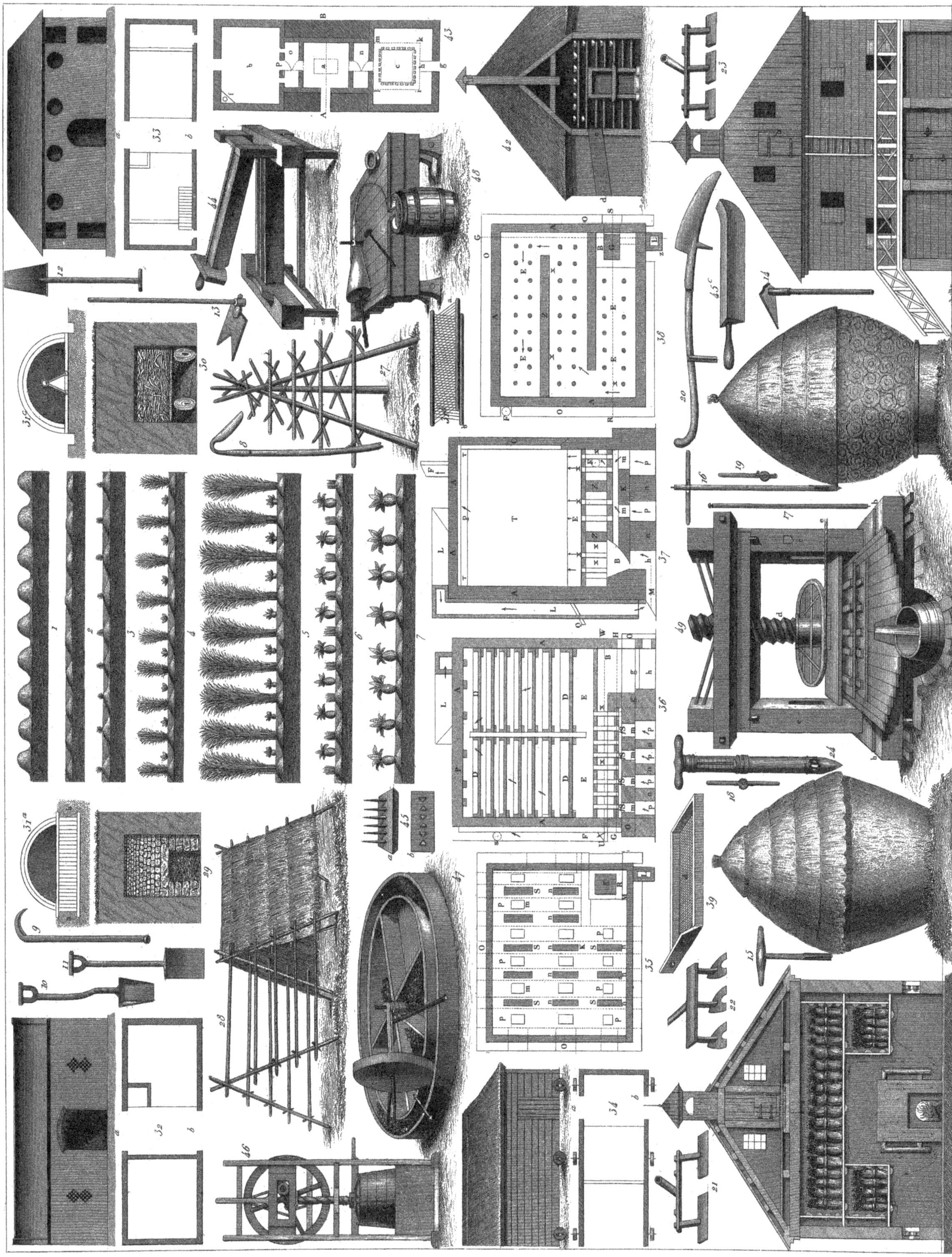

Plate 170 (Technology, Plate 30)

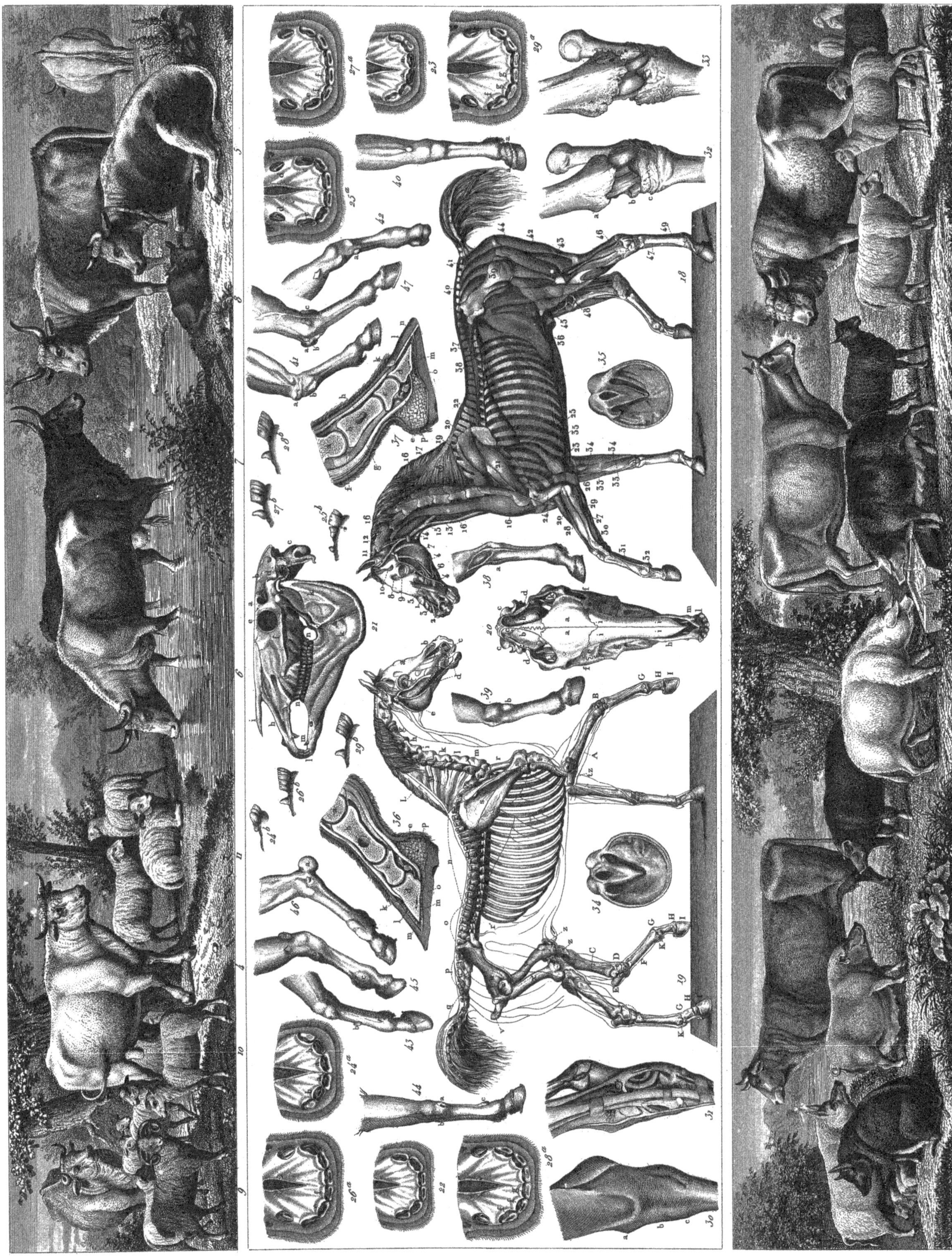

Plate 171 (Technology, Plate 21)

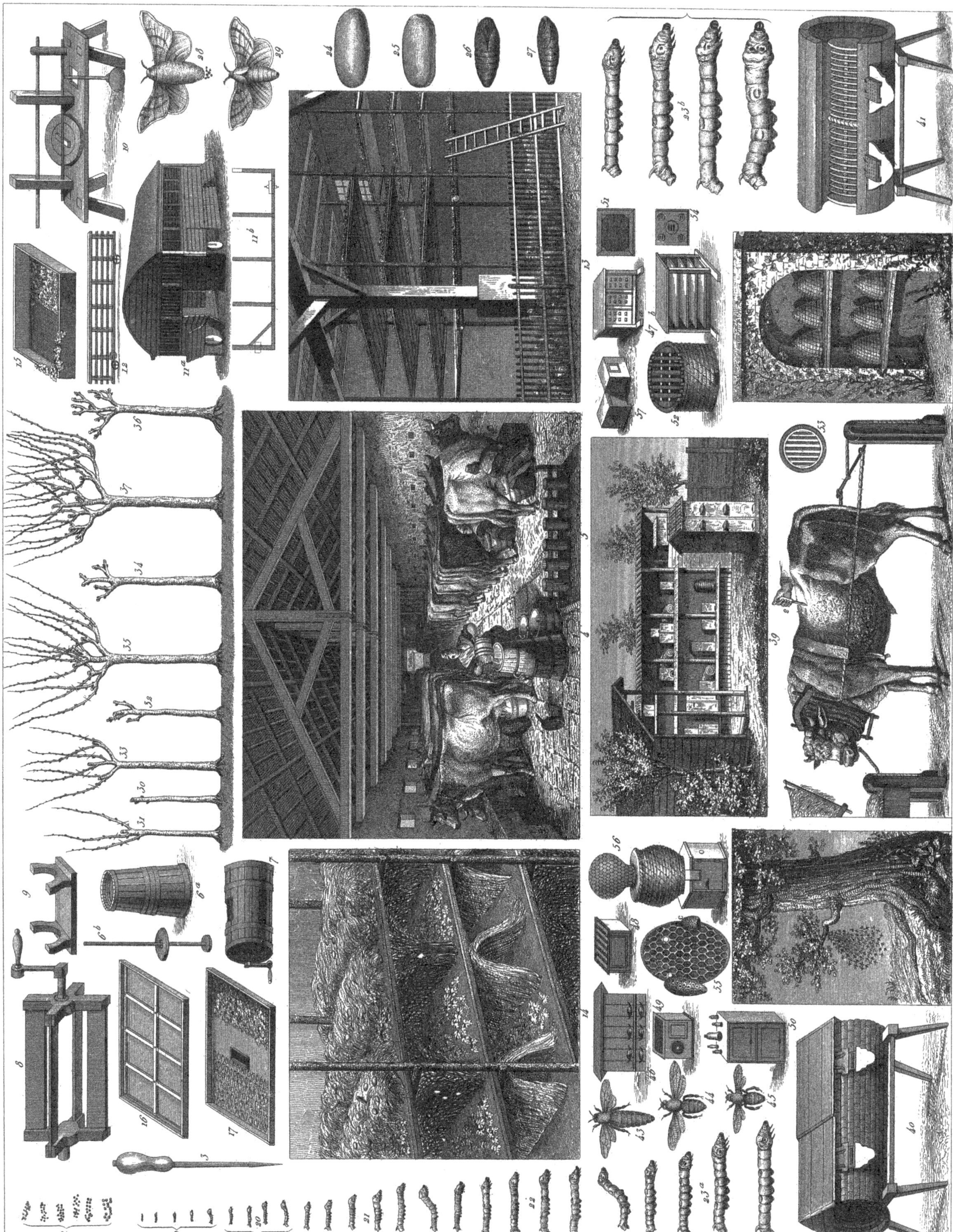

Plate 172 (Technology, Plate 32)

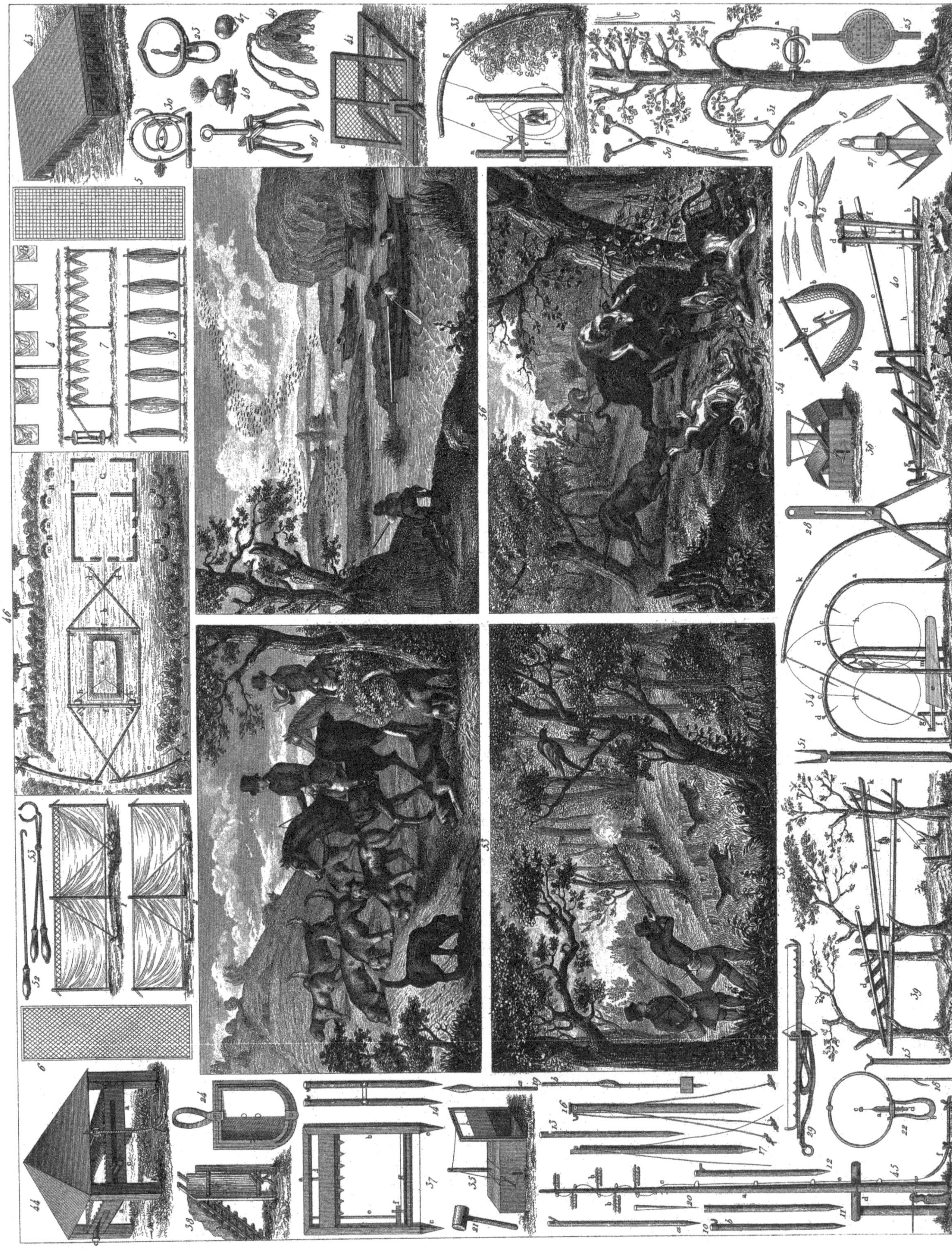

Plate 172 (Technology, Plate 32)

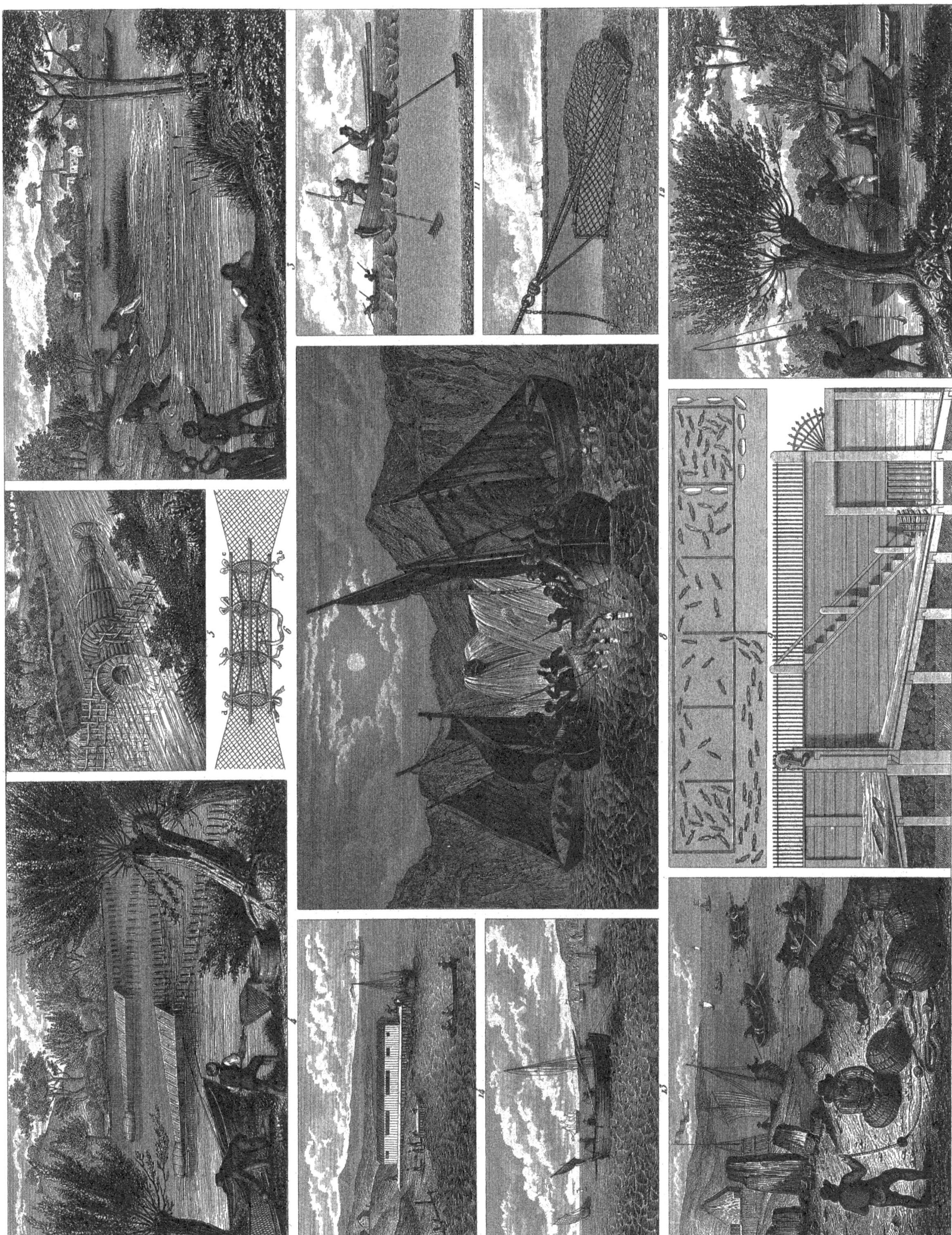

Plate 174 (Technology, Plate 34)

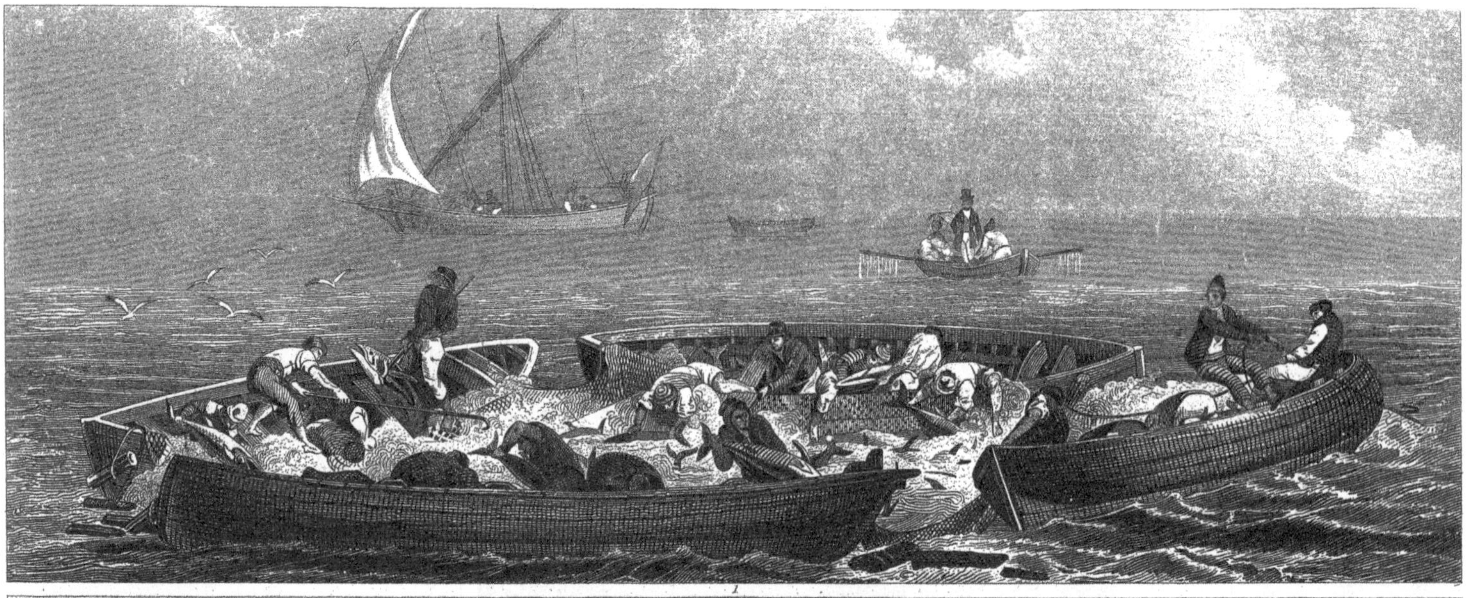

1

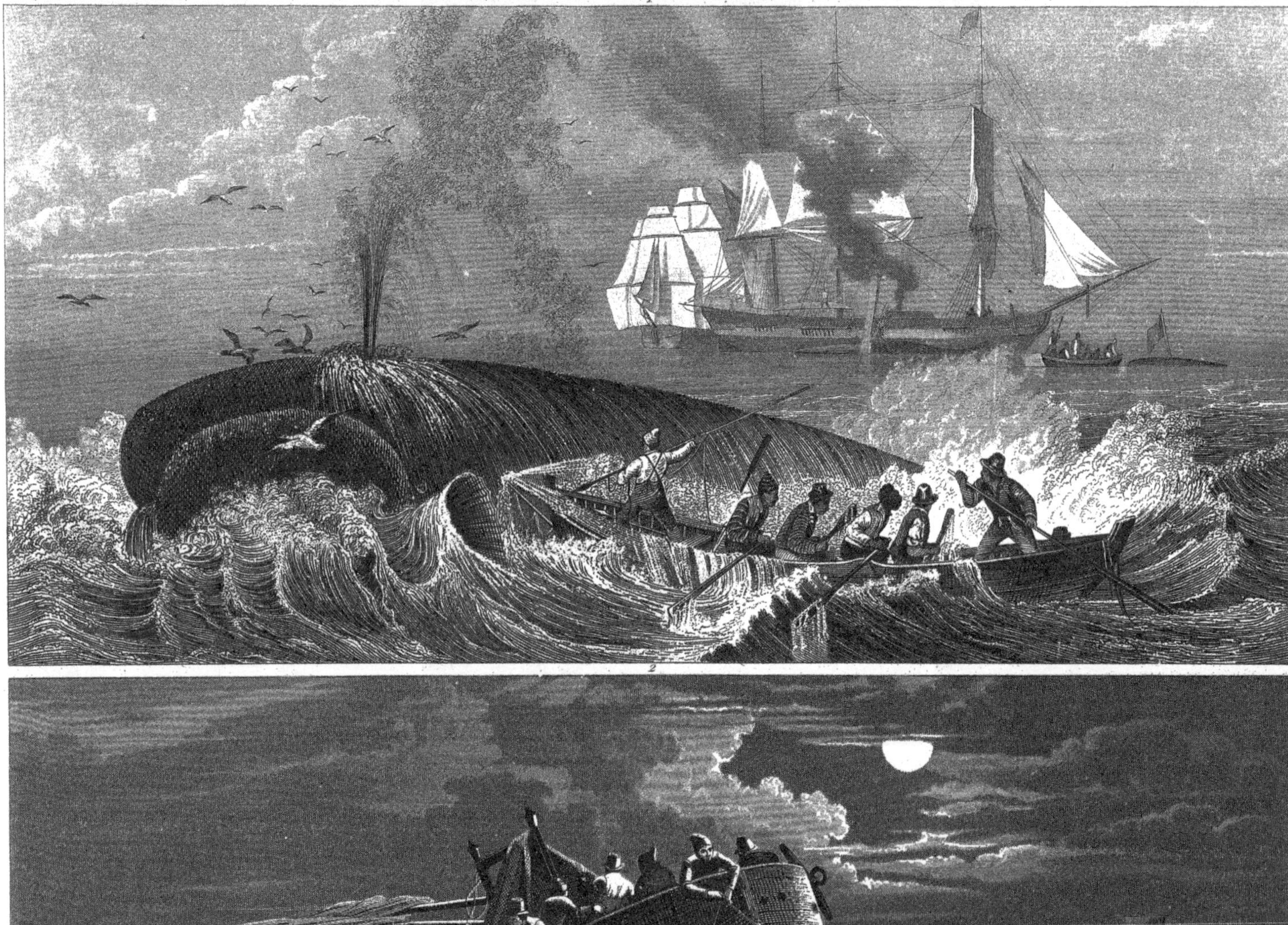

2

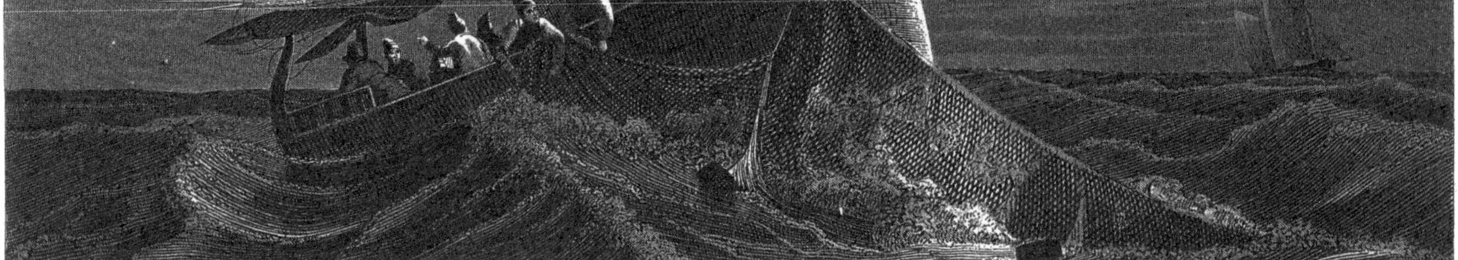

3

Plate 175 (Technology, Plate 35)